CHEMISTRY AND ANALYSIS OF HYDROCARBONS IN THE ENVIRONMENT

CURRENT TOPICS IN ENVIRONMENTAL AND TOXICOLOGICAL CHEMISTRY

This is a series of books based on papers from the two journals, *International Journal of Environmental Analytical Chemistry*, and *Toxicological and Environmental Chemistry*

ISSN: 0275-2581

Volume 1 ANALYTICAL ASPECTS OF MERCURY AND OTHER HEAVY METALS IN THE ENVIRONMENT
Edited by R. W. Frei and O. Hutzinger

Volume 2 RECENT ADVANCES IN ENVIRONMENTAL ANALYSIS
Edited by R. W. Frei

Volume 3 MUTAGENICITY TESTING AND RELATED ANALYTICAL TECHNIQUES
Edited by R. W. Frei and U. A. Th. Brinkman

Volume 4 MECHANISTIC ASPECTS OF THE THERMAL FORMATION OF HALOGENATED ORGANIC COMPOUNDS INCLUDING POLYCHLORINATED DIBENZO-*p*-DIOXINS
By G. G. Choudhry and O. Hutzinger

Volume 5 CHEMISTRY AND ANALYSIS OF HYDROCARBONS IN THE ENVIRONMENT
Edited by J. Albaigés, R. W. Frei and E. Merian

CHEMISTRY AND ANALYSIS OF HYDROCARBONS IN THE ENVIRONMENT

Edited by

J. ALBAIGÉS

Institute of Bio-Organic Chemistry
Barcelona

R. W. FREI

Analytical Chemistry Laboratory
The Free University, Amsterdam

E. MERIAN

International Association of
Environmental Analytical Chemistry, Therwil

GORDON AND BREACH SCIENCE PUBLISHERS

New York · London · Paris · Montreux · Tokyo

Gordon and Breach, Science Publishers, Inc.
One Park Avenue
New York, NY10016

Gordon and Breach Science Publishers Ltd.
42 William IV Street
London, WC2N 4DE

Gordon & Breach
58 rue Lhomond
75005 Paris

Gordon and Breach Science Publishers S.A.
Case Postale 161
CH-1820 Montreux 2

Gordon and Breach Science Publishers, Inc.
48-2 Minamidama
Oami Shirasato-machi
Sambu-gun
Chiba-ken
Japan 299-32

Library of Congress Cataloging in Publication Data
Main entry under title:

Chemistry and analysis of hydrocarbons in the environment.

(Current topics in environmental and toxicological chemistry; v. 5)
Contains most of the papers presented at a workshop, Barcelona, Nov. 1981, organized by the International Association of Environmental Analytical Chemistry (IAEAC) and the Societat Catalana de Ciències and Expoquimia.
Includes index.
1. Hydrocarbons—Analysis—Congresses. 2. Environmental chemistry—Congresses. I. Albaigés, J. II. Frei, R. W. (Roland W.) III. Merian, E. IV. International Association of Environmental Analytical Chemistry. V. Societat Catalana de Ciències Físiques, Químiques i Matemàtiques (Barcelona, Spain) VI. Expoquimia (1981: Barcelona, Spain) V. Series.

QD305.H5C45 1983 574′.01046 83-1603
ISBN 0-677-06140-4

Library of Congress catalog card number 83-1603. ISBN 0 677 06140 4. ISSN 0275-2581. Printed in Great Britain by Bell and Bain Ltd., Glasgow.

Contents

Preface

The ubiquitous occurrence of hydrocarbons in the environment is a problem we have been aware of for a long time. However, hydrocarbon sources are steadily changing, especially after the oil crisis, and thus creating new environmental impacts, which should be understood before we can deal with them. In addition, new or improved analytical procedures for the measurement of hydrocarbons in environmental samples are being developed and need to be adapted to requirements of new environmental policies.

This is the background which prompted the selection of the topic for a workshop, that was held in Barcelona in November 1981, organized by the International Association of Environmental Analytical Chemistry (IAEAC) and the Societat Catalana de Ciències and Expoquimia.

This book contains most of the papers presented at the workshop. They have already appeared in issues of the *International Journal of Environmental Analytical Chemistry* and the *Journal of Toxicological and Environmental Chemistry*. However, the collection in this volume will be more useful and more readily accessible to the researchers active in this area. The contributions have been classified into two main groups, one dealing with analytical methodologies recently designed for the determination of hydrocarbons and related compounds in air, water, sediments and including environmental fate and monitoring studies.

Since hydrocarbons are among the more widespread environmental organics, encompassing thousands of compounds of different origins, it is obvious that it was practically impossible to review all the aspects of the topic. Thus, as in other problem-oriented workshops organized by IAEAC, emphasis was placed on the discussion between participants (supported by papers presented in oral and poster sessions) in order to recognize the advances in the analytical chemistry and in the environmental understanding of hydrocarbons, as well as to assess where the gaps of our knowledge occur and where the priority areas for investigation are situated.

In the final discussion sessions, which were conducted by Milton L. Lee, Willie E. May and Bruce W. Tripp, three main questions have arisen: How does one decide on which compounds to monitor? What analytical techniques should be used or developed to make the measurements? How do we assure the quality of monitoring program and the data generated?

Since the main purpose of environmental studies is to identify and control the release of undesirable chemicals to the environment, it seems logical to first identify the hazardous compounds and then establish procedures to monitor this compound in environmental samples. In this respect, one has to bear in mind that, currently, hydrocarbon analysts are using a very narrow "analytical window". It should be remembered that petroleum is not the only source of hydrocarbons in the environment and that coal, shale oil and synfuels contain proportionally more toxic and mutagenic components. Consequently, future efforts should be directed toward the determination and monitoring of other biologically active components such as, e.g., the nitrogen-, sulfur-, and oxygen-containing PAH's which are largely present in these materials. Hence new analytical methods involving sample handling, fractionation, and detection must be developed for these more polar and often labile compounds. Selective pre-concentration, fractionation and detection methods are areas needing much future work.

The interest in new analytical methods for the study of the more polar and high molecular weight hydrocarbon fractions also stems from the actual economic interest in heavy crude oils or heavy fractions of conventional crude oils. These have rarely been studied in the past and modern HPLC and MS techniques should therefore be quite useful in this respect.

We also need to broaden our "hydrocarbon window" to include photochemical and metabolic products produced during environmental exposure. These reaction products can be biologically active, perhaps more so than the parent hydrocarbon. In addition, basic information on the physical-chemical form of hydrocarbons is important and frequently lacking. How these contaminants are distributed (adsorbed, dissolved, colloidal, etc.) has a definite influence on the biological availability.

Since it is impossible to select any particular technique as being suited for the analysis of all compounds in all matrices, laboratories must continue to use the most appropriate techniques for their specific applications. Therefore, steps should be taken to assure the quality of the data obtained. Standard methods are not generally accepted in research laboratories, and therefore, quality assurance standards or standard reference materials should be used for this purpose.

There is a problem with knowing how to handle and use all the data that are available today. Contemporary high resolution techniques can provide data on literally hundreds of components in a single sample. Much of this information is of little use and many of these sophisticated techniques are time consuming, costly, and cumbersome for routine monitoring of environmental pollutants. It was generally agreed that the maximum amount of information is desired for research studies concerning the health effects of various samples, but for routine monitoring, this data base should be reduced to a few compounds only.

However, monitoring is not simply an exercise in data accumulation. Any monitoring program needs to be designed to properly answer specific questions such as effects, degradation processes etc. Once the relevant questions are asked, decisions on the sampling scales and analytical scheme to be used will become important. In addition, any monitoring program should be coupled to concurrent research and should be flexible enough to be adjusted as new information becomes available. Data management and interpretation are also critical to any monitoring program.

All these arguments have prompted the organization of this workshop with the intention to bring together the experts in this area and to promote discussions between the various groups involved in this interdisciplinary field. In particular, it was intended to enhance collaboration between industry, government and university laboratories and between chemists and toxicologists. From the various comments received it seems that at least some of these goals have been realized in Barcelona.

At this point, we wish to thank all the attendants of the workshop, who provided the friendly atmosphere for a successful meeting. We hope that the papers presented here will promote awareness of the current problems in the field and will stimulate new or better approaches to future research in this area. We also like to express our deep appreciation to the Technical Secretary of Expoquimia, Mr. J. M. Otero, who kindly took charge of the many problems arising during the organization of the workshop.

Finally, the organizers acknowledge the financial support of the Department of Industry and Energy (Generalitat de Catalunya), ASERPETROL, CONCAWE and ENPETROL-Empresa Nacional del Petróleo.

J. Albaigés
R. W. Frei
E. Merian

Part I
ANALYTICAL METHODS

On the Determination of Benzo(a)pyrene in Atmospheric Aerosols Fractionated by a Low Pressure Impactor, by TLC Separation and *in situ* Spectrophotofluorometry at the Picogram Level†

ANTONIO H. MIGUEL‡

Instituto de Química, Universidade Federal do Rio de Janeiro, 21.941 Ilha do Fundão, Rio de Janeiro, Brasil

(*Received November 27, 1981*)

Urban aerosols were fractionated according to size by an eight-stage low pressure impactor with size resolution down to 0.05 μm diameter. Each stage of the impactor, containing picogram amounts of benzo(a)pyrene (BaP) and other polycyclic aromatic hydrocarbons (PAH) was ultrasonically extracted with benzene. After reduction to appropriate volumes, the extracts were spotted on a TLC plate and developed. The BaP spots were quantitated *in situ*, by finely focusing the TLC plate on the fluorescence excitation beam and measuring the resulting emission while scanning the plate. The regression lines passing thru the origins exhibited linear regression coefficients (R^2) between 0.9985 and 0.9996 for three-point BaP calibration data in the 100-1000 picogram range. The relative standard deviations varied between 3.4 and 12.9%, depending on the amount of BaP present. The detection limit for BaP is 5 picogram. The method allows the determination of typical atmospheric Bap size distributions with as little as 1.5 m^3 of sampled air.

KEY WORDS: Benzo(a)pyrene size distribution, low pressure impactor, TLC separation, *in situ* fluorescence, atmospheric aerosols.

---

†Presented at the Workshop on the Chemistry and Analysis of Hydrocarbons in the Environment, Barcelona, November 1981.

‡Present address: Departamento de Química, PUC, R. Marquês de S. Vicente, 225, Rio de Janeiro, 22.453, Brasil.

INTRODUCTION

Particle size and chemical composition of pollutants determine atmospheric extinction and transport, and deposition in the respiratory system. In addition, size data provide complimentary information regarding the source and mechanism of formation of toxic pollutant species.

Cascade impactors operating at atmospheric pressures have been used to fractionate PAH-containing ambient aerosols of aerodynamic diameter (d_p) larger than about 0.5 μm.[1–7] To obtain resolution below 0.5 μm, it is necessary to operate the impactors at reduced pressure.[8,9] However, low pressure impactors operate at flow rates two to three orders of magnitude smaller (typically 1 l/min.). This limitation demands large aerosol volumes—which result in increased observation "time windows"—and analytic methods with PAH sensitivities at the picogram level for 24 hr samples.

Particle bounce and overloading are two of the most common sources of error in the fractionation of aerosols. Particle bounce during sampling increases with increasing particle size and loading of the impaction surface.[10] According to Fuchs,[10] many investigators have not paid sufficient attention to this fact, and this resulted in large discrepancies in the published size distribution data of pollutant aerosols.

For cascade impactors with single nozzles (this study), there are two ways of reducing particle bounce: (i) by sampling small aerosol volumes and, (ii) by using coated impaction surfaces. The first condition imposes a high sensitivity demand on the analytic method used to determine the species of interest. The second requires that the presence of the coating material does not interfere with the detection method.

Currently, there are two relatively inexpensive methods for the determination of PAH at the picogram level, namely high performance (pressure) liquid chromatography with fluorescence detection (HPLC-F) and thin layer chromatography with *in situ* spectrophotofluorometry (TLC-SPF). For samples collected with multi-stage cascade impactors (this study), the latter has two major advantages: (i) it is much faster—blanks and samples can be developed simultaneously in less than 1.5 hrs as compared with about 6.5 hrs required in the HPLC-F method. (ii) it provides confirmatory excitation and emission spectra of the separated PAH at the 50 to 100 picogram level.

In the following sections we report on an improved TLC-SPF method capable of determining picogram amounts of BaP and other PAH with high precision. This high sensitivity allows the use of low aerosol volumes in the determination of submicron PAH at low atmospheric levels. An illustrative example is given, where the submicron size distribution of atmospheric BaP was determined in a sampled air volume as small as 1.44 m^3.

EXPERIMENTAL

Aerosol collection

Aerosols were segregated according to size by an eight-stage low pressure impactor built from high density aluminum stock (Alcan type 7075-T6) according to design and construction tolerances reported elsewhere.[11] The impactor samples at a rate of 1 l/min and has 50% efficiency cutoffs of 4.0, 2.0, 1.0, 0.5, 0.26, 0.12, 0.075, and 0.050 μm d_p. A Leybold-Heraeus type S4A vacuum pump was used.

Particle bounce was reduced by a layer of Vaseline^R about 2-μm thick by 3 mm diameter. This layer was obtained by applying 0.5 μl of a 2% solution of Vaseline^R in toluene—using a Hamilton no. 7001 SN microliter syringe—and allowing it to evaporate.

Aerosol extraction

At the end of the sampling period, using a Pasteur pipette, the particle deposit was removed by drop-wise addition of about 300 μl of spectrograde benzene, while tilting the impaction disc. The resulting suspension was collected in a 1 ml amber Reacti-Vial^{TM} (Pierce Chemical Co. no. 13097). After removal of all deposits, the vials were placed in an ultrasonic cleaner (Brosonic 220) and extracted for 30 min. The volume of each extract was then reduced to about 30 μl by flowing a stream of high grade nitrogen gas over their surface, at room temperature.

BaP analysis

Using 5 μl capillary pipettes, the extracts were spotted 0.7 cm apart and 1·5 cm from the bottom, on precoated TLC plates (21% acetylated cellulose, Schleicher and Schuel no. G1800/21 ac) cut into 20 × 4 cm pieces. The vials were rinsed three times with 5 μl of benzene also transferred to the same spot. BaP was separated from other PAH by development of the plates in the dark, in an equilibrated chamber containing 50 ml of a mixture of ethanol:toluene:water (17:4:4 (v/v). Plates containing BaP standards between 100 and 1000 picograms, spotted in the same manner with 5 μl pipettes were developed at the same time. The solvent front was allowed to travel 15 cm (~1 hr 20 min) and the plates were removed and allowed to dry.

A Perkin-Elmer model MPF-2A Fluorescence Spectrophotometer was used. A spectrodensitometer with slide mechanism built from 0.64 cm dia. recorder slide bars permitted vertical (normal to the excitation beam) and horizontal movement (in the plane of the beam) of the TLC plate. Motorized vertical movement of the TLC plate was provided by an electric clock motor which scanned the plate at a rate of 1 mm/min in the direction of solvent migration. The horizontal movement was manual. The entire unit was assembled on the platform of the module normally used in phosphorescence measurements (with the cam motor removed), which is equipped with optical lenses (Figure 1). The TLC plate is viewed at a 45° angle from both the excitation and emission slits.

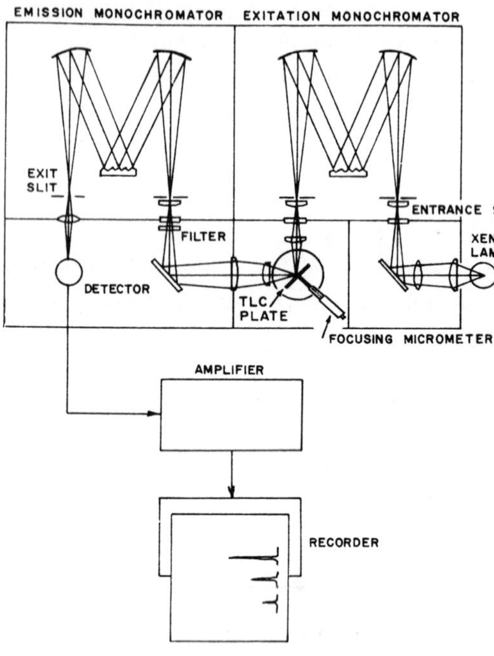

FIGURE 1 Schematic arrangement of the spectrophotofluorometer used, showing the TLC plate positioning and focusing micrometer.

After selection of the appropriate excitation and emission wavelengths and slits, the BaP spot is located and finely focused—using the straight micrometer shown in Figure 1—by optimizing the fluorescence emission intensity. Typical chromatograms are shown in Figure 2.

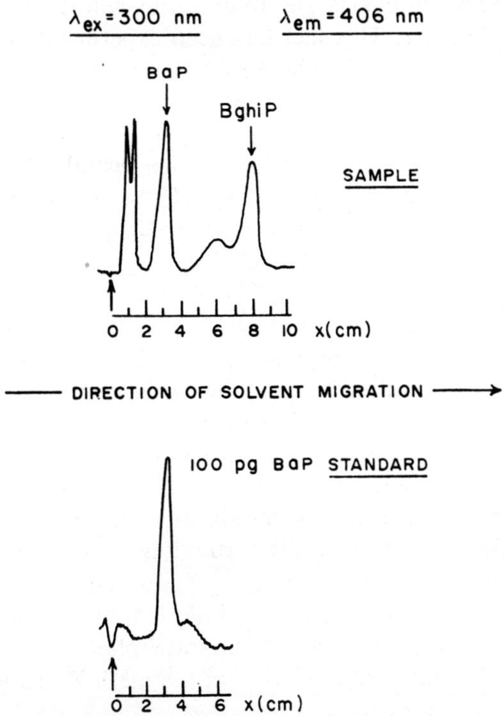

FIGURE 2 Typical TLC-SPF chromatograms. x = actual distance on the TLC plate.

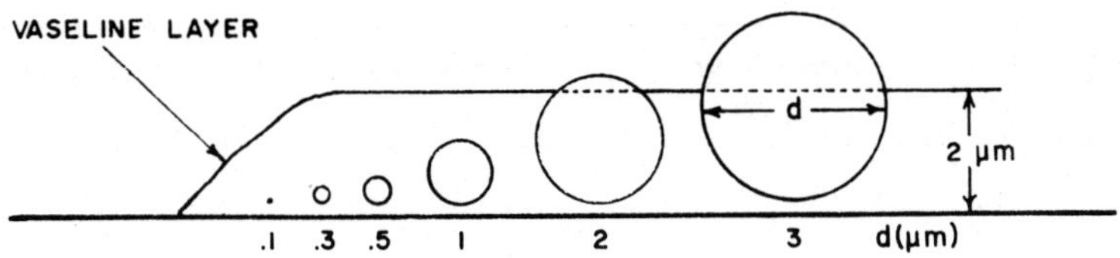

FIGURE 3 Representation of the sizes (μm) of deposited particles relative to a 2-μm thick layer of Vaseline coating.

RESULTS AND DISCUSSION

Collection efficiency

According to a model by Reischl and Johns,[12] the impaction area free of particle deposit decreases exponentially with the amount of aerosol sampled. Thus, for a coating layer 2-μm thick (Figure 3), particle bounce on the upper stages (cutoffs larger than 2 μm) will become critical only after monolayer coverage. This fact has been experimentally confirmed by current work of Hering and Friedlander.[13] Using an identical low pressure impactor, these investigators determined monolayer collection efficiencies of about 95% for polystyrene latex particles of 1.48 μm diameter collected on Vaseline coated surfaces. Collection efficiencies of submicron aerosols by the 2-μm Vaseline layer (Figure 3) should be greater than 95%, as a result of their embedding and retention by capillary forces.

TLC-SPF analysis

From the TLC-SPF chromatograms (Figure 2), either peak heights or areas of the spots can be used to plot the calibration curve and calculate the concentrations of the unknowns (provided that the same size pipettes are used for both sample and standard spotting). We found that, using either ordinate, the standard deviations of the calculated unknown concentrations were well within the standard deviation of the method.

The high precision of BaP analysis by this method is evidenced by correlation coefficients (R^2) between 0.9985 and 0.9996 (Table I) of linear regression lines passing thru the origins. Under the same operating conditions, calibration plots ran on separate plates also showed slope deviations within the standard deviation of the method (Table I).

TABLE I

BaP calibration data and linear regression coefficients $\lambda_{ex}=300$ nm, $\lambda_{em}=406$ nm, slits: ex = 14, em = 2.6 nm. Sensitivity = 3

BaP spotted, pg	Peak height,[a] cm	Peak height,[b] cm	Peak height,[c] cm
100	1.5	1.3	1.4
500	8.4	8.2	8.1
1,000	18.3	17.7	17.1
R^2	0.9985	0.9993	0.9996
intercept	−0.580	−0.669	−0.449
slope	0.0187	0.0183	0.0175

[a,b]Each set of standards applied on one plate.
[c]Each nominal standard (ran in duplicate) applied on three separate plates.

The relative standard deviations of several determinations varied between 3.4 and 12.9%, depending on the amount of BaP in the spot (Table II). The limit of detection—defined as a signal of amplitude twice the noise—is 5 picogram for BaP. Thus, to keep the RSDs near 5%, the

TABLE II

Reproductibility of the determinations as a function of BaP level[a]

pg found[b]	Ave. pg	Std. deviation	RSD, %
69, 61, 79	69.7	9.02	12.9
107, 103, 97	102.3	5.03	4.92
148, 139, 150	145.7	5.86	4.02
231, 228, 243	234.0	7.94	3.39

[a]1 μl aliquots of benzene extracts (20 min in ultrasound) of atmospheric particles collected on Hi-Vol glass fiber filters.

[b]Average of duplicate determinations.

amount of BaP collected on each stage of the impactor should be at least 20 times this limit. BaP in solvent and Vaseline blanks were below the detection limit.

The stability of BaP during fluorescence analysis was investigated by exposing a spot containing 500 picograms of this PAH to UV light of 300 nm under normal analytic conditions. After 30 min of continuous exposure, about 20% of the original BaP had photodecomposed or evaporated from the plate. In the TLC-SPF method described, a full BaP chromatogram is obtained in about 30 sec. During this period, about 3 picograms of BaP is lost. For this reason, the TLC plate should be scanned as fast as the recording device response permits. This fact may well be the limiting factor for achieving lower PAH detection limits using the method described.

Treatment of BaP aerosol size distribution data

There are several ways of reporting data on chemical composition as a function of particle size. For cascade impactors with wide size resolution ranges, such as the low pressure impactor used in this study, the best way is to report concentrations and mass distribution functions with respect to particle size (Table III). From such data, physically meaningful histograms can be obtained by plotting $\Delta M/M \cdot \Delta \log d_p$ vs $\log d_p$ (Figure 4). For instance, the modal nature of the aerosol is clearly shown by these plots. Figure 4 represents a distribution for which about 60% of the BaP is in what has been called[14] the "accumulation" mode ($0.05 < d_p < 0.26\,\mu$m) whereas about 30% is in the "coarse" mode ($0.26 < d_p < 2\,\mu$m). Similar distributions were found in Pasadena (California) aerosols,[8] where an

TABLE III

BaP concentration and mass distribution function with respect to particle size[a]

d_p range,[b] μm	ng BaP	ΔM, ng/m³	ΔM/MΔ log d_p
>4	0.13	0.090	0.184
2.0 –4.0	0.14	0.097	0.199
1.0 –2.0	0.13	0.090	0.184
0.5 –1.0	0.37	0.257	0.527
0.26 –0.5	0.21	0.146	0.317
0.12 –0.26	0.43	0.299	0.550
0.075–0.12	0.60	0.417	1.26
0.05 –0.075	0.33	0.229	0.804

[a]Aerosols sampled during a 24 hr period (Rio de Janeiro).

[b]The aerodynamic diameters refers to spherical particles of unit density. For other particle densities see ref. 11. The upper cutoff for the first stage was taken at 8 μm because particles larger than this size are not efficiently collected since sampling was not done isokinetically.

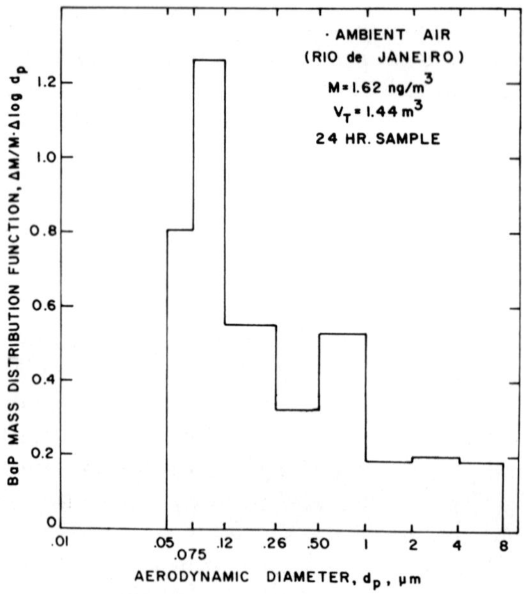

FIGURE 4 Distribution of BaP as a function of particle size. The distribution was normalized for M, the total BaP loading.

average of about 74 and 19% of the BaP was in the accumulation and coarse modes, respectively. These modal characteristics suggest similar sources of emission of this PAH into the atmosphere of the two cities.

Acknowledgements

The author wishes to express his gratitude to Celso Orsini (IF-USP, São Paulo) and Luiz C. Bouéres (Florida State University) for donation of the high density aluminum stock used in the construction of the low pressure impactor.

This research was supported in part by a grant from the National Research Council of Brasil (CNPq), FINEP and CEPG-UFRJ.

References

1. M. Kértész-Sáringer, E. Mészáros and T. Varkonyi, *Atmos. Environ.* **5,** 429 (1971).
2. R. Starkey and J. Warpinski, *J. Envir. Hlth* **36,** 503 (1974).
3. A. Albagli, H. Oja and L. Dubois, *Envir. Lett.* **6,** 241 (1974).
4. R. C. Pierce and M. Katz, *Envir. Sci. Technol.* **9,** 347 (1975).
5. M. Katz and R. C. Pierce, *Polynuclear Aromatic Hydrocarbons: Chemistry, Metabolism, and Carcinogenesis* (Raven Press, New York, 1976), R. I. Freudenthal and P. W. Jones, Eds., Vol. 1, pp. 413–429.
6. L. Van Vaeck and K. Van Cauwenberghe, *Atmos. Environ.* **12,** 2229 (1978).
7. M. Katz and C. Chan, *Envir. Sci. Technol.* **14,** 838 (1980).
8. A. H. Miguel and S. K. Friedlander, *Atmos. Environ.* **12,** 2407 (1978).
9. S. C. Stern, H. Zeller and A. I. Schekman, *Ind. Eng. Chem. Fundam.* **1,** 273 (1962).
10. N. A. Fuchs, *Fundamentals of Aerosol Science* (John Wiley and Sons, New York, 1978), Ed. by D. T. Shaw, Chap. 1, pp. 1-83.
11. S. V. Hering, S. K. Friedlander, J. J. Collins and L. W. Richards, *Envir. Sci. Technol.* **13,** 184 (1979).
12. G. P. Reischl and W. Johns, *Staub Reinhalt Luft* **38,** 55 (1978).
13. S. V. Hering and S. K. Friedlander, private communication (1981).
14. K. T. Whitby and G. M. Sverdrup, *California Aerosols: Their Physical and Chemical Characteristics.* (Particle Technology Laboratory Publication no. 347, 1978), Univ. of Minnesota.

Low Temperature Spectrofluorimetry of Complex Mixtures of PAH, Application to the Analysis of Isomeric PAH Extracted from Environmental Samples (Petroleum, Marine Sediments)†

P. GARRIGUES,‡ M. EWALD,¶‡ M. LAMOTTE,‡ J. RIMA,‡ A. VEYRES,‡ R. LAPOUYADE§ and J. JOUSSOT-DUBIEN‡

Laboratoires de Chimie Physique A‡ et de Chimie Organique,§ ERA du CNRS no. 167— Université de Bordeaux I, 351 Cours de la Libération 33405 Talence Cedex, France

and

G. BOURGEOIS

Centre d'Etude Structurale et d'Analyse des Molécules Organiques, Université de Bordeaux 1, 351 Cours de la Libération, 33405 Talence Cedex, France

(*Received December 10, 1981*)

Spectra obtained by High Resolution Spectrofluorimetry (HRS) in *n*-alkane matrices are more resolved and more structured than those obtained in nitrogen matrix. Analysis of petroleum and marine sediment fractions obtained by a two steps HPLC procedure allow us to identify and quantify the five isomers of the monomethylphenanthrene by HRS in *n*-hexane at 15 K. Some results are compared with those obtained by GC-MS analysis.

KEY WORDS: Polycyclic-aromatic-hydrocarbons (PAH), monomethyl-phenanthrene isomers (MP), monomethyl-chrysene isomers (MC), petroleum, marine sediment, high-resolution-spectrofluorimetry (HRS), high-performance-liquid-chromatography (HPLC), gas-chromatography-mass-spectrometry (GC-MS).

---

†Presented at the Workshop on the Chemistry and Analysis of Hydrocarbons in the Environment, Barcelona, November 1981.

¶To whom correspondence should be addressed.

INTRODUCTION

During the last ten years, two methods with high resolution spectrofluorimetric detection at low temperature ($T \ll 77$ K) have been developed: the *n*-alkane frozen solutions of PAH (Shpolskii matrices)[1] and the matrix isolation spectroscopy in inert gas matrix.[2] Our purpose was to choose the best method for the analysis of isomeric PAH in environmental samples. Qualitative studies on synthetic mixtures show that the Shpolskii matrices method is the most convenient one. This method was applied to the analysis of monomethylphenanthrenes (MP) in petroleum and marine sediment fractions.

Quantitative determinations are made by combined internal-standard and standard addition methods. Results are briefly compared with those of GC-MS analysis and further developments are under way to obtain quantitative determination with standard deviations for the two techniques (HRS and GC-MS).

EXPERIMENTAL

Authentics of the monomethylphenanthrenes have been synthetized in the laboratory,[3] except the 2-methylphenanthrene (2 MP) (K and K Fine Products). Purity of these compounds have been tested by GC-MS. Authentics of methyl-chrysene have been provided by Dr. A. Colmsjö of Arrhenius Laboratory (Stockholm, Sweden). Results presented in this work have been obtained with a crude petroleum from North Sea (marine origin) and with a marine sediment from Cariaco Trench (Orgon II Cruise).[4,5] Two other petroleums[6] and three different marine sediments were examined which lead to the generality of the informations presented here.

(a) Fractionation

The chromatographic procedure has been described in previous papers.[7,8] After adsorption chromatography of crude samples on Florisil (SiO_2 and MgO) with pentane elution to eliminate aliphatics and very polar compounds, two HPLC steps allow us to isolate first the triaromatic compounds on μ-silica-NH_2, secondly the monomethylated triaromatics on μ-silica-C_{18}. Elution of the aromatic compounds was monitored at 254 nm with UV-VIS monochromatic absorption detector (LDC Spectromonitor III).

(b) High resolution spectrofluorimetry (HRS)

A home-made high resolution spectrofluorimeter has been previously described.[9] For the frozen solution method, quartz tubes containing *n*-alkane solutions of PAH were attached to the cold head of a closed cycle helium cryogenerator (CTI Cryodyne 21S) operating at 15 K. Preliminary fast freezing of solutions at 77 K in liquid nitrogen is necessary to obtain sharp emission spectra. For matrix isolation spectroscopy, PAH were sublimed and mixed with N_2 gas before trapping on a cold surface attached to the head of the cryogenerator.[2]

The light of the excitation source (Xenon lamp XBO Osram, 450 W) was dispersed by a monochromator (Jobin Yvon HR 20) and focused on the analyzed samples. Fluorescence emission was observed at 90° through a high resolution monochromator (Jobin Yvon HR 1000, resolution 0.8 Å).

RESULTS AND DISCUSSION

(a) Qualitative experiments on artificial mixtures

We have compared emission spectra obtained for artificial mixtures of isomeric PAH in *n*-alkane frozen solutions and in nitrogen matrix.

The first mixture was composed of the five monomethylphenanthrene isomers. In Shpolskii matrices, very sharp emission bands (bandwidth about 0.1 nm) are observed. Each compound can be detected by a specific emission band for selective excitation wavelength, by comparison with emission spectra of authentic compounds. With an excitation wavelength centered at 298.5 nm (bandwidths about 2.5 nm) (Figure 1), characteristic emission bands belonging to 9 MP, 4 MP and 3 MP are observed. In nitrogen matrix on the contrary, only one very intense broad band (bandwidth about 1.2 nm) is observed for any excitation wavelength, indicating clearly that monomethylphenanthrene individual emissions cannot be distinguished by this method.

A mixture of the six monomethylchrysene isomers was also analyzed. Each isomer is well detected in *n*-heptane frozen solutions by characteristic sharp fluorescence peaks (Figure 2). In nitrogen matrix, some isomers are only detected as shoulders on more intense fluorescence bands. A change in excitation wavelength conditions, does not allow a better observation of these compounds.

In conclusion, emission spectra of PAH in *n*-alkane matrices are more resolved and more structured than those obtained in nitrogen matrix and allow the identification of each compound in a complex mixture of isomers.

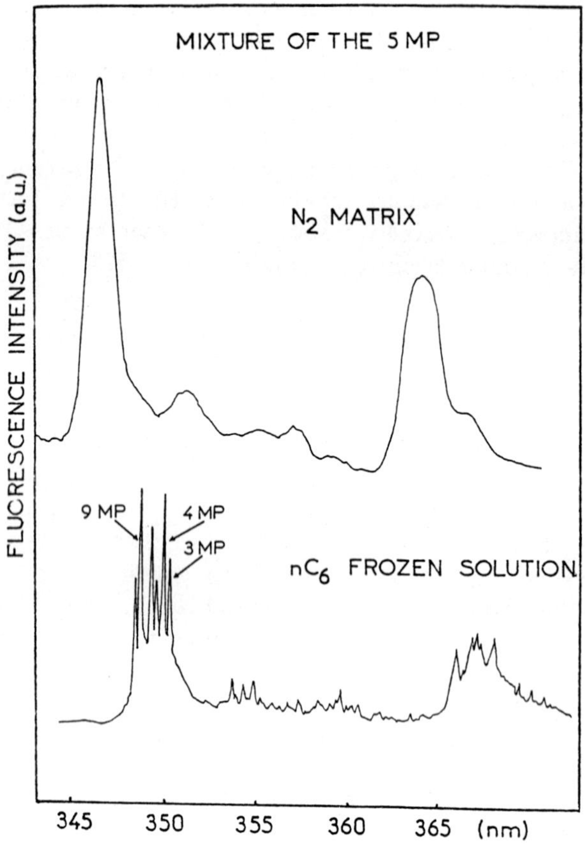

FIGURE 1 Fluorescence spectra at 15 K of monomethylphenanthrenes (MP) in a synthetic mixture:

— in nitrogen matrix: 200 ng of each isomer, excitation wavelength at 292 nm, molecular ratio $N_2/MP > 10^6$.

— in nC_6 frozen solution: $c = 2.10^{-7}$ M for each isomer, excitation wavelength at 298.5 nm.

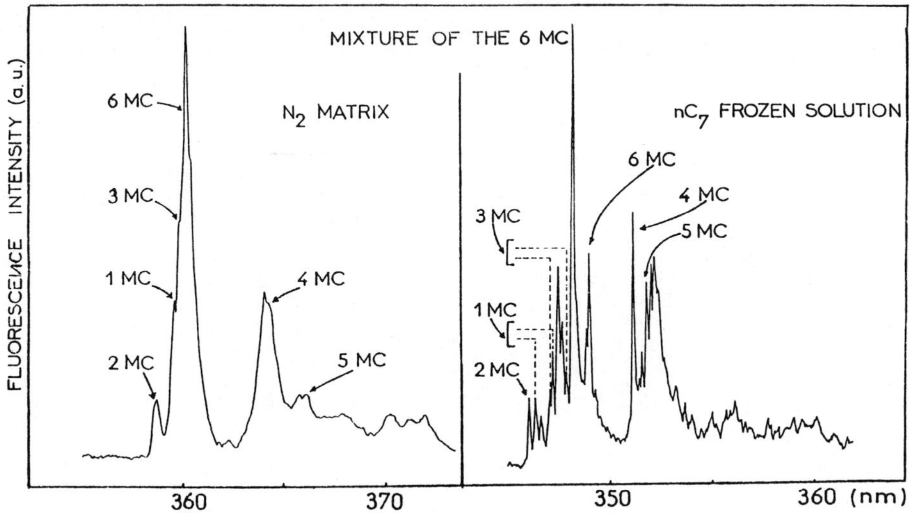

FIGURE 2 Fluorescence spectra at 15 K of the six monomethylchrysenes (MC) in a synthetic mixture:

— in nitrogen matrix: about 100 ng for each isomer, excitation wavelength at 261 nm, molecular ratio $N_2/MC > 10^6$.

— in nC_6 frozen solution: $c \sim 2.10^{-6}$ M for each isomer, excitation wavelength at 266 nm.

(b) Identification of monomethylphenanthrenes in natural samples

We have analyzed the monomethylphenanthrenes contained in fractions of petroleum and of marine sediment, obtained by a two step HPLC procedure. The optimum conditions for the characterization of each isomer were first defined by using a synthetic mixture with a judicious choice of the excitation wavelength and by recording either the fluorescence (for 9 MP, 4 MP and 3 MP) or the phosphorescence (for 2 MP and 1 MP). The spectra of natural samples compared with those of synthetic mixtures indicate the small amount of 4 MP both in petroleum and in marine sediment fractions (Figure 3). The fact that natural samples emission spectra are similar to those of synthetic mixtures, (i.e. good signal/noise ratio, no detectable interference from others solutes), points out the selectivity of the chromatographic separations and of the spectrofluorimetric analysis.

(c) Quantitative determination in the petroleum fraction

Due to different kinds of solute aggregations in Shpolskii matrices

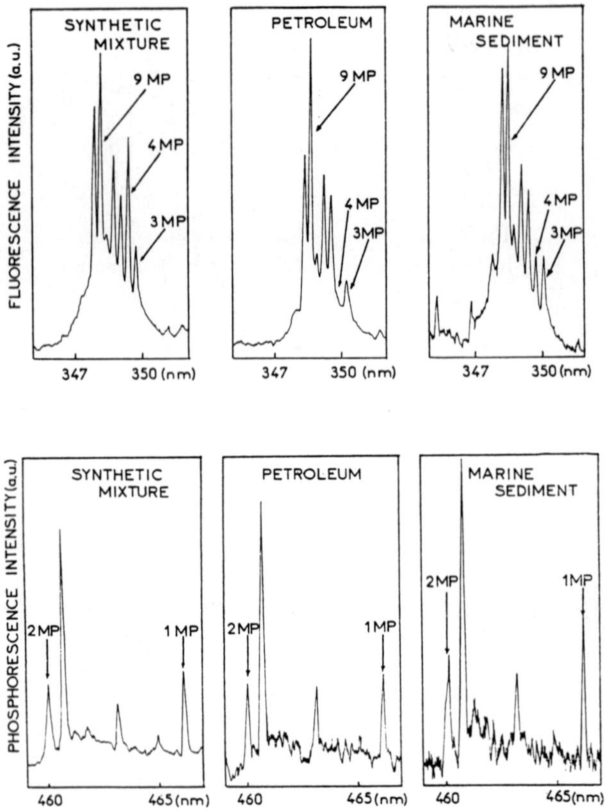

FIGURE 3 Emission spectra of monomethylphenanthrenes (MP) in a synthetic mixture (each MP isomer at 2×10^{-7} M) and in petroleum and marine sediment fractions, frozen in polycristalline *n*-hexane at 15 K. Excitation at 299 nm for fluorescence spectra. Excitation at 296.5 nm for phosphorescence spectra. Each isomer is identified by one characteristic emission peak.

occurring at high concentration ($c > 10^{-5}$M), fluorescence intensity depends linearly in concentration only below $c = 10^{-6}$ M.[10,11]

Quantitative determinations have been carried out by the combined standard addition-internal standard calibration procedure.[12]

This method has been discussed for the analysis of monomethylphenanthrenes extracted from natural samples.[13]

This method has been extended here to the quantitative analysis of a North Sea petroleum fraction. The standard addition calibration curve is calculated for each isomer by a mean square method and the concentration of each compound in the sample is determined by extrapolation. Analyses have been also carried out by GC-MS which allows an identification of each monomethylphenanthrene.[14] Preliminary results on the relative distribution of MP isomers obtained respectively by these two methods are in good agreement (Figure 4). Works are in progress to determine concentration of each isomer with standard deviations in several natural samples analyzed both by HRS and GC-MS.

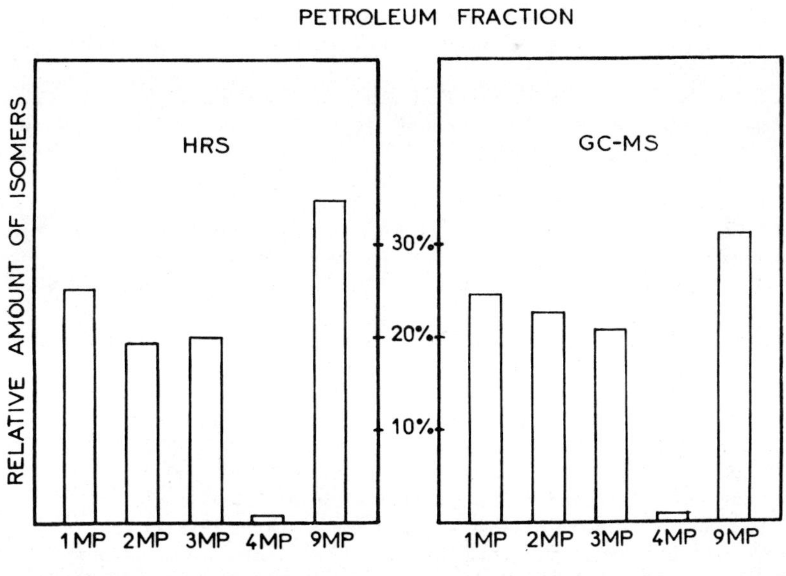

FIGURE 4 Relative distribution of monomethylphenanthrene isomers in North Sea petroleum fraction, determined by HRS and by GC-MS (fragmentographic detection at $m/z = 192$). (Standard deviations on the results obtained with several different samples are under examination and will be published later, including a discussion on the relative accuracy of the two methods HRS and GC-MS.)

In conclusion, results presented in this paper demonstrate the capability of the high resolution spectrofluorimetry in Shpolskii matrices at 15 K in solving identification and quantification problems occurring in complex mixtures of isomeric PAH. Further quantitative applications of this technique will be done on biogeochemical markers of the phenanthrene and chrysene series, derivatives of triterpenoid compounds, which have been also identified by HRS in marine sediments.[8,14]

Acknowledgements

Miss J. Bellocq is acknowledged for technical assistance.

Mr. J. L. Oudin from Compagnie Française des Pétroles and Mr. D. Jonathan from Société Nationale Elf Aquitaine (Production) are acknowledged for financial support and the gift of petroleum samples. Marine sediments were obtained during the Orgon cruises sponsored by the "Centre d'Etudes de Géochimie Marine" (CEGMA), which is acknowledged for partial support.

References

1. E. V. Shpolskii, *Sov. Phys. Usp* **5,** 522 (1962).
2. E. L. Wehry and G. Mamantov, *Anal. Chem.* **51,** 643 A (1979).
3. R. Lapouyade, A. Veyres, N. Hanafi, A. Couture and A. Lablache-Combier, *J.* Org. *Chem.* in press.
4. A. Saliot, M. J. Tissier, A. Fevrier, D. Tusseau, M. Ewald and C. Boussuge, *Géochimie Organique des Sédiments Marins Profonds, Orgon II, Atlantique Nord-Est Brésil,* Ed. CNRS, Paris, pp. 233–238 (1978).
5. M. Ewald, *Géochimie Organique des Sédiments Marins Profonds, Orgon II, Atlantique Nord-Est Brésil,* Ed. CNRS, Paris, pp. 285–292 (1978).
6. M. Ewald, M. Lamotte, P. Garrigues, J. Rima, A. Veyres and R. Lapouyade, *Proc Xth International Meeting on Organic Geochemistry.* Bergen (Norway) 1981.
7. S. A. Wise, S. N. Chesler, H. S. Hertz, L. R. Hilpert and W. E. May, *Anal. Chem.* **49,** 2306 (1977).
8. M. Ewald, M. Lamotte, F. Redero, M. J. Tissier and P. Albrecht, *Phys. Chem. Earth* **12,** (Adv. Org. Geochemistry 1979), Eds. A. G. Douglas, J. R. Maxwell, Pergamon Press, pp. 275–279 (1980).
9. P. Garrigues, M. Lamotte, M. Ewald and J. Joussot-Dubien, *C.R. Acad. Sci., Ser. II* **293,** 567 (1981).
10. J. Rima, M. Lamotte and A. M. Merle, *Nouv. J. Chim.* **5,** 605 (1981).
11. Y. Yang, A. P. D'Silva and V. A. Fassel, *Anal. Chem.* **53,** 894 (1981).
12. T. Ya Gaevaya and A. Ya Khesina, *J. Anal. Chem. USSR* **29,** 1913 (1974).
13. J. Rima, M. Lamotte and J. Joussot-Dubien (to be published in *Anal. Chem.* (1982)).
14. M. Ewald, A. Moinet, J. Bellocq, P. Wehrung and P. Albrecht, "*Geochimie Organique des Sédiments Marins Profonds, Orgon IV, Golfe d'Aden, Mer d'Oman,*" Ed. CNRS Paris, pp. 405–414 (1981).

Analysis of Multiring Aromatic Hydrocarbons by HPLC Using a Reverse-Phase Radial Compression Column System†

ROBERT ALEXANDER, KIMBERLEY M. CUMBERS and ROBERT I. KAGI

Petroleum Geochemistry Group, School of Applied Chemistry, Western Australian Institute of Technology, Kent Street, BENTLEY 6102, Western Australia

(*Received November 26, 1981*)

KEY WORDS: Aromatic hydrocarbons, Liquid chromatography, Effluent waters, Marine Biota, Soot.

The performance of a radially compressed C_{18} reverse-phase column system for analysis of multiring aromatics has been assessed in terms of efficiency, resolution and analysis time, and these parameters are compared with those reported for other column systems. The optimum conditions for analysis were established using nine standard aromatic compounds. These conditions are aslo shown to be suitable for analysis of complex mixtures of aromatic compounds obtained from petroleum refinery effluent waters, marine biota and combustion residues (soot).

INTRODUCTION

Analysis of aromatic compounds in samples of effluent waters and marine biota is of continuing interest.[1,2] HPLC techniques have proven particularly useful for the analysis of polynuclear aromatics (PNA's) because of the speed, sensitivity and specificity which can be achieved, particularly when used in conjunction with UV fluorescence detectors.[3,4]

---

†Presented at the workshop on the Chemistry and Analysis of Hydrocarbons in the Environment, Barcelona, November 1981.

Separations using normal-phase column packing materials have been widely used,[5,6] however the more recently-developed bonded-phase packings offer significant practical advantages to the analyst in terms of reduced activation cycles and speed of analysis. In this paper we report the use of a radially-compressed non-polar column for the simultaneous, rapid analysis of aromatics ranging from dinuclear compounds through to six-ring PNA's.

EXPERIMENTAL

HPLC analysis

All measurements were performed using a Varian 8500 dual pump LC system equipped with a VARISCAN variable wavelength UV detector. The column system consisted of a WATERS ASSOCIATES Radial Compression Module equipped with a RADIAL PAK C_{18} Cartridge.

Sample volumes were usually 1.5 μl, and the optimum flow rate was 1.5 ml min^{-1}. The initial composition of the solvent was 62% acetonitrile in water. This was continued for ten minutes, then the acetonitrile content was increased at 4% per minute for four minutes, followed by constant composition for five minutes, and a final increase to 100% acetonitrile at the initial rate of 5% per minute for four minutes and finally at a rate of 2% per minute.

GC-MS analysis

All measurements were made using a Hewlett-Packard 5985B system equipped with a 25 m × 0.25 mm ID, fused-silica capillary column coated with SP2100 liquid phase. The column was programmed from an initial temperature of 60°C, to a final temperature of 270°C at 4°C/min. Hydrogen was used as carrier gas at a flow velocity of 30 cm sec^{-1}.

Standards

All compounds used as standards were of analytical grade and were purchased from Unilab (naphthalene), BDH (phenanthrene), KOCH-LIGHT (pyrene, fluoranthrene, chrysene, benzo(α)pyrene, benzo(g,h,i)perylene), SIGMA (perylene, dibenz(a,h)anthracene).

Sample preparation

Samples of oil refinery effluent water, marine biota and soot were extracted with dichloromethane using established procedures.[7,8] The concentrated organic extract was then subjected to column chromatography using silicic acid and the fraction eluting from the column with pentane-ether (20:1) was retained. This fraction was further purified by medium pressure liquid chromatography using a SILICA 60 MERCK column. The compounds which had retention times between those of naphthalene and perylene eluted from the column with hexane-dichloromethane (7:1) solvent were retained for further HPLC analysis.

RESULTS AND DISCUSSION

In order to provide a direct comparison of the performance of the RADIAL PAK C_{18} radial compression system with that reported recently for other column systems,[9] performance parameters were measured under isocratic conditions with perylene and benzo(α)pyrene as substrates. Table I contains the retention time (tr), plate number (N) and HETP (H) data for the two substrates, together with resolution values (Rs) determined using acetonitrile-water (70:30) as solvent. Comparison of the performance of the radial compression system with that reported by Amos for two other reverse-phase columns[9] shows that its performance is intermediate between that of the other two. The radial compression system does however meet the minimum requirements specified by this author for analysis of PNA's, in that resolution is greater than 1.5 and the longer retention time is less than 30 minutes.

TABLE I

Performance parameters for reverse-phase columns using isocratic conditions, (a) $Rs=2\ (tr_2 - tr_1)/w_2+w_1$ where tr_1 and tr_2 represent the retention times and w_1 and w_2 the base peak width of the two compounds and (b) data from Reference 9.

Column	Perylene			Benzo(α)pyrene			Rs^a
	t_r(min)	N (plates)	H(mm)	t_r(min)	N (plates)	H(mm)	
Radial Pac C_{18}	16.0	2200	0.054	18.3	2400	0.050	1.7
μ Bondapack C_{18}[b]	22.7	4303	0.058	25.0	4303	0.058	1.6
HC-ODS-Sil-x-1[b]	9.3	1697	0.15	13.7	1764	0.14	4.0

Samples often contain PNA's because of contamination by material of petrogenic origin. Such samples usually contain di- and trinuclear aromatic compounds in addition to the higher ring homologues. To analyse the complete range of aromatic compounds, the analysis conditions were optimised using a solvent gradient. Figure 1 shows the traces obtained from chromatographs of a set of standard aromatic compounds ranging from two to six aromatic rings using a UV detector at three different wavelengths. From these traces it is apparent that efficiency

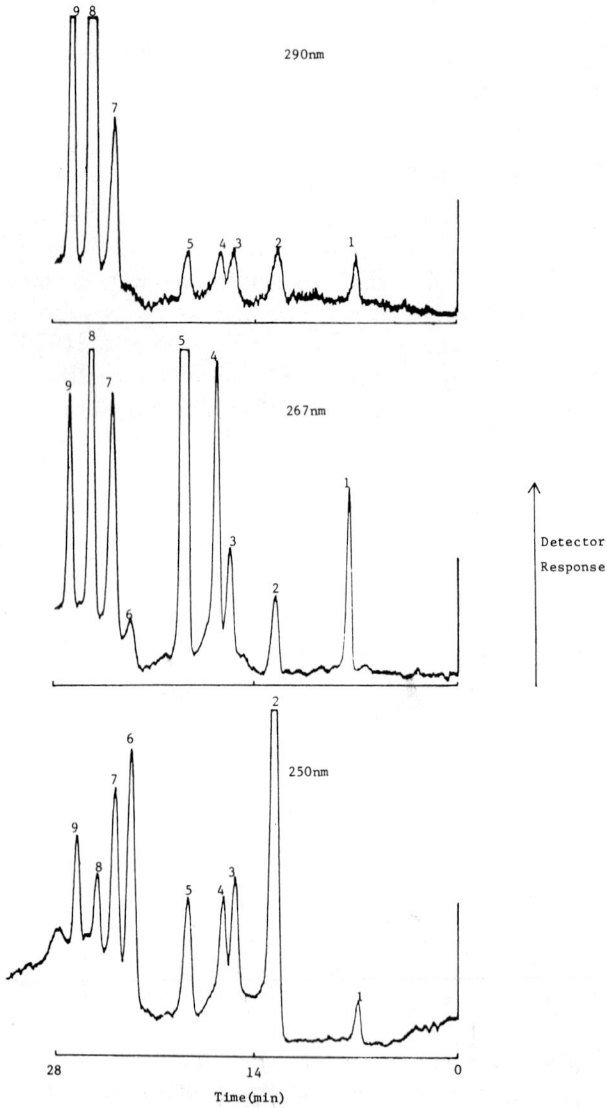

FIGURE 1 HPLC traces showing separation of aromatic standards using gradient elution and the detector response at three wavelengths. Numbers represent naphthalene (1), phenanthrene (2), fluoranthrene (3), pyrene (4), chrysene (5), perylene (6), benzo(α) pyrene (7), dibenz(a, h)anthracene (8) and benzo(g, h, i)perylene (9).

and resolution are maintained at suitable levels throughout the analysis, that the analysis can be completed in under thirty minutes, and that the most suitable wavelength for detection of these compounds is 267 nm.

Figure 2 shows traces obtained from the aromatic fraction of extracts

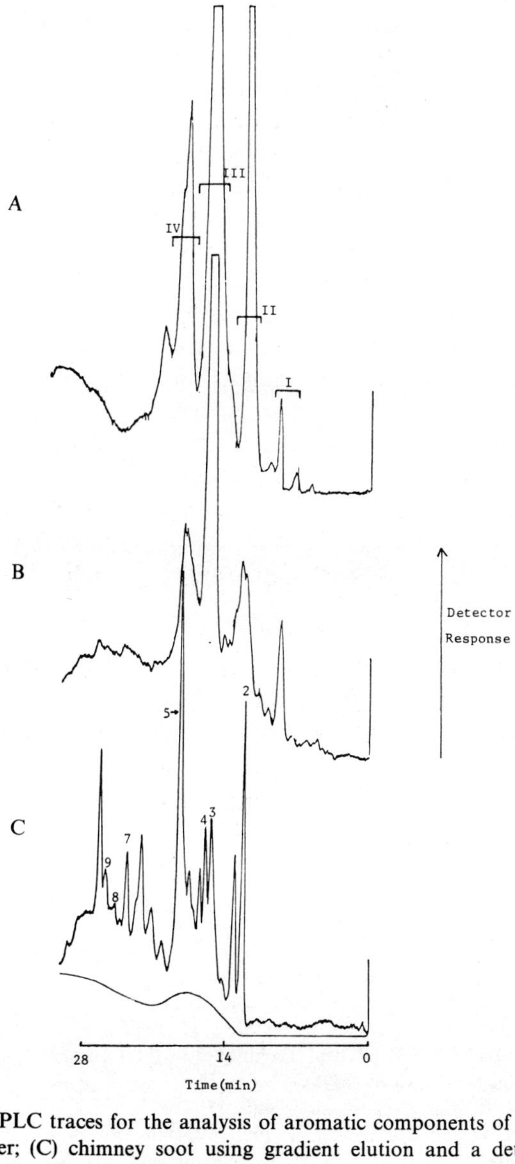

FIGURE 2 HPLC traces for the analysis of aromatic components of extracts of (A) mullet; (B) effluent water; (C) chimney soot using gradient elution and a detection wavelength of 267 nm. Peaks are numbered as described for Figure 1.

from fish (A), refinery effluent water (B) and chimney soot (C). The refinery effluent waters originate from a petroleum refinery and are discharged into a semi-enclosed bay. The fish sample was obtained from a specimen of mullet (*Aldrichetta foresteri*) captured within the effluent mixing zone adjacent to the effluent outfall. Traces A and B both show characteristics of petrogenic aromatic hydrocarbons. Peaks with retention times indicated by I coincide with those of naphthalene and methylnaphthalenes while peak II retention times coincide with dimethynaphthalenes and phenanthrene. These assignments were verified by collecting the fractions eluting over these time intervals and subjecting them to further analysis by capillary GC-MS. This technique provided further evidence that peak III represents both trimethylnaphthalenes and methylphenanthrenes, and peak IV dimethylphenanthrenes. This order of elution is also consistent with that reported by other workers.[10] Such a distribution of aromatic compounds is typical of crude oil[11] and suggests that the mullet has been contaminated by petroleum contained in the effluent waters. Trace C shows the results obtained from analysis of a sample obtained from chimney soot. Assignment of peaks was carried out by by coinjection with an authentic standard: the numbers on this trace represent the same compounds as those in Figure 1. One major difference between Trace C and Trace A and Trace B is that Trace C contains much sharper peaks. This is the result of the differences in composition of aromatic hydrocarbons derived from combustion sources from those of petrogenic origin.

Usually, the petrogenic compounds contain a greater variety of alkylsubstituted aromatic rings than do typical combustion-derived aromatics which contain predominantly unsubstituted aromatic compounds.[12,13] The resultant narrower retention time intervals in the HPLC trace for the combustion-derived sample may prove to be of value in deciding between possible sources of contamination by aromatic compounds.

References

1. B. K. Afghan and A. W. Wolkoff, *J. Liq. Chromatogr.* **4,** 99 (1981).
2. D. W. Grant and R. B. Meiris, *J. Chromatogr.* **203,** 293 (1981).
3. K. Ogan, E. Katz and W. Slavin, *Anal. Chem.* **51,** 1315 (1979).
4. R. C. Lao and R. S. Thomas, Analytical Techniques in Environmental Chemistry, Ed. J. Albaiges, p. 97, Pergamon, Oxford (1978).
5. R. K. Sorrell and R. Reding, *J. Chromatogr.* **185,** 655 (1979).
6. J. P. Durane and N. Petroff, *J. Chromatogr.* **190,** 85 (1980).
7. R. J. Law, Oily Water Discharges. Eds., C. S. Johnston and R. J. Morris, p. 167, Applied Science, London (1980).

8. D. G. Shaw and B. A. Baker, *Environ. Sci. Technol.* **12,** 1200 (1978).
9. R. Amos, *J. Chromatogr.* **204,** 469 (1981).
10. J. Chmiclowiec, J. E. Beshai and A. E. George, *Fuel,* **59,** 838 (1980).
11. B. P. Tissot and D. H. Welte, *Petroleum Formation and Occurrence* p. 351, Springer-Verlag, Berlin (1978).
12. J. L. Lake, C. Norwood, C. Dimock and R. Bowen, *Geochim. Cosmochim. Acta.* **43,** 1847 (1979).
13. R. E. Laflamme and R. A. Hites, *Geochim. Cosmochim. Acta.* **42,** 289 (1978).

Capillary Column Gas Chromatography of Environmental Polycyclic Aromatic Compounds†

MILTON L. LEE, DANIEL L. VASSILAROS, and DOUGLAS W. LATER
Department of Chemistry, Brigham Young University, Provo, Utah 84602

(*Received November 26, 1981*)

The methodologies are described for isolating clean fractions of polycyclic aromatic compounds from diverse environmental samples such as air particulate matter, sediments, and fish tissue. The common step in all procedures is the separation of the polycyclic aromatic compound fraction into well-defined chemical classes by adsorption chromatography on an alumina column. These procedures greatly facilitate the detailed characterization of the polycyclic aromatic hydrocarbons, sulfur heterocycles, and nitrogen heterocycles by capillary column gas chromatography.

KEY WORDS: Capillary column, gas chromatography, polycyclic aromatic compounds, air particulates, sediment, fish.

INTRODUCTION

It has long been known that chemical agents present in fossil fuels and produced during combustion of organic matter cause skin carcinomas in man. In addition, exposure of workers to vapors of synfuels (such as coal-derived liquids), coke-oven emissions, and coal processing emissions has been correlated with the increased incidence of cancers of other organ systems.[1] The polycyclic aromatic compounds (PAC) have been identified as major contributors to this activity. One sub-class of PAC, the

†Presented at the Workshop on the Chemistry and Analysis of Hydrocarbons in the Environment, Barcelona, November 1981.

polycyclic aromatic hydrocarbons (PAH), has been the most extensively studied in the past, but research to identify and assess the health effects of environmental PAC containing heteroatoms of nitrogen, oxygen, and sulfur is presently being pursued in many laboratories. These heterocyclic species have been shown in many cases to be the most significant environmental hazards. For example, the amino polycyclic aromatic hydrocarbons (APAH) have recently been shown to be the determinant mutagens in synthetic fuels,[2–4] and the nitro polycyclic aromatic hydrocarbons (NPAH) have been identified as direct-acting mutagens in carbon blacks[5] and diesel exhaust.[6,7] Several recent studies have also corroborated the observation that the polycyclic aromatic sulfur heterocycles (PASH) may be the most persistent of the PAC in the environment.[8–10] The importance of this observation is emphasized in light of recent tests [1,11] that show the PASH to be significantly more toxic than either the PAH or PANH (polycyclic aromatic nitrogen heterocycles).

The main pathways for human uptake of environmental PAC are ingestion, inhalation, or absorption through the skin. PAC are transported from the pollution source to man through the air and water. PAC can be inhaled or absorbed from air as a vapor, aerosol (such as in a coal conversion facility), or more commonly, as molecules adsorbed on particulate matter originating from an emission source. PAC enter the aqueous environment from a variety of sources including oil pollution, fallout from air pollution, effluents from industries and sewage treatment plants, and storm drain runoff. Man is most likely to encounter PAH in water by ingestion of contaminated food (such as fish) or drinking water.

In order to assess the effects of PAC on the environment and man, detailed characterization of PAC mixtures in various sample types must be done. Analysis of the distribution of these compounds in air particulates, water, sediments, and biological materials is important. The ideal analytical methodology would be one that could be uniformly applied to all sample types with comparable high quality results. Unfortunately, the vastly different sample matrices preclude the ideal situation, and modifications in procedures must be made for specific sample types. Nevertheless, it is still desirable to standardize as much of the analytical procedure as possible.

In this paper, the chemical class separation of PAC found in three different matrices (air particulate matter, sediment, and fish tissue) is described. The common aspects of the analytical methodology are the class separation of PAC by alumina column adsorption chromatography and the final separation of components by capillary column gas chromatography.

EXPERIMENTAL

Sample fractionation

Air particulate matter. A 15-g quantity of urban air particulate matter which was collected in the Washington, D.C., area over a 12-month period by the National Bureau of Standards, was extracted in a Soxhlet apparatus for 12 hrs with 250 ml of 1:3 methanol/benzene (v:v). The Soxhlet thimble and apparatus were pre-extracted with the same solvent mixture for 4 hrs before actual sample extraction.

The extract was concentrated to approximately 5 ml using a rotary evaporator, and a 1-ml portion was pre-adsorbed onto 3 g of neutral aluminum oxide (Brockman Activity 1, 80-200 mesh, Fisher No. A950), transferred to the top of a 6-g alumina column (10 cm × 11 mm i.d.), and eluted with 100 ml of chloroform containing 0.75% ethanol. The eluate was then reduced to a volume of 1 ml by rotary evaporation. This fraction was then transferred to a 22 mm i.d. column containing 60 g of Bio-Beads SX-12 (200–400 mesh, Bio-Rad), and eluted with methylene chloride. The first 55 ml which contained principally aliphatic compounds were discarded, and the next 100 ml which contained the PAC were collected and reduced in volume to 1 ml. It was found that this two-step alumina-gel column chromatographic procedure produced a relatively clean PAC fraction.

The chemical class fractionation of the PAC fraction was accomplished as described previously[13] for coal liquids on neutral aluminum oxide. Two fractions were collected for analysis in this work: the neutral PAC fraction containing PAH and PASH, and a more polar fraction containing mostly PANH. A schematic diagram of the fractionation procedure is given in Figure 1.

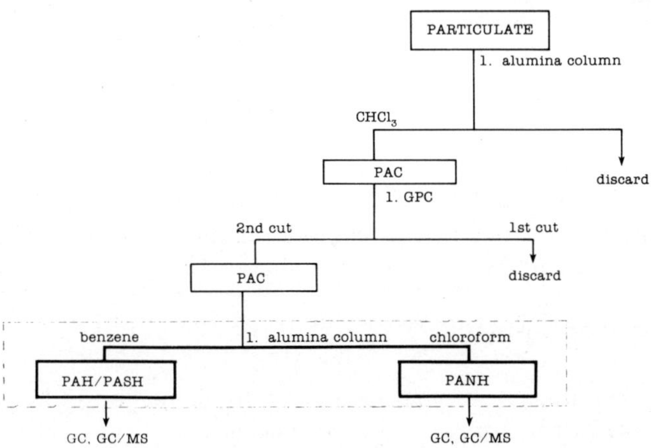

FIGURE 1 Fractionation scheme for air particulate matter.

Fish Tissue. Composited whole fish samples of a brown bullhead catfish (*Ictalurus nebulosus*) from the Black River in Ohio and a striped bass (*Morone saxatillis*) from the Sacramento River in California were provided by the U.S. Fisheries and Wildlife Service, Columbia National Research Laboratory, Columbia, MO. These samples were analyzed as previously described[14] and shown in Figure 2. Again, two fractions were collected from each sample for analysis: a PAH/PASH fraction and a PANH fraction.

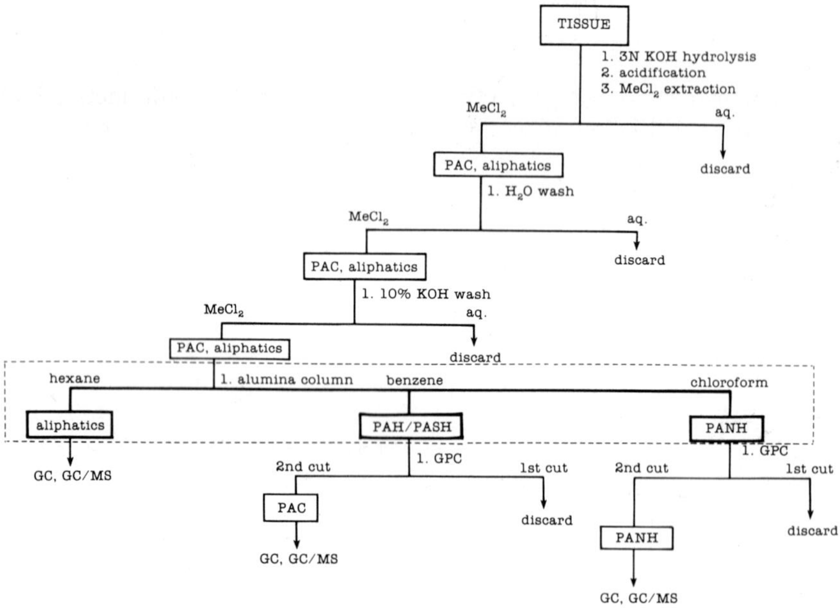

FIGURE 2 Fractionation scheme for fish tissue.

River Sediment. A river sediment sample from the Black River, also provided by the U.S. Fisheries and Wildlife Service, was extracted and fractionated according to the same scheme as shown in Figure 2, except the acidification step following the KOH hydrolysis and the acqueous washes following the methylene chloride extraction were omitted. Both PAH/PASH and PANH fractions were obtained.

Gas chromatography

Capillary column gas chromatography with flame ionization detection (FID) was performed using a Hewlett–Packard 5880 gas chromatograph with splitless vaporization injection. A flame photometric detector (FPD) on a Perkin-Elmer Sigma III gas chromatograph was used to obtain chromatograms of the PASH. Fused silica capillary columns coated with SE-52 were prepared in this laboratory.

Mass spectral data of resolved components were obtained using a Hewlett–Packard 5982A quadrupole mass spectrometer which was operated in the electron impact mode at 70 eV electron energy. Spectra were acquired and processed with a Hewlett–Packard 5934A data system.

RESULTS AND DISCUSSION

As discussed earlier, no single analytical procedure is applicable to all sample types. In comparing Figures 1 and 2, it can be seen that the common step in the fractionation schemes is the alumina column separation of the PAC fraction into specific chemical classes. The virtues of this procedure have been discussed in detail elsewhere and applied to the characterization of PAC in coal liquids.[13] The main advantage is that clean, well-defined chemical classes of PAC are separated from each other for further chemical analysis. The various other steps in each fractionation scheme are designed to solve particular separation problems characteristic of the sample matrix analyzed. For instance, the two-step alumina and gel column clean-up was found essential to remove polar organic materials from the air particulate extracts before clean class separations could be obtained from the second alumina column. In contrast to this, the base hydrolysis, aqueous wash, and gel permeation steps were essential for separating the PAC from lipids and oils which were present in large amounts in the fish tissue.[14] When analyzing samples that are composed almost entirely of PAC, such as coal liquids, no fractionation steps are required prior to the alumina column class separation.[13]

Figures 3–5 show representative chromatograms of the PAH/PASH fractions isolated from air particulates, sediment, and fish tissue, respectively. All three fractions are almost entirely free of non-PAC compounds. This is essential for identifying PAC by retention measurements alone,[15] comparing chromatographic profiles, or facilitating identification by gas chromatography-mass spectrometry. Compounds identified in these fractions are listed in Table I. It is interesting to note the relatively high concentrations of PASH and alkylated PAC in the Black River fish. The ratio of phenanthrene concentration to dibenzothiophene concentration in the fish was 9:1 as compared to a ratio of 19:1 in the sediment (a sample of which was taken in the same vicinity as where the fish was caught). These results support other reports of selective bioconcentration of PASH in aquatic organisms.[8–10]

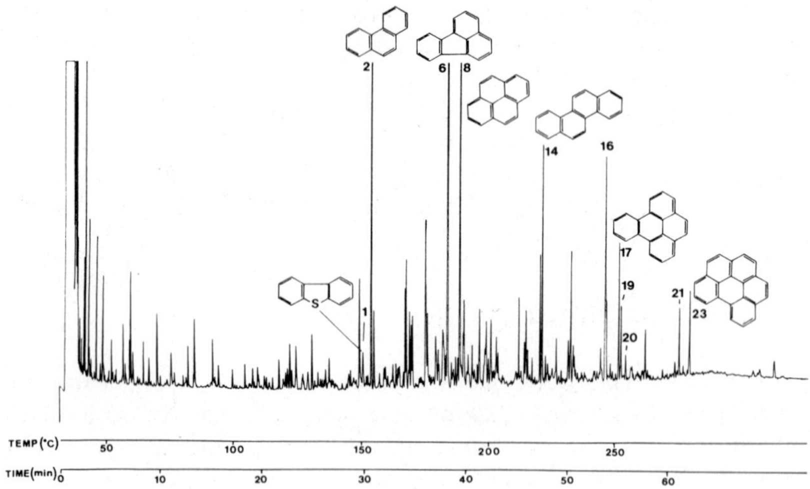

FIGURE 3 Capillary column gas chromatogram (FID) of PAH/PASH fraction from Washington air particulate matter. Conditions: 25 m × 0.20 mm i.d. fused silica capillary column coated with SE-52 (0.17 μm film thickness); oven temperature held at 40°C for 2 min, and then programmed at 4°C/min to 260°C; hydrogen carrier gas.

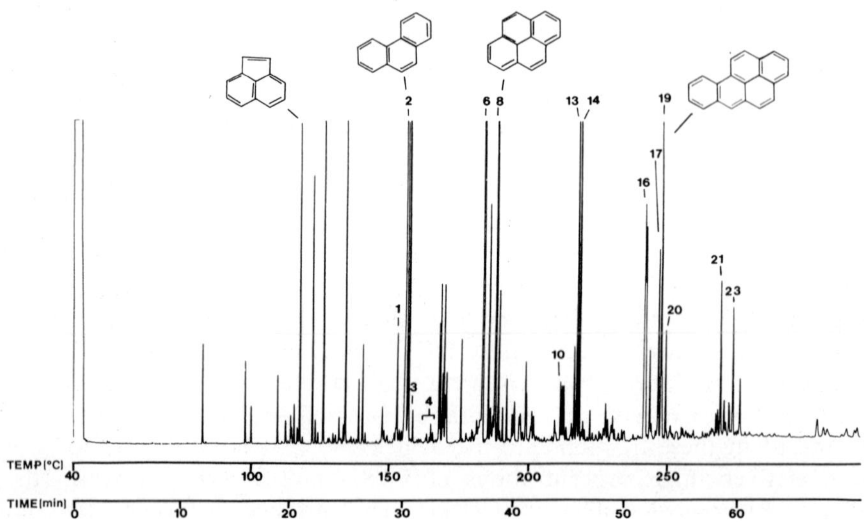

FIGURE 4 Capillary column gas chromatogram (FID) of PAH/PASH fraction from a Black River sediment. Conditions: 25 m × 0.20 mm i.d. fused silica capillary column coated with SE-52 (0.17 μm film thickness); oven temperature held at 40°C for 2 min, and then programmed at 4°C/min to 260°C; hydrogen carrier gas.

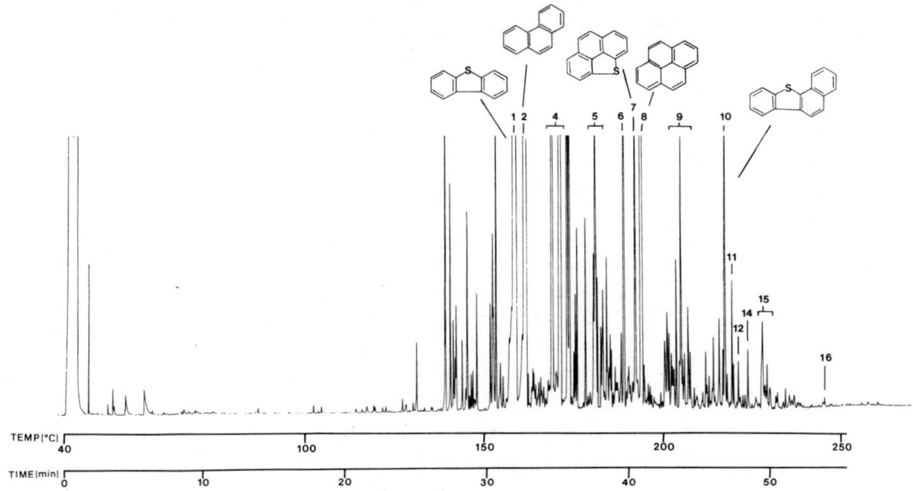

FIGURE 5 Capillary column gas chromatogram (FID) of PAH/PASH fraction from Black River bullhead catfish. Conditions: 20 m × 0.30 mm i.d. fused silica capillary column coated with SE-52 (0.34 μm film thickness); oven temperature held at 40°C for 2 min, and then programmed at 4°C/min to 260°C; hydrogen carrier gas.

Sulfur-selective chromatographic detection of the PASH in the neutral PAC fractions provides clearer comparisons of the distribution of sulfur-containing compounds in these fractions (Figures 6–8). A rather high relative concentration of alkylated PASH can be observed in the Sacramento River fish.

Representative chromatograms of the PANH fractions from air particulates, sediment, and fish tissue are shown in Figures 9–11. Compounds identified are listed in Table II. In all three samples, various carbazoles and azaarenes have been identified. It is also interesting to note the presence of several PAC containing keto-functional groups in the sediment and fish samples. The isoquinolines appear to be the major PANH components in the air particulate extract.

The efficient and reproducible methods described here for analyzing PAC in various sample matrices are important for proper utilization of high resolution gas chromatographic techniques. This is ultimately important for studies leading to the further understanding of the possible human health effects of environmental PAC.

TABLE I
Compounds identified in the PAH/PASH fractions of Washington air particulate, Black River Sediment, and Sacramento and Black River fish.

Peak #[a]	Compound Name
1	Dibenzothiophene
2	Phenanthrene
3	Naphtho[2,3-*b*]thiophene
4	Methyldibenzothiophenes
5	C_2-Dibenzothiophenes
6	Fluoranthene
7	Phenanthro[4,5-*bcd*]thiophene
8	Pyrene
9	Methylphenanthro[4,5-*bcd*]thiophenes
10	Benzo[*b*]naphtho[2,1-*d*]thiophene
11	Benzo[*b*]naphtho[1,2-*d*]thiophene
12	Benzo[*b*]naphtho[2,3-*d*]thiophene
13	Benz[*a*]anthracene
14	Chrysene/Triphenylene
15	Methyl 4-ring thiophenes
16	Benzo[*b*] & [*k*]fluoranthenes
17	Benzo[*e*]pyrene
18	Peri-condensed 5-ring thiophenes
19	Benzo[*a*]pyrene
20	Perylene
21	Indeno[1,2,3-*cd*]pyrene
22	cata-condensed 5-ring thiophenes
23	Benzo[*ghi*]perylene

[a]Peak numbers refer to peaks labelled in Figures 3–8.

TABLE II
Compounds identified in the PANH fractions of Washington air particulate, Black River Sediment, and Sacramento and Black River fish.

Peak #[a]	Compound Name
1	9H-Fluoren-9-one
2	Acenaphthoquinone
3	Benzo[*h*]quinoline
4	Acridine
5	Phenanthridine (Int. Std.) and Benzo[*f*]quinoline
6	Carbazole
7	Methylbenzoquinolines
8	Methylcarbazoles
9	Anthraquinone
10	Azafluoranthenes and Azapyrenes
11	Benzo[*def*]carbazole
12	Naphthoquinolines
13	Benzo[*a*]carbazole
14	7H-benz[*de*]anthracen-7-one
15	Benzocarbazole

[a]Peak numbers refer to peaks labelled in Figures 9–11.

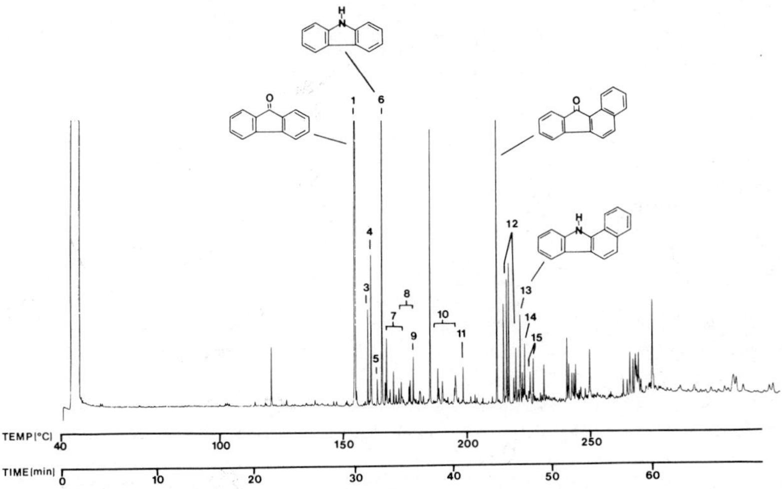

FIGURE 10 Capillary column gas chromatogram (FID) of PANH fraction from Black River sediment. Conditions: 25 m × 0.20 mm i.d. fused silica capillary column coated with SE-52 (0.17 μm film thickness); oven temperature held at 40°C for 2 min, and then programmed at 4°C/min to 260°C; hydrogen carrier gas.

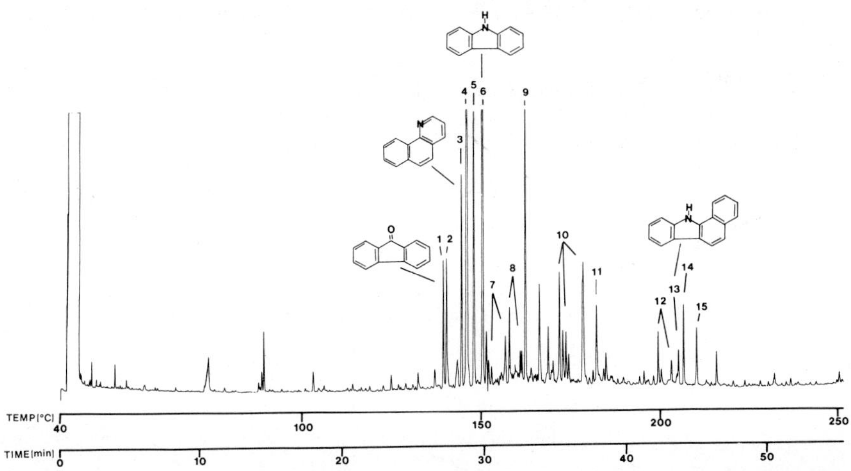

FIGURE 11 Capillary column gas chromatogram (FID) of PANH fraction from Black River bullhead catfish. Conditions: 25 m × 0.20 mm i.d. fused silica capillary column coated with SE-52 (0.17 μm film thickness); oven temperature held at 40°C for 2 min, and then programmed at 4°C/min to 260°C; hydrogen carrier gas.

Acknowledgement

The authors thank Stephen A. Wise of the National Bureau of Standards, Washington, D.C., for providing the sample of Washington air particulates. The authors also thank Larry Ludke, Tom May, Mike Ribick, and Paul Baumann of the U.S. Fisheries and Wildlife Service, Columbia National Fisheries Research Laboratory, Columbia, MO, for providing the fish and sediment samples.

References

1. H. M. Braunstein, E. D. Copenhaver, and H. A. Pfuderer, *Environmental, Health, and Control Aspects of Coal Conversion: An Information Overview* (Oak Ridge National Laboratory, Oak Ridge, TN, 1977), Vol. 2.
2. B. W. Wilson, R. A. Pelroy, and J. T. Cresto, *Mut. Res.* **79,** 193 (1980).
3. M. R. Guerin, C.-h. Ho, T. K. Rao, B. R. Clark, and J. L. Epler, *Environ. Res.* **23,** 42 (1980).
4. B. W. Wilson, M. R. Petersen, R. A. Pelroy, and J. T. Cresto, *Fuel* **60,** 289 (1981).
5. H. S. Rosenkranz, E. C. McCoy, D. R. Sanders, M. Butler, D. K. Kiriazides, and R. Mermelstein, *Science* **209,** 1039 (1980).
6. D. Schuetzle, F. S. C. Lee, T. J. Prater, and S. Tejada *Intern. J. Environ. Anal. Chem.* **9,** 93 (1981).
7. X. B. Xu, J. P. Nachtman, S. M. Rappaport, E. T. Wei, S. Lewis, and A. L. Burlingame, *J. Appl. Toxicol.* **1,** 196 (1981).
8. M. Ogato, Y. Miyake, K. Fuhisawa, and Y. Yoshida, *Bull. Environ. Contam. Toxicol.* **25,** 130 (1980).
9. F. Berthou, Y. Gourmelun, Y. Dreano, and M. P. Friocourt, *J. Chromatogr.* **203,** 279 (1981).
10. D. L. Vassilaros, D. A. Eastmond, W. R. West, G. M. Booth, and M. L. Lee, *Proc. 6th Intl. Symp. on Polynuclear Aromatic Hydrocarbons*, Columbus, Ohio, 1981.
11. G. M. Booth, Brigham Young University, unpublished results.
12. M. L. Lee, M. V. Novotny, and K. D. Bartle, *Analytical Chemistry of Polycyclic Aromatic Compounds* (Academic Press, New York, 1981), 446 pp.
13. D. W. Later, M. L. Lee, K. D. Bartle, R. C. Kong, and D. L. Vassilaros, *Anal. Chem.* **53,** 1612 (1981).
14. D. L. Vassiiaros, P. W. Stoker, G. M. Booth, and M. L. Lee, *Anal. Chem.* **54,** 106 (1982).
15. M. L. Lee, D. L. Vassilaros, C. M. White, and M. Novotny, *Anal. Chem.* **51,** 768 (1979).

Analysis of Complex Mixtures of Aromatic Hydrocarbons. Relations between Retention Index and Molecular Structure†

J. BERMEJO, J. S. CANGA and O. M. GAYOL

Instituto Nacional de Carbón y sus Derivados "Francisco Pintado Fe" (C.S.I.C.), Oviedo, Spain

(*Received November 27, 1981*)

Relationships between Kováts indices on nonpolar stationary phases and the structure of alkyl- and alkenylbenzenes, defined by the molecular connectivity or connectivity index, have been established.

Almost all components of aromatic naphthas are included in the different linear equations which relate both empirical indices. The constant terms of the straight lines depend on the nature and position of the substituents in the benzene ring. Substituents in *para* position do not influence each other, allowing the study of the effect of its nature on the ratio $\delta I/\Delta\chi$.

Equations for different series of compounds, including *n*-alkyl- and 1,4-di-*n*-alkylbenzene homologous series, are calculated in order to support its identification in aromatic naphthas.

KEY WORDS: Gas chromatography, Kováts indices, molecular connectivity, molecular structure, aromatic hydrocarbons.

INTRODUCTION

Molecular connectivity is a concept intended for the quantification of the topology of molecules. It was introduced by Randić[1] and widely described by Kier and co-workers[2–4] through the topological branching index, which shows correlations with several molecular properties. This topological branching index, also called connectivity index, "χ", has been used by Kaliszan and co-workers[5–8] to investigate the relationships with

†Presented at the Workshop on the Chemistry and Analysis of Hydrocarbons in the Environment. Barcelona, November 1981.

the chromatographic behaviour of several types of organic compounds, finding good correlations with chromatographic parameters such as the Kováts index, "I", particularly when nonpolar stationary phases are used.

It is known that "I" values are related with the interactions between the solute molecules and the stationary phase. When saturated hydrocarbons are chromatographed on nonpolar stationary phases, only dispersion forces are significant. In this case, the molecular connectivity is able to describe the chromatographic behaviour more exactly than any other empirical parameter.[9] In addition connectivity indices are very easy to calculate.

Aromatic naphthas are complex industrial products whose principal components are alkylbenzenes. Since we deal with aromatic hydrocarbons which are compounds with low polarity, it is supposed that if they are chromatographed on nonpolar stationary phases, in order to minimize the polar interactions, it will be possible to obtain a considerable help in the identification of such components from the relationships between chromatographic retention and connectivity.

The present paper deals with our investigation concerning the relationships between Kováts indices and connectivity indices of alkylbenzenes on squalane at 100°C and alkenylbenzenes on methylsilicone (SE-30) at 65°C.

EXPERIMENTAL

The connectivity index was calculated according to the Kier and Hall's[4] equation $\chi = \Sigma(\delta i \cdot \delta j)^{-1/2}$ where δi and δj are the number of nonhydrogen bonds of the bonded "*i*" and "*j*" carbons. For example, connectivity of isopropylbenzene is calculated as follows:

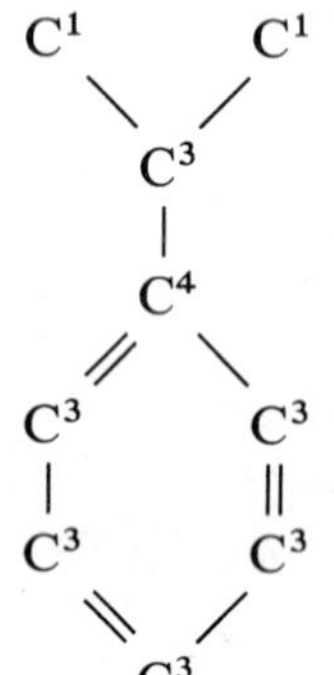

$$\chi = 1/(1.3)^{1/2} + 1/(1.3)^{1/2} + 1/(4.3)^{1/2} + 1/(4.3)^{1/2} + 1/(4.3)^{1/2}$$
$$+ 1/(3.3)^{1/2} + 1/(3.3)^{1/2} + 1/(3.3)^{1/2} + 1/(3.3)^{1/2} = 3.353$$

"*I*" values on squalane at 100°C are available in the literature[10–13] although, when it was possible, data from Engewald and Wennrich[11] were preferably used in order to secure homogeneity of the results. All data about alkenylbenzenes were taken from Svob and Deur-Siftar[12] who obtained them on methylsilicone (SE-30) at 65°C.

RESULTS

Data about alkylbenzenes on squalane and alkenylbenzenes on methylsilicone (SE-30) are presented in Figure 1. This figure shows that there are different linear relationships between "*I*" and "χ" in both nonpolar phases.

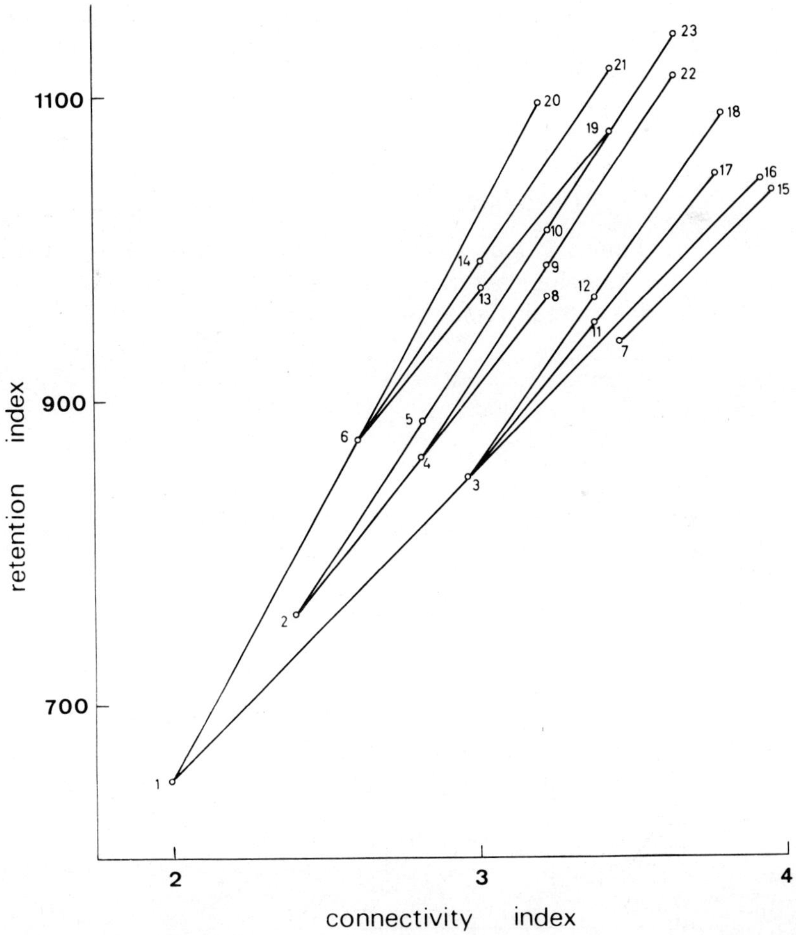

FIGURE 1 Plot of Kovàts indices "*I*" against connectivity indices "χ" for alkylbenzenes on squalane at 100°C and alkenylbenzenes on methylsilicone (SE-30) at 65°C. Numbers correspond to the components quoted in Tables I–IV.

The constant terms of the straight regression lines depend on the nature of the substituents and their relative positions in the benzene ring. The only lines where benzene is included appear defined only by three points belonging to the series: benzene, R-benzene and 1,4-RR-benzene. The linearity of these points in the plot "I" versus "χ", states that the values $\delta I/\Delta\chi$ ($\delta I = I_2 - I_1$ and $\Delta\chi = \chi_2 - \chi_1$) do not vary when a second substituent is introduced in the *para position. On the contrary when a second substituent enters in an ortho* position the ratio $\delta I/\Delta\chi$ increases, whereas a very small decrease is observed if the substituent occurs in a *meta* position.

For studying the dependency between $\delta I/\Delta\chi$ values and the nature of the substituent, the series of 1,4-disubstitutions, in which the substituents are not influenced each other, are appropriated. The corresponding equations and their respective constants are shown in Table I. For alkylbenzenes the $\delta I/\Delta\chi$ values increase in the following order:

$$R = \text{methyl} > \text{ethyl} > n\text{-} > \text{tert-} > \text{iso-} > \text{sec-alkyl}$$

As it is well known,[11] the value of I_{CH_2} (index difference per methylen group) from the homologous series of n-alkylbenzenes present a minimum for the second and the third addition of a methylene group (Figure 2). On

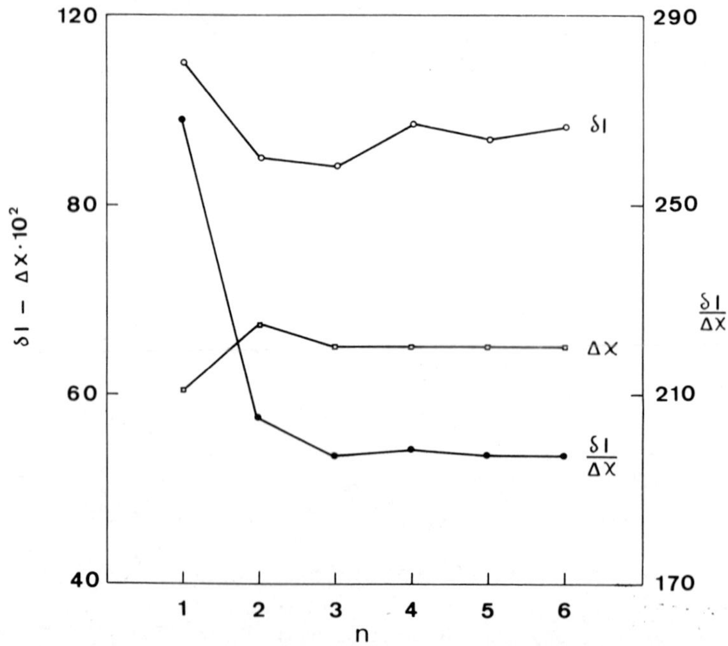

FIGURE 2 Dependence of index increments per CH_2-group (on squalance at 100°C) (δI), connectivity increments ($\Delta\chi$) and $\delta I/\Delta\chi$, on carbon number (n) of alkyl benzene substituent.

TABLE I

Equations $I=f(\chi)$ for series benzene, R-benzene and 1,4-RR-benzene

Components (see Figure 1)	I	χ	$I=a\chi+b$
benzene (1)	654.8	2.000	$a=362.4$
vinylbenzene (6)	874.9	2.606	$b=-69.9$
1,4-divinylbenzene (20)	1095.9	3.217	$r=1.00000$
benzene (1)	650.5	2.000	$a=260.4$
methylbenzene (2)	760.1	2.409	$b=131.6$
1,4-dimethylbenzene (4)	864.6	2.822	$r=0.99986$
benzene (1)	650.5	2.000	$a=200.7$
ethylbenzene (3)	850.0	2.970	$b=250.7$
1,4-diethylbenzene (16)	1040.5	3.943	$r=0.99990$
benzene (1)	650.5	2.000	$a=191.0$
n-propylbenzene (7)	938.0	3.470	$b=270.8$
1,4-di-*n*-propylbenzene	1212.5	4.943	$r=0.99990$
benzene	650.5	2.000	$a=185.9$
isopropylbenzene	908.4	3.353	$b=280.8$
1,4-diisopropylbenzene	1154.1	4.709	$r=0.99989$
benzene (1)	650.5	2.000	$a=192.8$
n-butylbenzene (15)	1036.8	3.970	$b=267.0$
1,4-di-*n*-butylbenzene	1410.8	5.943	$r=0.99995$
benzene	650.5	2.000	$a=189.8$
tert-butylbenzene	971.7	3.659	$b=272.9$
1,4-di-tert-butylbenzene	1281.3	5.323	$r=0.99993$
benzene	650.5	2.000	$a=172.7$
sec-butylbenzene	990.2	3.892	$b=309.3$
1,4-di-sec-butylbenzene	1304.3	5.785	$r=0.99974$
benzene	650.5	2.000	$a=179.8$
isobutylbenzene	991.3	3.825	$b=295.1$
1,4-diisobutylbenzene	1307.7	5,655	$r=0.99975$

the other hand, the introduction of a methyl group in the benzene ring increases the "χ" value by 0.409, the first methylene group by 0.561 and the second and following by 0.500. Consequently, the $\delta I/\Delta\chi$ value decreases when the second methylene group enters, becoming constant after the propylbenzene.

The order within each group of the alkylbenzene isomers is explained because of the different variation of their retention indices (*n*- > iso- > sec- > tert-) and their connectivities (*n*- > sec- > iso > tert-).

Effects of the substituents in ortho position

When a second group enters in an *ortho* position it originates an increase in the $\delta I/\Delta\chi$ ratio respective of the value determined by the first substituent, whatever the nature of both groups is.

In Table II it is shown that the compound series methyl-, 1,2-dimethyl-, 1,2,3-trimethyl-, 1,2,3,4-tetramethyl- and pentamethylbenzene defines a straight line, in a plot $I \overset{\text{vs}}{—} \chi$, with a regression coefficient >0.9999, in spite of that the third methyl group enters in an *ortho* or *meta* substituted position. The line belonging to the 1,4-dimethylbenzene and its trimethyl- and tetramethyl- derivatives obtained by consecutive introduction of a methyl group in a carbon atom whose *meta* and *ortho* position are occupied, has the same slope as the preceding line. Both series show again that the substituents have very little influence on *meta* and *para* positions (Table II).

When the methyl group enters in a position flanked by other methyl groups (*ortho* and *ortho* substituted position) the slope of the equation $I = f(\chi)$ increases as a logic consequence of the double influence of the two methyl groups in *ortho* positions. This happens whatever the entering group and the substituent groups are, although the effect is greater when the substituent groups are methyl groups than when they are other alkyl or alkenyl groups.

The plot of the series of trisubstituted benzenes, where two of the substituents are methyl and the other is a vinyl or ethyl group, shows two straight lines almost parallels to the ordinate axis because of the small difference in the "χ" values which these isomers have (Figure 3). The "I" values for the dimethyl-ethyl isomers, fall in the expected order according to the slopes of the lines in Table II, which take on the greatest value when a methyl group is introduced in a position which is *ortho* methylated or flanked by methyl groups.

In the group of the dimethyl-vinyl isomers there are some deviations from the expected order. The most important belong to the 1,3-dimethyl-2-vinylbenzene whose location in the last place of the series has to be considered as an anomaly.

TABLE II

Equations $I = f(\chi)$ for series originated when an R-group enters in an ortho substituted position or in an ortho flanked position. Indices from References 10 and 12.

Components (see Figure 1)	I	χ	$I = a\chi + b$
methylbenzene (2)	756.6	2.409	$a = 301.7$
1, 2-dimethylbenzene (5)	882.9	2.827	$b = 30.3$
1, 2, 3-trimethylbenzene (10)	1011.1	3.243	$r = 0.99998$
1, 2, 3, 4-tetramethylbenzene (23)	1133.9	3.660	
pentamethylbenzene	1260.1	4.077	$r = 0.99938$
1, 4-dimethylbenzene (4)	860.8	2.822	$a = 301.0$
1, 2, 4-trimethylbenzene (9)	985.5	3.239	$b = 11.2$
1, 2, 3, 5-tetramethylbenzene (22)	1111.8	3.656	$r = 1.00000$
ethylbenzene (3)	846.9	2.970	$a = 269.0$
1-methyl-2-ethylbenzene (12)	963.9	3.388	$b = 49.6$
1,2-dimethyl-3-ethylbenzene (18)	1071.2	3.804	$r = 0.99972$
vinylbenzene (6)	874.9	2.606	$a = 290.7$
1 methyl-2-vinylbenzene (14)	989.0	3.024	$b = 114.9$
1, 3-dimethyl-2-vinylbenzene (21)	1117.9	3.442	
1, 3, 5-trimethylbenzene (8)	969.0	3.234	$a = 352.0$
1, 2, 3, 5-tetramethylbenzene (22)	1113.1	3.656	$b = -171.7$
pentamethylbenzene	1260.1	4.077	$r = 0.99987$
hexamethylbenzene	1415.0	4.500	
1, 3-dimethylbenzene (4)	864.8	2.822	$a = 351.8$
1, 2, 3-trimethylbenzene (10)	1012.9	3.243	
1-methyl-3-ethylbenzene (11)	947.9	3.383	$a = 331.4$
1, 2-dimethyl-3-ethylbenzene (18)	1087.4	3.804	
1-methyl-3-vinylbenzene (13)	974.0	3.019	$a = 291.0$
1, 2-dimethyl-3-vinylbenzene	1097.1	3.442	

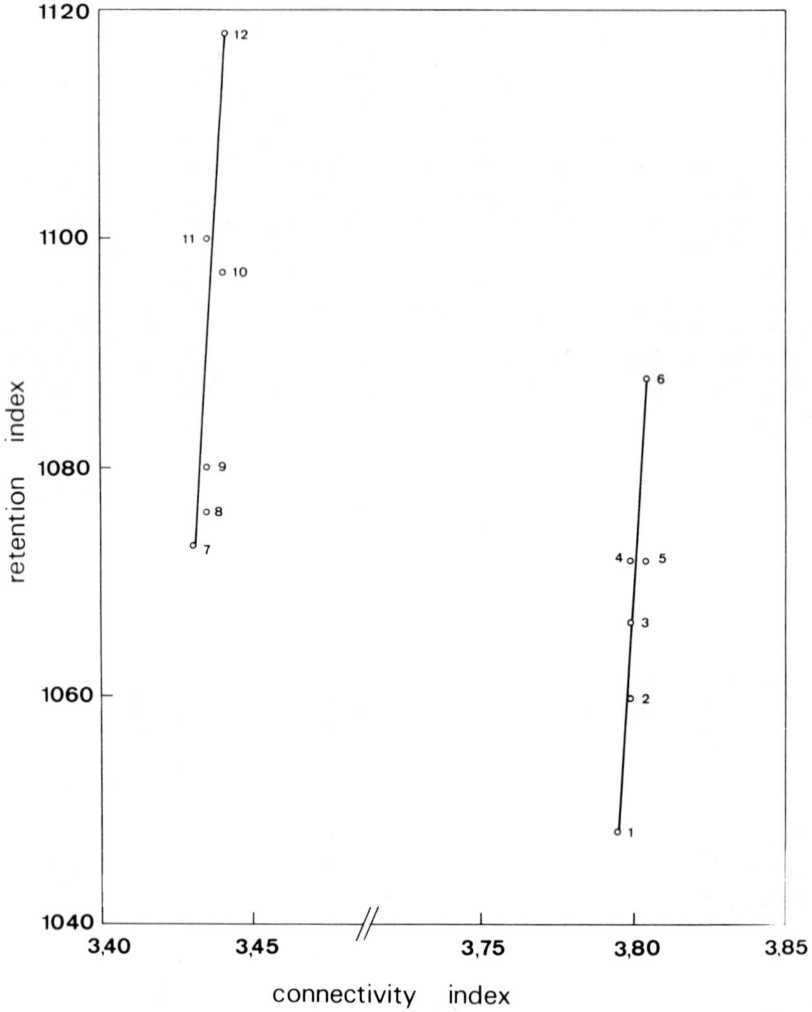

FIGURE 3 Relationships between "*I*" and "χ" in the series of dimethyl-ethylbenzenes and dimethyl-vinylbenzenes.

(1) 1, 3-dimethyl-5-ethylbenzene;
(2) 1, 4-dimethyl-2-ethylbenzene;
(3) 1, 3-dimethyl-4-ethylbenzene;
(4) 1, 2-dimethyl-4-ethylbenzene;
(5) 1, 3-dimethyl-2-ethylbenzene;
(6) 1, 2-dimethyl-3-ethylbenzene;
(7) 1, 3-dimethyl-5-vinylbenzene;
(8) 1, 4-dimethyl-2-vinylbenzene;
(9) 1, 3-dimethyl-4-vinylbenzene;
(10) 1, 2-dimethyl-3-vinylbenzene;
(11) 1, 2-dimethyl-4-vinylbenzene;
(12) 1, 3-dimethyl-2-vinylbenzene.

Homologous series of straight chain mono- and 1,4-dialkylbenzenes

As a consequence of the variation of "I" and "χ" shown in Figure 2, the correlation decreases when toluene and ethylbenzene are included. However, when the series start from the *n*-propylbenzene, the regression coefficient is >0.9999 for all series shown in Table III.

TABLE III

Equations $I=f(\chi)$ for homologous series mono- and 1,4-*n*-dialkylbenzenes

Components (see Figure 1)	I	χ	$I=a\chi+b$
n-propylbenzene (7)	938.0	3.470	$a=195.0$
n-butylbenzene (15)	1036.8	3.970	$b=261.6$
n-pentylbenzene	1133.0	4.470	$r=1.00000$
n-hexylbenzene	1231.0	4.970	
1-methyl-4-propylbenzene	1040.2	3.883	$a=195.4$
1-methyl-4-butylbenzene	1138.8	4.383	$b=281.8$
1-methyl-4-pentylbenzene	1236.3	4.883	$r=1.00000$
1-methyl-4-hexylbenzene	1333.4	5.383	
1-ethyl-4-propylbenzene	1126.4	4.443	$a=195.8$
1-ethyl-4-butylbenzene	1227.1	4.943	$b=257.3$
1-ethyl-4-pentylbenzene	1322.1	5.443	$r=0.99994$
1-ethyl-4-hexylbenzene	1421.1	5.943	
1,4-di-*n*-propylbenzene	1212.5	4.943	$a=198.3$
1-propyl-4-butylbenzene	1311.0	5.443	$b=232.1$
1-propyl-4-pentylbenzene	1410.8	5.943	$r=1.00000$

The I_{CH_2} values originated in the sequence toluene-ethylbenzene-*n*-propylbenzene are the same for all 1,4-dialkylbenzene series, for example, for 1-methyl-4-*n*-butyl-, 1-ethyl-4-*n*-butyl- and 1-*n*-propyl-4-*n*-butylbenzene.

Table IV shows other series of alkyl- and alkenylbenzenes with *meta* substitutions, which give also straight lines with very good regression coefficients, therefore they can give support in the identification of components of aromatic naphthas.

TABLE IV

Equations $I=f(\chi)$ for series of *meta* substitutions. Indices from Reference 12

Components (see Figure 1)	I	χ	$I=a\chi+b$
methylbenzene (2)	758.0	2.409	$a=254.1$
1, 3-dimethylbenzene (4)	863.2	2.822	$b=146.0$
1, 3, 5-trimethylbenzene (8)	967.6	3.234	$r=1.00000$
ethylbenzene (3)	847.7	2.970	$a=243.3$
1-methyl-3-ethylbenzene (11)	948.6	3.383	$b=125.1$
1, 3-dimethyl-5-ethylbenzene (17)	1048.2	3.794	$r=1,00000$
propylbenzene	936.1	3.470	$a=235.6$
1-methyl-3-propylbenzene	1033.4	3.883	
isopropylbenzene	908.0	3.353	$a=228.6$
1-methyl-3-isopropylbenzene	1002.4	3.766	
tert-butylbenzene	972.9	3.659	$a=214.9$
1-methyl-3-tert-butylbenzene	1055.8	4.072	$b=184.6$
1, 3-dimethyl-5-tert-butylbenzene	1150.2	4.484	$r=0.99927$
vinylbenzene (6)	874.0	2.606	$a=240.0$
1-methyl-3-vinylbenzene (13)	974.0	3.019	$b=249.4$
1, 3-dimethyl-5-vinylbenzene (19)	1072.9	3.431	$r=1.00000$
allylbenzene	929.3	3.080	$a=239.0$
1-methyl-3-allybenzene	1028.0	3.493	
isoallylbenzene	1004.2	3.109	$a=237.5$
1-methyl-3-isoallylbenzene	1102.3	3.522	

CONCLUSIONS

The multiple relationships found between the chromatographic retention indices on nonpolar stationary phases and the molecular structure, defined by the molecular connectivity, for an extensive series of alkyl- and alkenylbenzenes, the very good correlations between both parameters found for all the series, and the easy calculation of the connectivity indices, make possible to set up a method for the identification of these compounds in environmental or industrial samples, namely in aromatic naphthas.

This method is less time-consuming and has advantages in material requirements and precision over other identification methods based on ΔI, δI or homomorphic factors, and their relations with the polarity of the stationary phase.[11,13,14]

References

1. M. Randić, *J. Am. Chem. Soc.* **97,** 6609 (1975).
2. W. J. Murray, L. H. Hall and L. B. Kier, *J. Pharm. Sci.*, **64,** 1978 (1975).
3. L. H. Hall, L. B. Kier and W. J. Murray, *J. Pharm. Sci.* **64,** 1974 (1975).
4. L. B. Kier and L. H. Hall, *Molecular Connectivity in Chemistry and Drug Research.* Academic Press, New York, N.Y., 1976.
5. R. Kaliszan, *Chromatographia,* **10,** 529 (1977).
6. R. Kaliszan and H. H. Lamparczyk, *J. Chromatogr. Sci.* **16,** 246 (1978).
7. R. Kaliszan. *Chromatographia* **12,** 171 (1979).
8. A. Radecki, H. Lamparczyk and R. Kaliszan, *Chromatographia* **12,** 595 (1979).
9. M. Gassiot-Matas and G. Firpo-Pamies, *J. Chromatogr.* **187,** 1 (1980).
10. L. Soják and J. A. Rijks, *J. Chromatogr.* **119,** 505 (1976).
11. W. Engewald and L. Wennrich, *Chromatographia,* **9,** 540 (1976).
12. V. Svob and D. Deur-Siftar, *J. Chromatogr.* **91,** 677 (1974).
13. J. Bermejo, S. R. Moinelo and C. Suárez Canga, *Publicacion INCAR no. 6.* Octubre, 1980.
14. S. D. West and C. H. Randall, *J. Chromatogr., Sci.* **13,** 5 (1975).

Analytical Standards and Methods for the Determination of Polynuclear Aromatic Hydrocarbons in Environmental Samples†

W. E. MAY, S. N. CHESLER, H. S. HERTZ and S. A. WISE

Organic Analytical Research Division, Center for Analytical Chemistry, National Measurement Laboratory, National Bureau of Standards, Washington, DC 20234, U.S.A.

(*Received March 21, 1982*)

Standard reference materials (SRM's) have been produced, certified, and issued by the United States National Bureau of Standards (NBS) since 1905. NBS currently issues more than 1000 SRM's of various types, including nuclear materials, rubber, clinical and environmental trace metal standards. The most recent addition to this group is a series of environmental trace organic materials with certified concentrations of selected polynuclear aromatic hydrocarbons (PAH), phenols, and N-heterocyclic compounds. Until recently, trace organic SRM's were non-existent due to the lack of analytical methodology necessary for certification. Details concerning the analytical methods developed and used for certification of the concentrations of several PAH in SRM's 1580 (Organics in Shale Oil), 1644 (Generator Columns for PAH in water), 1647 (PAH in Acetonitrile), and 1649 (Urban Particulate Matter) are given along with some suggested uses for these SRM's.

KEY WORDS: Gas chromatography/mass spectrometry, generator columns, high performance liquid chromatography, polynuclear aromatic hydrocarbons, priority pollutants, shale oil, standard reference materials, urban particulate matter.

---

†Paper presented at the Workshop on the Chemistry and Analysis of Hydrocarbons in the Environment. Barcelona, November 1981.

INTRODUCTION

We have previously reported on the state-of-the-art in the frontier area of quantitative trace organic analysis.[1] In recent years, the Organic Analytical Research Division of the National Bureau of Standards (NBS) has been involved in a number of collaborative analytical studies whose purpose was to further assess the accuracy of data obtained from the trace organic analysis of environmental samples. The results of these studies revealed the existence of biases among laboratories involved in the measurement of polycyclic aromatic hydrocarbons (PAH), phenols, and nitrogen containing PAH compounds in environmental samples.[2-4] It has been demonstrated, particularly in the trace metals area, that the use of appropriate quality assurance standards or Standard Reference Materials (SRM's) can enhance the accuracy of analytical measurements.[5] However, until recently, SRM's for environmental trace organic analyses were nonexistent due to the lack of the analytical methodology necessary for certification.

At NBS three modes of measurements have historically been used to assure that certified values of SRM properties are accurate. These are: (a) definitive methods, (b) reference methods, (c) two or more independent and reliable methods.

In the first mode, measurement of the property is made using a method of analysis based on "first principles". A definitive method is one in which all major significant parameters have been related directly, or by a solid chain of evidence, to the base or derived units of the International System of Units (SI). Direct determination by mass is an example of a definitive method. At NBS, isotope dilution mass spectrometry (ID/MS) has been used as a definitive method for the measurement of organic and inorganic constituents at trace levels in several matrices.[5] In this procedure, all chemical manipulations are done on a weight basis. All mass spectrometric determinations involve ratios, not absolute measurement of the individual isotopes. Therefore, no known instrumental corrections or errors are involved. However, ID/MS has not been investigated at NBS as a definitive method for environmental trace organic analysis mainly due to the unavailability of appropriate internal standards for many of the important analytes.

Reference methods (as defined by NBS) are those methods whose accuracy have been substantiated by comparison with a definitive method or by analysis of a previously certified SRM. These methods are generally more cost and time-effective than definitive methods, and usually do not require highly specialized equipment. Since reference methods require the existence of either a definitive method and/or an appropriate

SRM, environmental trace organic reference methods (as defined by NBS) are also non-existent.

The third possible mode for certification, and the one currently being used for the certification of environmental trace organic SRM's, is measurement by at least two totally independent and reliable methods. This approach is based on the rationale that the likelihood of two independent methods being biased by the same amount in the same direction is small. When the analytical results agree, we can state with some assurance that they are indeed accurate.

The interlaboratory comparison or "round-robin" mode is a subset of mode C and has also been used at NBS for certification of some SRM's. However, this procedure is normally used only for the purpose of reissuing an SRM, with the old certified SRM's being used by NBS as a quality control check on the participating laboratories. The use of this mode in an uncontrolled manner would yield results of questionable accuracy and poor precision as shown by the data given in Table I from an interlaboratory comparison of the determination of several PAH in a shale oil sample.

In this paper, details concerning the analytical methods developed and used for certification of the concentrations of several PAH in SRM's 1580 (Organics in Shale Oil), 1644 (Generator Columns for PAH in Water), 1647 (Priority Pollutant PAH in Acetonitrile), and 1649 (Urban Particulate Matter) are presented along with some suggestions for the use of these materials.

SRM 1580, ORGANICS IN SHALE OIL

The shale oil for this SRM is from a 150-ton retort for *in situ* simulated combustion of oil shale, operated by the Laramie Energy Technology Center, Laramie, Wyoming. The shale oil was obtained by Oak Ridge National Laboratory (ORNL) in November 1975, where it underwent centrifugation to separate water and sludge from the oil.

At NBS, the centrifuged sample was filtered through fine filter paper and mixed in a 20-liter, Teflon-stoppered, glass bottle by rolling for 40 hours. Samples were aliquoted into 2 mL amber glass ampoules. Although not intended to be representative of all shale oils, SRM 1580 provides a typical specimen of this matrix for use in developing analytical methods.

Randomly selected ampoules were analyzed in the certification process. Each analyst examined at least six ampoules and made a minimum of three replicate measurements from each ampoule. No correlation was found between the measured values and the ampouling sequence. The PAH measured in this sample are pyrene, fluoranthene, benzo[a]pyrene, benzo[e]pyrene, and perylene.

TABLE I

Interlaboratory comparison of shale oil analyses (Concentrations in $\mu g/g$).

		Laboratory									
Compound	NBS[a]	2	3[b]	3[c]	3[d]	4	5	6	7	8	9
fluoranthene	61 ± 8	102	247	75		80	108	126	122	116	380
pyrene	102 ± 5	155	161	138	147	150	168	177	141	185	620
benzo[a]pyrene	22 ± 2	—		19	13	—			3.3		
benzo[e]pyrene	21 ± 1								1.3		

[a]Results reported by NBS represent the mean of values obtained by GC, GC/MS, and HPLC uncertainties represent the standard deviation of the mean.
[b]Quantification by GC/MS.
[c]Quantification by HPLC.
[d]Quantification by laser excited spectroscopy.

Certification

Two methods of analytical sample preparation were used. Dilution of the sample in methylene chloride was the only sample preparation step necessary prior to analysis by gas chromatography/mass spectrometry (GC/MS). Samples to be analyzed by high performance liquid chromatography (HPLC) were prepared by HPLC fractionation. Samples containing approximately 14 mg of shale oil were injected onto a 300 mm × 8 mm aminosilane (NH_2) column using a loop injector. The mobile phase was composed of two to five percent methylene chloride in hexane. Standards of the compounds to be measured were injected prior to the shale oil to determine the appropriate elution volumes for fraction collection. The fractions were collected in centrifuge tubes and reduced to 50–500 μL by passing nitrogen over the sample.

The GC/MS determinations were performed using selected ion monitoring and a standard addition technique for quantification. The polynuclear aromatic hydrocarbons of interest were separated on a 30 m SE-52 wall coated open tubular column interfaced to a quadrupole mass spectrometer.

Liquid chromtographic analysis was performed using an instrument equipped with a gradient pumping system, loop injector, and a spectrofluorimetric detector. A digital integrator and strip chart recorder were used for data acquisition. As mentioned previously, PAH were initially isolated from the shale oil matrix by chromatography on an aminosilane column using hexanemethylene chloride as the mobile phase. The analytes were then separated from the PAH by chromatography on an octadecylsilane column (C-18) with a water-acetonitrile mobile phase. Further selectivity was obtained by monotoring the chromatographic effluent fluorometrically with excitation and emission wavelengths optimized for detection of each analyte. External standard and standard addition methods were used for quantification.

A summary of the results obtained using the two analytical methods are given in Table II. Further details concerning these methods can be found elsewhere.[3,4] The certified values for five PAH along with values for three phenols and one N-heterocyclic compound are given in Table III.

Uses

SRM 1580 is intended primarily for evaluating the reliability of analytical methods used for the determination of trace level organic compounds in an oil matrix. In addition to PAH, several phenols and N-heterocycles have been certified in SRM 1580. These compounds are representative of acidic, basic, and neutral organic species which are isolated in most extraction schemes.

TABLE II

Summary of results of analytical methods used in certification.

Compound	Conc.[a] (μg/g)	Sample preparation technique	Analytical technique
fluoranthene	55 .±5	Direct Injection	GC/MS
	53 ±2	HPLC	HPLC
pyrene	101 ±5	Direct Injection	GC/MS
	107 ±8	HPLC	HPLC
benzo[a]pyrene	20 ±1	Direct Injection	GC/MS
	23 ±1	HPLC	HPLC
benzo[e]pyrene	17 ±1	Direct Injection	GC/MS
	20 ±3	HPLC	HPLC
perylene	2.8±0.5	Direct Injection	GC/MS
	3.9±0.6	HPLC	HPLC

[a]Uncertainty is the standard deviation of a single measurement.

TABLE III

Certified values of PAH in SRM 1580.

Compound	Concentration (μg/g)[a]
fluoranthene	54 ±10
pyrene	104 ± 9
benzo[a]pyrene	21 ± 6
benzo[e]pyrene	18 ± 8
perylene	3.3± 1.1
phenol	407 ±50
o-cresol	385 ±50
2,6-dimethylphenol	175 ±30
5,6-benzoquinoline	16 ± 4

[a]The estimated uncertainty listed for a constituent is based on judgment and represents an evaluation of the combined effects of method imprecision, possible systematic errors among methods, and material homogeneity. The uncertainty given represents the best estimates of possible systematic errors and the 95% confidence level of the means.

As efforts to develop alternate fuels intensify, the number of laboratories required to perform single species quantitative analyses will increase dramatically, and new measurement techniques and methods will continue to evolve. It must be emphasized that SRM 1580 will not assure accurate analysis for PAH, phenols and N-heterocyclic compounds in complex environmental matrices; however this SRM can serve as a benchmark against which the accuracy of emerging analytical techniques can be evaluated. SRM 1580 can also serve as an intercalibration material on which laboratories can evaluate the comparability of methods for compounds and classes of compounds that have not yet been certified.

SRM 1644, GENERATOR COLUMNS FOR POLYNUCLEAR AROMATIC HYDROCARBONS

SRM 1644 consists of separate generator columns for anthracene, benz(a)anthracene, and benzo(a)pyrene. These columns are 50 cm × 0.6 cm stainless steel tubes packed with 80–100 mesh quintus quartz (sea sand) and coated with 0.5% (w/w) of the PAH compound of interest. These columns were prepared commercially under the technical direction of NBS.

Saturated solutions are generated by pumping HPLC grade water through the generator column at flow rates between 0.1 and 5 mL/min. Since the acqueous solubility is a well-defined thermodynamic quantity, saturated solutions are standard solutions.

Certification

Two independent analytical techniques were used to certify the concentration of anthracene, benz(a)anthracene and benzo(a)pyrene in the acqueous effluents from their respective generator columns. The first technique involved the quantitative extraction of the PAH from the aqueous stream by an "extractor column" packed with a C-18 bonded phase; the use of an acetonitrile-water eluent to transfer components from the extractor column to an analytical C-18 column for separation of the analyte from non-analyte components; and detection of the analyte by absorbance at 254 nm. Details concerning this method have been reported elsewhere.[6–8] The second technique involved the use of a "standard addition" technique for "on stream" analysis. The effluent from the generator column was mixed with PAH standards dissolved in acetonitrile, and the resultant analyte concentration in the mixed solvent determined by fluorescence.

These methods were used to measure the concentrations of anthracene, benz(a)anthracene and benzo(a)pyrene in the aqueous effluents from their respective generator columns at temperatures between 10 to 30°C for 20 columns randomly selected from the SRM pool of 240 columns for each compound. The data obtained were combined and fitted to an empirical expression of the form $\ln[\text{Conc}] = A + B(1/T) + C(1/T)^2$ by least squares. In this equation, ln [Conc] is the natural logarithm of the concentration, T is the absolute temperature, and A, B, and C are constants. The derived equations for anthracene, benz(a)anthracene, and benzo(a)pyrene were used to calculate the certified concentrations at one degree intervals between 10 and 30°C for each compound. These certified concentrations are given in Table IV. The certification of generator columns for Indeno(1,2,3-cd)pyrene is now in progress. These columns will provide aqueous solutions of this PAH at certified concentrations between 0.05 and 0.3 μg/kg over the 10–30°C temperature range.

Uses

SRM 1644 is intended to provide accurate concentrations of anthracene, benz(a)anthracene and benzo(a)pyrene in water. Because the acqueous solubility of PAH in general and the above in particular are very low (43, 9.1, 1.6 and 0.2 μg/kg) at 25°C, respectively, such solutions are difficult to prepare and impossible to preserve for long periods of time. The use of this SRM allows the user to circumvent these problems by generating the desired solution as needed.

TABLE IV

Temperature °C	Concentration and its uncertainty,[a] μg/kg		
	Anthracene	Benz(a)anthracene	Benzo(a)pyrene
10	16.6 ± 0.7	3.38 ± 1.2	0.59 ± 0.06
11	17.6 ± 0.6	3.60 ± 1.1	0.63 ± 0.05
12	18.7 ± 0.6	3.83 ± 0.91	0.67 ± 0.04
13	19.8 ± 0.5	4.09 ± 0.79	0.71 ± 0.04
14	21.1 ± 0.5	4.36 ± 0.68	0.76 ± 0.03
15	22.4 ± 0.5	4.65 ± 0.59	0.81 ± 0.03
16	23.8 ± 0.5	4.96 ± 0.54	0.87 ± 0.03
17	25.4 ± 0.5	5.29 ± 0.55	0.93 ± 0.03
18	27.0 ± 0.5	5.65 ± 0.60	0.99 ± 0.03
19	28.8 ± 0.5	6.04 ± 0.68	1.06 ± 0.03
20	30.7 ± 0.6	6.45 ± 0.77	1.13 ± 0.03
21	32.8 ± 0.6	6.90 ± 0.87	1.21 ± 0.04
22	35.0 ± 0.6	7.38 ± 0.94	1.30 ± 0.04
23	37.4 ± 0.6	7.90 ± 1.0	1.39 ± 0.04
24	39.9 ± 0.6	8.45 ± 1.0	1.49 ± 0.04
25	42.7 ± 0.6	9.05 ± 1.0	1.59 ± 0.04
26	45.7 ± 0.8	9.69 ± 1.0	1.71 ± 0.04
27	48.9 ± 1.0	10.4 ± 1.1	1.83 ± 0.05
28	52.4 ± 1.3	11.1 ± 1.2	1.96 ± 0.05
29	56.1 ± 1.7	11.9 ± 1.3	2.11 ± 0.07
30	60.1 ± 2.2	12.8 ± 1.7	2.26 ± 0.09

[a]The uncertainties are 99 percent (Working–Hotelling) confidence bands for the entire regression curve. The difference between the true and certified concentration should be less than the stated uncertainty at the 99 percent confidence level.

SRM 1647, PRIORITY POLLUTANT POLYNUCLEAR AROMATIC HYDROCARBONS (IN ACETONITRILE)

A 12.0 liter solution containing the 16 PAH on the U.S. Environmental Protection Agency's list of priority pollutants was prepared for NBS by a commercial vendor. Approximately 11 of 12 liters of this solution were dispensed (1.2 mL ampoule) into 9000 ampoules before shipment to NBS.

Samples representing the early, middle and final stages of ampouling were analyzed by HPLC immediately upon arrival at NBS. No significant differences in concentration for any of the 16 compounds was found. However, visual inspection of the residue remaining in the 12-L flask after the ampoules had been filled revealed the presence of some undissolved crystalline material. HPLC analysis of an aliquot of this residual solution revealed that the two compounds that did not dissolve completely were chrysene and dibenz(a),h)anthracene.

Certification

Three analytical methods were used in the certification of SRM 1647. The gravimentric values (corrected for compound purity) supplied by the commercial vendor were used for fourteen of the sixteen compounds. Gravimetric values for chrysene and dibenz(a,h)anthracene were rejected as a result of the preliminary HPLC determinations made on the ampouled and residual solution. HPLC separation on a polymeric C-18 column, using an acetonitrile-water mobile phase and UV (254 nm) determine response factors for each of the sixteen compounds. Six mixture were measured. Four external standard solutions were used to determine responcse factors for each of the sixteen compounds. Six replicate measurements were made on each of the external standard solutions and nine ampoules randomly selected from the SRM pool. Figure 1 is a typical chromatogram from the HPLC analysis of SRM 1647.

Gas chromatography on a fused silica SE-52 capillary column was used for the determination of eight of the sixteen compounds. Two standard solutions were used to obtain compound responses relative to 1-methylpyrene and m-tetraphenyl which were used as internal standards. Four replicate analyses were made on each of the two calibration solutions and two ampoules taken from the SRM pool.

Gas chromatography was also used to determine the purity of each of the crystalline materials used in this SRM. These purity values were used directly in the gravimetric method and in the calculation of the concentration of each of the 16 PAH in the standard solutions used in the

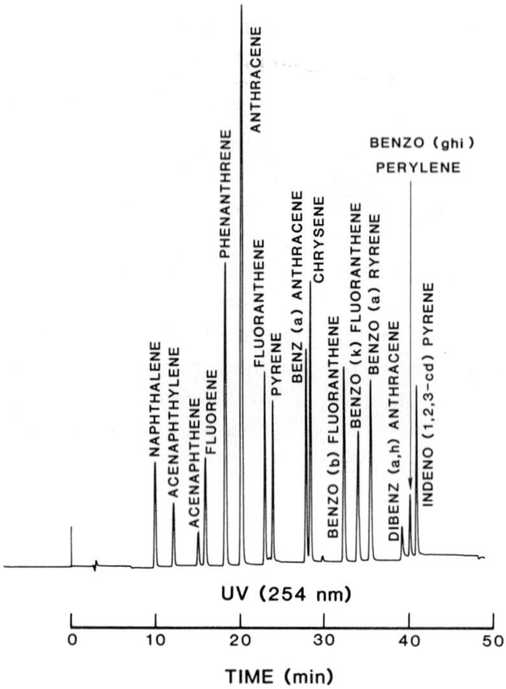

FIGURE 1 HPLC separation of 16 priority pollutant PAH.

GC and HPLC methods. The certified values along with a summary of the results obtained for the three analytical methods are given in Table V.

Uses

SRM 1647 is intended primarily for calibrating chromatographic systems used in the determination of priority pollutant PAH. This material can also be useful in PAH recovery studies. For example, because of its miscibility with water, SRM 1647 can be used to fortify aqueous samples with known amounts of these PAH. This SRM was developed at the request of the U.S. Environmental Protection Agency to support method 610 for PAH.[9]

SRM 1649, URBAN PARTICULATE MATTER

NBS has recently completed a preliminary analysis of an urban air particulate matter sample as an SRM to be certified for trace organic

TABLE V

Summary of results by the analytical methods used for certification of SRM 1647 (PAH in Acetonitrile).

	Concentration (μg/mL)			
	Gravimetry	HPLC	GC	Certified[a]
1. Naphthalene	22.5	22.4 ±0.5		22.5 ±0.2
2. Acenaphthylene	19.0	19.2 ±0.5		19.1 ±0.2
3. Acenaphthene	20.8	21.2 ±0.4		21.0 ±0.4
4. Fluorene	4.89	4.96±0.18		4.92±0.10
5. Phenanthrene	5.00	5.12±0.18		5.06±0.10
6. Anthracene	3.25	3.33±0.10		3.29±0.10
7. Fluoranthene	9.99	10.3 ±0.5		10.1 ±0.2
8. Pyrene	9.82	9.85±0.58		9.84±0.10
9. Benz(a)anthracene	4.99	5.12±0.14	4.97±0.06	5.03±0.10
10. Chrysene		4.69±0.15	4.68±0.06	4.68±0.10
11. Benzo(b)fluoranthene	5.11	5.13±0.21	5.09±0.06	5.11±0.10
12. Benzo(k)fluoranthene	5.00	5.06±0.15	4.99±0.10	5.02±0.10
13. Benzo(a)pyrene	5.28	5.32±0.13	5.31±0.19	5.30±0.10
14. Benzo(ghi)perylene	4.00	4.09±0.30	3.99±0.14	4.01±0.10
15. Dibenz(a,h)anthracene		3.73±0.12	3.63±0.07	3.68±0.10
16. Indeno(1,2,3-cd)pyrene	4.07	4.11±0.15	4.02±0.06	4.06±0.10

[a]The estimated uncertainty given for each compound is based on judgment, and represents an evaluation of the combined effects of method imprecision and possible systematic errors among methods.

constituents. SRM 1649 will consist of 10 gram quantities of this material with certified concentrations for several PAH. The particulate material for this SRM was collected in Washington, D.C. in a baghouse especially designed for this purpose. The sample was collected over a period in excess of 12 months and, therfore, is a time-integrated sample. The particulate material was removed from the filter bags by a specially designed vacuum cleaner and combined into a single lot. This material was screened through a fine mesh sieve to remove most of the fibers and other extraneous material from the bags. The sieved material was then thoroughly mixed in a V-blender, bottled, and sequentially numbered.

Certification

Randomly selected bottles have been selected for the determination of 13 PAH by at least one of two independent analytical schemes. Sample aliquots of 1 g are being extracted in a Soxhlet extractor for 48 h with a cycle time of about 20 min. Samples being prepared for GC analysis are extracted with 450 mL of a 1:1 mixture of benzene/methanol, whereas samples for HPLC analysis are extracted with a similar volume of methylene chloride. An internal standard solution of 1-methylpyrene (for GC analysis) or 7-methylfluoranthene (for LC analysis) is added to the particulate samples prior to extraction. Prior to GC analysis the extract is concentrated in a rotary evaporator, re-dissolved in cyclohexane, and liquid-liquid partitioned between N,N-dimethylformamide (DMF) and water as described by Bjørseth.[10] After the liquid-liquid partition the total PAH fraction is isolated by normal-phase HPLC on an aminosilane column. The PAH fraction is collected, concentrated, and analyzed by GC on a 30 m × 0.25 mm i.d. SE-52 fused silica column.

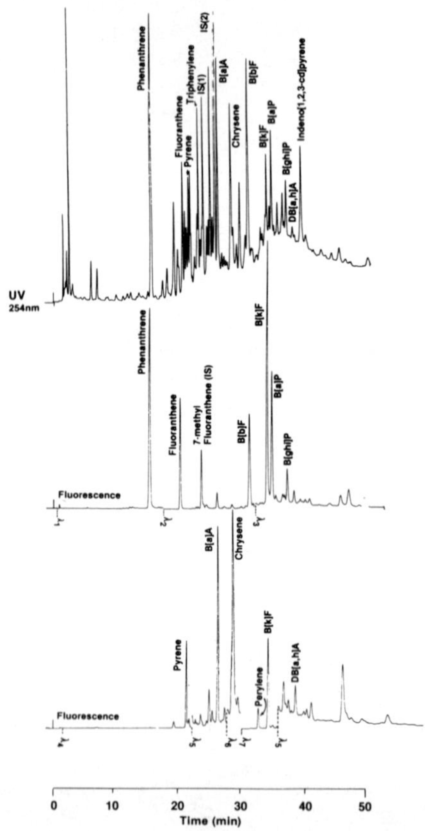

FIGURE 2 HPLC analysis of Washington Urban Dust Sample. The upper chromatogram is the UV detection at 254 nm; the middle and lower chromatograms are with fluorescence detection with excitation and emission conditions given in Table VI. (IS) = internal standard, B(b)F = benzo-(b)fluoranthene, B(k)F = benzo(k)fluoranthene, B[ghi]P = benzo(ghi)perylene and DB[a,h]A = dibenz(a,h)anthracene.

TABLE VI

Fluorescence condition for HPLC determination of PAH in air particulate extract.

	Wavelengths (nm)		
	Excitation	Emission	PAH determined
λ_1	250	360	phenanthrene
λ_2	285	450	fluoranthene, 7-methylfluoranthene (I.S.), benzo[b]fluoranthene
λ_3	295	400	benzo[k]fluoranthene, benzo[a]-pyrene, benzo[ghi]perylene
λ_4	335	385	pyrene
λ_5	285	390	benz[a]anthracene, dibenz[a,h]-anthracene
λ_6	270	360	chrysene
λ_7	400	440	perylene, benzo[k]flouranthene

TABLE VII

Preliminary results of the determination of selected PAH on Washington urban particulate matter (μg/g).[a]

	GC	HPLC
Phenanthrene		4.8 ± 0.3
Fluoranthene	7.3 ± 0.2	7.4 ± 0.7
Pyrene	7.2 ± 0.4	6.3 ± 0.4
Benzo[a]anthracene	2.4 ± 0.1	2.5 ± 0.3
Chrysene	4.6 ± 0.2[b]	3.7 ± 0.2
Benzo[b]fluoranthene		6.2 ± 0.5
Benzo[k]fluoranthene		2.0 ± 0.2
Benzo[a]pyrene	2.9 ± 0.3	2.6 ± 0.4
Benzo[e]pyrene	3.3 ± 0.2	4.1 ± 0.3
Perylene	0.8 ± 0.1	0.8 ± 0.1
Benzo[ghi]perylene	4.7 ± 0.2	3.9 ± 0.8
Indeno[1,2,3-cd]pyrene	3.3 ± 0.3	3.4 ± 0.4
Dibenz[a,h]anthracene		0.5 ± 0.1

[a]These values are *not* certified. Research is continuing to resolve discrepancies between methods. Certification will be contingent upon the agreement between methods for each compound after all analytical measurements have been made.

[b]Includes triphenylene, which co-elutes with chrysene.

Sample clean-up for the LC chloride analyses consists of evaporation of the methylene chloride to near dryness, solvent exchange to cyclohexane, and liquid-liquid partition between cyclohexane and nitromethane.[11] The nitromethane solution is concentrated to approximately 2 mL and analyzed by reversed phase HPLC on a polymeric C-18 column, with an acetonitrile/water gradient elution. A fluorescence detector is being used which is capable of changing excitation and emission wavelengths as a function of time during the chromatographic run. The excitation and emission wavelengths used for detection of each of the 13 PAH determined are listed in Table VI. Preliminary data for SRM 1649 obtained using the GC and HPLC methods on a limited number of samples are given in Table VII. Figures 2 and 3 are GC and HPLC chromatograms representing the analysis of extracts of candidate SRM 1649. Further details concerning these methods will be reported elsewhere.[12,13]

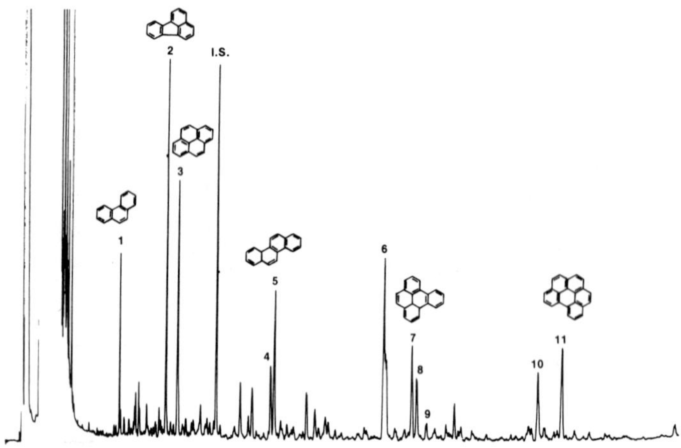

FIGURE 3 Gas chromatographic separation of fraction from Washington Urban Dust Sample. Peaks identified are (1) phenanthrene, (2) fluoranthene, (3) pyrene, (I.S.) 1-methylpyrene, (4) benz(a)anthracene, (5) chrysene/triphenylene, (6) benzofluoranthenes, (7) benzo(e)pyrene, (8) benzo(a)pyrene, (9) perylene, (10) indeno(1,2,3-cd)pyrene, and (11) benzo(ghi) perylene.

Uses

This particulate material will be useful for the validation of analytical methods for the determination of PAH on particulate matter samples. In addition the availability of this large quantity of homogeneous particulate material will provide analysts with adequate material for use in intercalibration and biological studies.

SUMMARY

The SRM's discussed in this paper are intended to serve either or both of two functions; calibration of analytical instrumentation or validation of analytical procedures for determination of individual PAH in complex samples. SRM 1647, Priority Pollutant PAH in Acetonitrile, is intended for use in the calibrating chromatographic instrumentation used in PAH analyses. SRM 1644, Generator Columns for PAH in Water, can be used for producing aqueous solutions for anthracene, benz(a)anthracene and benzo(a)pyrene. SRM's 1580, Organics in Shale Oil and 1649, Urban Particulate Matter, can be used for evaluating the reliability of analytical methods used for determining individual PAH in complex matrices. For example, several research groups have used SRM 1580 to validate new analytical procedure for determining PAH in various oil shale, coal, and petroleum derived samples.[14–19]

In our own laboratory, SRM 1580 has been used to validate an on-line multidimensional HPLC approach for the determination of PAH in oil matrices.[20] This system (shown in Figure 4) utilizes an aminosilane column in the normal phase mode for isolation of a selected fraction from a complex mixture and a C-18 column in the reversed phase mode with spectrofluorimetric detection for analytical separation and selective detection, as does the off-line sequential HPLC method that we have described previously.[3,4,21–23] In addition, this system employs a "concentrator column" (diamino) for interfacing the normal and reversed phase systems. The desired fraction is valved through, and trapped on, the

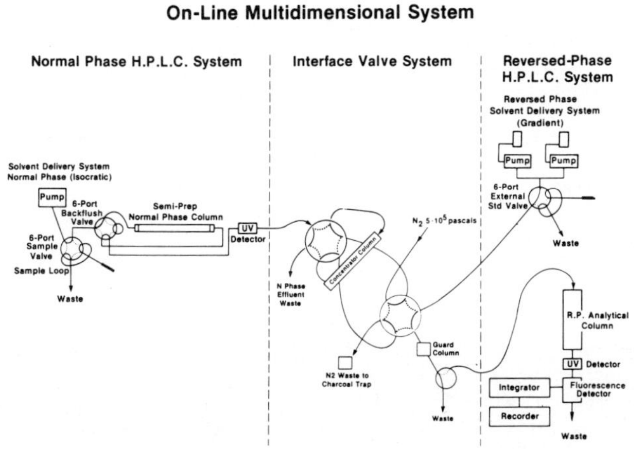

FIGURE 4 Schematic diagram of the on-line multidimensional HPLC system.

"concentrator column". The normal phase solvent is evaporated using an inert gas purge. The concentrator column is then valved into the reversed-phase analytical system and the fraction transferred to the analytical column by gradient elution focusing. SRM 1580 was used to validate this approach as shown by the data given in Table VIII.

The methods developed and/or used at NBS for certification of these SRM's can, with minor modifications, be employed for the routine determination of PAH in real-world samples.

TABLE VIII

Determination of selected PAH in three oil samples by multidimensional HPLC.

Compound	Multidimensional HPLC (on-line)[a]	Multidimensional HPLC (off-line)[a]	Certified[c]
	SRM 1580 (Shale Oil) Concentration in μg/g)		
Fluoranthene	53.8 ±3.5(3)[b]	53 ±2(9)	54 ±10
Pyrene	109.7 ±7.1(3)	107 ±8(10)	104 ±18
Perylene	2.32 ±0.15(5)	3.9 ±0.6(11)	3.4±2.2
Benzo[a]pyrene	20.7 ±2.2(8)	23 ±1(8)	21 ± 6
	SRC II Coal Liquid (Concentration in mg/g)		
Fluoranthene	3.15 ±0.04(3)	3.3 ±0.16(3)	
Pyrene	6.65 ±0.05(3)	6.0 ±0.2(3)	
Perylene	0.024±0.002(3)		
Benzo[a]pyrene	0.133±0.008(5)	0.134±0.007(3)	
Benzo[e]pyrene	0.153±0.006(3)	0.143±0.005(3)	
Benzo[k]fluoranthene	0.062±0.001(5)		
	Wilmington Crude Oil (Concentration in μg/g)		
Fluoranthene	2.4 ±0.3(3)	3 ±1(3)	
Pyrene	14.8 ±2.2(3)	14 ±2(3)	

[a]Uncertainty is 1 standard deviation.
[b](*n*) indicates the number of replicate measurements.
[c]Certified values derived from a combination of HPLC and GC/MS values.

Acknowledgments

The research summarized in this paper represents the efforts of many staff members of the NBS Organic Analytical Research Division.

The authors wish to acknowledge the efforts of J. M. Brown-Thomas, S. Bowie, W. Cuthrell, F. R. Guenther, L. R. Hilpert, R. Parris, K. Richie, W. Sonnefeld, R. A. Velapoldi and P. A. White for making many of the measurements leading to the certification of the SRM's mentioned in this manuscript. Partial financial support from the Office of Health and

Environmental Research of the U.S. Department of Energy and the Environmental Monitoring and Support Laboratory (Cincinnati, Ohio), U.S. Environmental Protection Agency is gratefully acknowledged.

References

1. H. S. Hertz, W. E. May, S. A. Wise and S. N. Chesler, *Anal. Chem.* **50** (4), 428A–436A (1978).
2. H. S. Hertz, L. R. Hilpert, W. E. May, S. A. Wise, J. M. Brown, S. N. Chesler and F. R. Guenther, In: *Measurement of Organic Poll. in Water and Wastewater*, ASTM STP 686, C. E. Van Hall, ed., 291–301 (1978).
3. W. E. May, J. Brown-Thomas, L. R. Hilpert and S. A. Wise, *Chemical Analysis and Biological Fate: Polynuclear Aromatic Hydrocarbons*, M. Cooke and A. Dennies, ed. (Battelle Press, Columbus, Ohio) 1–16 (1981).
4. H. S. Hertz, J. M. Brown, S. N. Chesler, F. R. Guenther, L. R. Hilpert, R. M. Parris, W. E. May and S. A. Wise, *Anal. Chem.* **52** (11), 1650–1657 (1980).
5. G. Uriano and C. Gravatt, *CRC Critical Reviews in Analytical Chemistry*, October 1977, 361–411.
6. W. E. May, S. P. Wasik and D. H. Freeman, *Anal. Chem.* **50,** 1, 175–179 (1978).
7. W. E. May, J. M. Brown, S. N. Chesler, F. R. Guenther, H. S. Hertz and S. A. Wise, *Polynuclear Aromatic Hydrocarbons*, P. Jones and P. Leber, ed., (Ann Arbor Science Publishers, Inc., Ann Arbor, MI) 411–418 (1979).
8. W. E. May, *Petroleum in the Marine Environment* **7,** 143–192 (1980).
9. *Federal Register* **44** (233), 69514 (December 3, 1979).
10. A. Bjørseth, *Anal. Chim. Acta* **94,** 21–27 (1977).
11. M. Novotny, M. L. Lee and K. D. Bartle, *J. Chromotogr. Sci.* **12,** 606–612 (1974).
12. S. A. Wise, S. L. Bowie, S. N. Chesler, W. F. Cuthrell, W. E. May and R. E. Rebbert, *Polynuclear Aromatic Hydrocarbons: Phys. and Biol. Chem.* (Battelle Press, Columbus), in press.
13. W. E. May, W. F. Cuthrell and S. A. Wise, *The quantitation of selected polycyclic aromatic hydrocarbons from air particulate matter by liquid chromatography*, manuscript in preparation.
14. B. A. Tomkins, R. R. Reagan, J. E. Caton and W. H. Greist, *Anal. Chem.* **53,** 1217–1222 (1981).
15. Y. Yang, A. P. D'Silva, V. A. Fassel and M. Iles, *Anal. Chem.* **52,** 1350–1351 (1980).
16. W. D. Spall, *Anal. Chem.*, submitted for publication.
17. Y. Yang, A. P. D'Silva and V. A. Fassel, *Anal. Chem.* **53,** 2187–2209 (1981).
18. T. Vo-Dinh and P. R. Martinez, *Anal. Chim. Acta* **125,** 13–19 (1981).
19. C. S. Woo, A. P. D'Silva and V. S. Fassel, *Anal. Chem.* **52,** 159–164 (1980).
20. W. J. Sonnefeld, W. H. Zoller, W. E. May and S. A. Wise, *Anal. Chem.* **54,** 723–727 (1982).
21. S. A. Wise, S. N. Chesler, H. S. Hertz, L. R. Hilpert and W. E. May, *Anal. Chem.* **49** (14), 2306–2310 (1977).
22. S. A. Wise, S. N. Chesler, H. S. Hertz, L. R. Hilpert and W. E. May, *Carcinogenesis 3: Polynuclear Aromatic Hydrocarbons*, P. W. Jones and R. I. Freudenthal, eds. (Raven Press, New York, NY), 175–182 (1978).
23. J. M. Brown, S. A. Wise and W. E. May, *J. Environ. Sci. Health* **A15**(6), 613–623 (1980).

Determination of Total Hydrocarbons in Water†

S. ROTTERI

CONCAWE, Babylon—Kantoren A, Koningin Julianaplein 30, 2595 AA Den Haag, Netherlands

(Received November 26, 1981)

It is important to the oil refining industry to obtain accurate and reliable data on the levels of hydrocarbons in aqueous discharges. However, considerable difficulties are presented in placing data from different locations on a comparable basis, because of the number and diversity of procedures used. The analytical principle involved may differ, for example infrared spectroscopy as opposed to gravimetric determination. Other aspects of the procedures, such as standardisation and calculation, may also vary. A CONCAWE task force of experts from the oil refining industry has identified certain key technical elements to the determination of hydrocarbons in water. The task force concluded that infrared spectroscopy is the best available analytical technique and that any procedure used should include a solvent extraction step, separation of non-polar material and measurement at three wavelengths with the use of typical absorptivities for the calculation of oil-in-water values.

KEY WORDS: Hydrocarbons, oil-in-water, analysis, determination, infrared

INTRODUCTION

The oil refining industry has made substantial reductions in recent years in the total amount of oil and other components of crude oil discharged in aqueous effluents.[1,2] This information has been obtained from sources within the industry and such surveys have clearly shown that a variety of methods are used to determine oil-in-water (i.e. dissolved and suspended hydrocarbons) and other parameters. These methods are used routinely by refinery laboratories and, while the variation in analytical procedure does not affect the overall trend, it is clearly desirable that analytical methods are placed on a comparable basis in order to obtain more accurate

†Presented at the Workshop on the Chemistry and Analysis of Hydrocarbons in the Environment, Barcelona, November 1981.

estimations of the amounts of oil discharged. It must be emphasised that this report is concerned with the overall discharge of oil-in-water and not with the determination of individual trace substances derived from crude oil. Techniques for such, more detailed, analyses are under current investigation by a separate CONCAWE group of experts.

Accordingly, CONCAWE established a special task force of experts, with experience in analysis of oil-in-water in refinery discharges, with the objective of identifying the key technical elements of an analytical procedure for determination of oil-in-water. CONCAWE had previously recommended[3] the use of techniques based on infrared (IR) spectroscopy and the present task force confirmed this recommendation. The survey was therefore limited to IR based methods, including at least thirteen different procedures in current use in Western European refineries. As CONCAWE is not a standardisation body the study excluded the possibility of developing a new standard procedure.

TECHNICAL ASPECTS OF DETERMINATION OF HYDROCARBONS IN WATER

Considerable difficulties are presented in selecting a single method, because of the variable nature of oil and because of the vast range of components, from gaseous hydrocarbons to bituminous compounds and from paraffinic to aromatic hydrocarbons. Further, these components can vary both within different crude oils and, particularly, within oily wastes as a result of different processing conditions. This situation is illustrated in Figure 1 which shows the differences in the IR spectra of toluene, cyclohexane and *n*-heptane. Thus it was accepted at the outset of the study that no approach could be identified which could give an absolute measure of oil-in-water. After careful consideration of a number of published procedures the following technical elements were identified.

SOLVENT AND SOLVENT EXTRACTION PROCEDURE

The choice of solvent is affected by several aspects including: the recovery of hydrocarbons from the water sample; the possibility of loss of solvent by evaporation with the consequent concentration of hydrocarbons; and the requirements for health and safety in the workplace. Two solvents are generally used: carbon tetrachloride which is less volatile than the other, Freon 113, the use of which has been recommended on grounds of its lower toxicity.[4]

IR TRACES FOR PURE HYDROCARBONS

(25 mg in 100 ml CARBON TETRACHLORIDE)

from Palmer (1980)

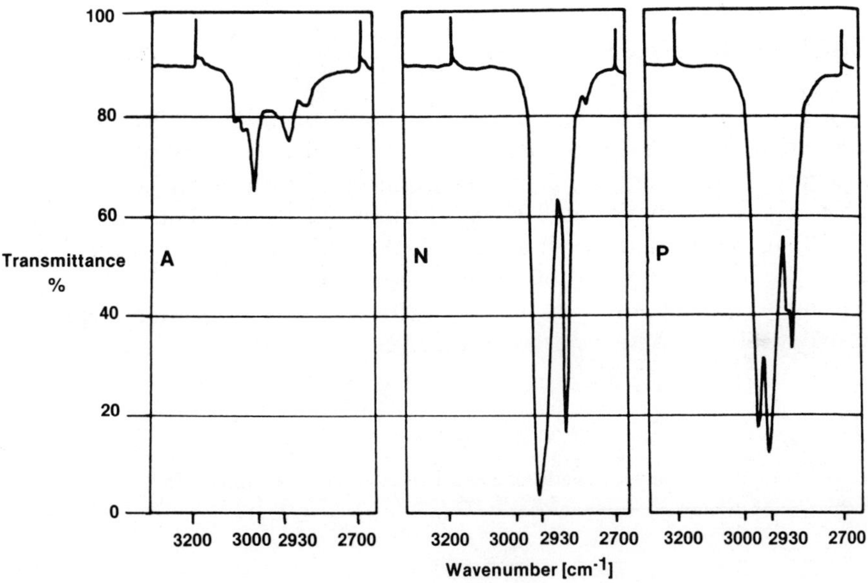

A = TOLUENE (AROMATIC)
N = CYCLOHEXANE (NAPHTHENE)
P = n-HEPTANE (PARAFFIN)

FIGURE 1

11/81

REMOVAL OF INTERFERING COMPOUNDS

This aspect is of particular importance where contamination may arise from compounds not originating from the oil refining industry. These compounds are generally polar in nature. The separation of the non-polar hydrocarbons from the polar compounds is generally achieved by use of an appropriate adsorbent either by direct addition of the adsorbent to the solvent, following extraction, or by elution through a column. Experience has shown that it is not possible to carry out this step without simultaneously adsorbing some hydrocarbons (higher molecular weight aromatics). Such compounds generally comprise only a minor proportion of the hydrocarbons in any aqueous sample and any reduction in the value of oil-in-water is much less than the relatively massive distortion that can arise from extraneous material which is removed by this step.

ABSORPTION WAVELENGTHS

The use of infrared spectrophotometry is based on the identification and quantitative measure of absorption by particular chemical groups at characteristic wavelengths. For hydrocarbons, the characteristic groups are (Table I): CH (aromatic) at 3.30 μm (3030 cm^{-1}); $>CH_2$ (methylene) at 3.42 (2924); and CH_3 (methyl) at 3.38 (2958). All of these groups are also present in a vast number of other non-hydrocarbon compounds and, accordingly, it is necessary to include extraction and separation steps in the analytical procedure, as described above.

TABLE I

Absorption wavelengths of characteristic groups in hydrocarbons

	μm	cm^{-1}	Characteristic
CH_2	3.30	3030	Low molecular weight aromatics
$>CH_2$	3.42	2924	Long straight chain saturated hydrocarbons and cycloalkanes
CH_3	3.38	2958	Saturated hydrocarbons

Of the three chemical groups indicated CH is most characteristic of low molecular weight aromatic hydrocarbons. In the higher aromatic homologues the C atoms are partially or wholly substituted. The $>CH_2$ group is mostly present in long straight chain saturated hydrocarbons and in cycloalkanes, and CH_3 is the terminal group of saturated hydrocarbons, although it can be present as a substituent in other compounds.

Consequently the measure of absorbance at three wavelengths, or the overall IR spectrum including these wavelengths, can be used both as a quantitative measure of the hydrocarbons in the sample and as a qualitative evaluation of its composition.

REFERENCE MATERIAL

The choice of reference material in an analytical procedure is of particular importance where a broad group of compounds is to be collectively determined. "Known oils", such as those recovered from the oil separators in a water treatment system, are often used to give measurement of the concentration of oil in aqueous effluents. The relevance of the use of "known" oil may change with the pattern of crude used and with the different processes operating within a single refinery. Nevertheless, reasonable knowledge of the amount of oil discharged from a refinery can be obtained, providing the "known" oil used is a representative long-term composite sample. However, the composition of all such separated oils (also known as "slop" oils) is different from the mixture of hydrocarbons in the water. This is particularly true where further treatment stages, such as air-flotation and biotreatment, are included, owing to selective removal and degradation and the differing distribution of individual hydrocarbons between the oily and aqueous phases. In addition, it may be difficult to compare analytical results obtained by different refinery laboratories as "known" oils will necessarily differ from one refinery to another.

The use of a "reference oil", consisting of a defined mixture of hydrocarbons such as benzene, isooctane and cetane (*n*-hexadecane), does enable comparison of data obtained by different laboratories, including checking of procedures by control authorities. Although this enables the same reference mixture to be used consistently and at different locations, the mixture selected may be quite different from the oil being measured, with a consequent loss of accuracy. The use of such mixtures has no apparent advantage over the use of typical specific absorption coefficients (typical absorptivities)[5] as described below.

CALCULATION OF RESULTS BASED ON TYPICAL ABSORPTIVITIES

In recent years, particular attention has been paid to the development of accurate procedures for direct calculation of results rather than the use of known oils or mixture of pure compounds. The use of typical absorptivities for measurement of CH, $>CH_2$ and CH_3 is generally accepted as the best approach. The absorptivities are defined as "typical" in that they are intended to be of general application, irrespective of the model of IR apparatus used and operating conditions, that is of the resolution of the bands. The simplicity of application of such typical figures counterbalances the minor accuracy. None of the IR procedures directly measures C atoms fully substituted by groups other than H atoms as there is no IR absorption. Such fully substituted H atoms are widely present in hydrocarbons. In order to take their presence into account, suitable correction factors have been developed.[5] This procedure can be refined by calculating the absorptivities for CH, $>CH_2$ and CH_3 for the individual IR apparatus and operating conditions.[6] While this may require an extra effort at the beginning in each laboratory, it would improve the accuracy and allow a sounder comparison of data from different sources.

The calculation procedure using such absorptivities is based on the single spectra of three single compounds (toluene, *n*-hexadecane, pristane), carefully selected in order to give the best overall fit to a broad range of petroleum products without the use of the correction factors mentioned above. In order to allow for the relatively weak absorption of the CH group, a suitable correction should be calculated for the $>CH_2$ and CH_3 shoulder.

CONTINUOUS ANALYSERS

An important recent trend has been the development of apparatus that allows continuous recording of oil-in-water levels. Although such analysers have many practical drawbacks this approach could eventually allow a reduction in the routine laboratory work and the compilation of a much more substantial data base on a cost-effective basis. At the moment such methods are based on IR stream analysers which measure only $>CH_2$ absorption. Consequently, these instruments may be of use for monitoring performance of treatment systems within a single refinery but do not allow compilation on a broader basis of accurate data on the amount of oil discharged by oil refineries.

CONCLUSIONS

The CONCAWE task force has concluded that infrared spectroscopy provides the best basis currently available for the measurement of oil-in-water. Any method that is used should incorporate a solvent extraction step, separation of non-polar material by elution through an adsorbent column and measurement of the hydrocarbons at three wavelengths characteristic of the three predominant groups CH, $>CH_2$ and CH_3. For comparative purposes, the most practical approach to the calculation of oil-in-water values is to use typical absorptivities for these groups, with incorporation of a factor to allow for those groups not measured in the infrared. Continuous stream analysers have only specialised application at present and the development of suitable instrumentation for stream analysis of oil-in-water on a comparative basis across the industry remains an objective for future technical innovation.

References

1. CONCAWE. Survey on quality of refinery effluents in Europe—comparison between 1969–1974 performance. CONCAWE report 4/77 (1977).
2. CONCAWE. The environmental impact of refinery effluents—CONCAWE's assessment. CONCAWE report 1/80 (1980).
3. CONCAWE. Methods for the analysis of oil in water and soil. CONCAWE report 9/72 (1972).
4. I. N. Sax, *Dangerous Properties of Industrial Materials*. 5th ed. Reinhold, New York, 471, 696 (1979).
5. H. Hellman, *Vom Wasser* **50**, 231–245.
6. P. J. Whittle, W. A. McCrum and M. W. Horne, *Analyst* **105,** 673–684.
7. T. Palmer, Sampling and analysis of oil water effluents, in "Oily Water Discharges". Ed. C. S. Johnston and R. J. Morris, *Applied Science*, London (1980).

A Comparative Study of Chemical and Microbiological Monitoring of Pollutant Hydrocarbons in Urban Aquatic Environments†

A. M. SOLANAS and R. PARÉS
Department of Microbiology, Faculty of Biology, University of Barcelona (Spain)

and

C. MARFIL and J. ALBAIGÉS‡
Institute of Bio-Organic Chemistry (CSIC). Jorge Girona Salgado, Barcelona 34 (Spain)

(*Received February 4, 1982*)

Conventional chemical and microbiological methods—aromatics by UV-fluorescence and the number of oil-degrading microorganisms, respectively—have been used for the monitoring of pollutant hydrocarbons in three different aquatic systems: two rivers, one harbour and three marine coastal areas. An evaluation of the first year data of such study is presented.

Relative populations of total heterotrophic microorganisms and those of degrading hydrocarbons correlate satisfactorily with hydrocarbon concentrations in marine areas, where chronic pollution situations occur, whereas unreliable results were obtained in the river systems. The water temperature seems to have a positive influence on the response of microorganisms to oil pollution.

KEY WORDS: Oil pollution, marine pollution, aromatic hydrocarbons, oil-degrading microorganisms, UV-fluorescence.

INTRODUCTION

Anthropogenic hydrocarbons are ubiquitous in nature and primarily in aquatic environments as a consequence of the accidental or intentional discharges of petroleum products into either fresh waters or sea waters.

The ecological impact produced in these environments by acute oil spills

---

†Presented at the Workshop on the Chemistry and Analysis of Hydrocarbons in the Environment, Barcelona, November 1981.

‡To whom correspondence should be addressed.

is actually well documented both from the chemical and microbiological standpoints. In such situations, the identification of pollutant sources and the ultimate fate of hydrocarbons are questions of major concern.[1]

On the contrary, less attention has generally been paid to areas receiving chronic inputs (estuaries, rivers, etc.) in which the conditions, although apparently less severe, are not necessarily less harmful.[2] The emphasis in these cases should be placed in long term monitoring surveys. Questions of interest are the intercomparability of the information provided by the different methodologies used in different places and, particularly, the assessment of to what extent the parameters measured reflect the historical exposure of the environment to hydrocarbons. In this respect indigenous microorganisms play an important role in determining the capability of the systems for handling these pollutants.

During the last decade scientific efforts have gone toward the development of analytical techniques sensitive enough for pollution baseline studies and capable of making distinction between natural and anthropogenic hydrocarbons.[3] On the other hand, significant progress has been made on the microbiology and ecology of the latter mainly on the occasion of large marine oil spills or simulated acute discharges.[5,6] However, the information available from chronic or apparently unpolluted situations and especially from waters nearby urban areas, including distributions of hydrocarbons and oil degrading microorganisms, is rather scarce.[7,8] Therefore, a long term monitoring of petroleum hydrocarbons by conventional chemical and microbiological methods was undertaken in the vicinities of the highly populated area of Barcelona. Three different systems were selected, namely a harbour, rivers and marine coastal sites. The first year data will be presented and evaluated below.

EXPERIMENTAL

Sampling

The water sampling was carried out at a series of stations located in the surroundings of Barcelona as indicated in Fig. 1. Shoreline water samples were collected in stations 1, 4 and 6, river samples were obtained from stations 2 and 5, and the harbour samples from station 3.

These stations were visited fortnightly during the period of October 1980–September 1981 with weighted dark glass bottles. The bottles were immediately sunk to a depth of 20 cm, where they filled, as a small buoy prevented them from sinking further.[9]

Bottles for hydrocarbon analysis (2·5 l of capacity) were thoroughly cleaned with spectro-grade carbon tetrachloride before use. Immediately after sampling, 40 ml of the same solvent was added to prevent degradation and then were stored at 4°C until analysis. This was performed within the first 24 hours.

Water samples for microbial enumeration were collected asseptically in sterile 250 ml glass bottles and used for analysis directly.

Hydrocarbon analysis

No separation was made between particulate and soluble hydrocarbons. Approximately 2 litres of water were extracted twice with 40 ml of spectrograde carbon tetrachloride. The joined extracts were dried over Na_2SO_4 and evaporated.

The resulting residue was dissolved in 5 ml of spectro-analyzed *n*-hexane and then analyzed by UV-fluorescence (excitation at 300 nm) with a Perkin–Elmer MPF-3 spectrophotometer. Quantitation was performed on the emission spectra at 330 nm, using a Kuwait diesel-oil reference sample (b.p. 250–350°C).

Alternatively, control extracts were fractionated in a 30 cm × 1 cm i.d. silica-alumina column (1:1), 4% water deactivated. One column volume of *n*-hexane-methylene chloride (1:1) was eluted and analyzed for hydrocarbons as described previously. No significant differences were found in the results when compared with the direct quantitation of the extracts.

Microbiological methods

The quantitative determination of microorganisms was carried out using the most-probable number technique (MPN) in three tubes series. MPN procedure have been suggested as a substitute for plate count procedures for enumerating oil-degrading microorganisms.[10,11]

The liquid medium used for enumeration of the total heterotrophic microorganisms (in g/l) consisted of 2.0 g glucose, 5.0 g casamino acids and 1.0 g yeast extract dissolved in sterilised artificial marine water (Biolife D-1141-505) or distilled water, according to the sample origin. Tubes were incubated one week at 22°C, at which time growth, as evidenced by turbidity, was scored.

Hydrocarbon-utilizing microorganisms were also enumerated. The basal medium was that of Bushnell-Haas Broth (Difco 0578–01),[12] supplemented with 3% ClNa, when the sample consisted of seawater. Two sets were used, one to which no carbon source was added, as a control, and the other with 1% sterilized kerosene as the sole source of carbon and energy.[13] Incubation was for 14 days at 22°C and 120 r.p.m. in an orbital shaker, after which relative growth was scored. A kerosene, containing 35% alkanes, 50% naphtenes and 15% aromatics was chosen as substrate, because it provides an adequate range of hydrocarbons to which the microbes could respond and is more representative of the products that might pollute the sampling area.

RESULTS AND DISCUSSION

The area of Barcelona, we have selected for study, is adequate for stating the problem we are dealing with, as the comparability of the information given by chemical and microbiological methods regarding hydrocarbon chronic pollution. It is well representative of an urban-industrial area, inhabited by 2.5 million persons, with a suspected chronic incidence on

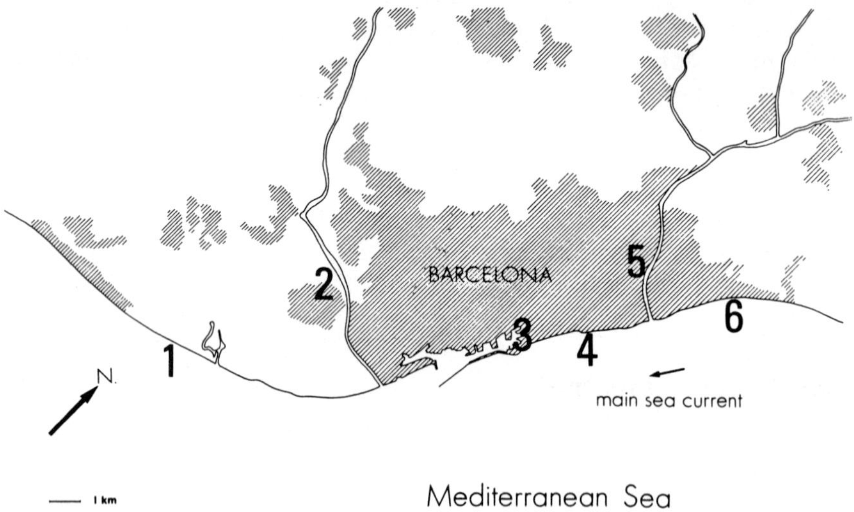

FIGURE 1 Location of sampling sites: (1) Gavá beach, (2) Llobregat river, (3) Barcelona harbour, (4) Barceloneta beach, (5) Besós river, (6) Badalona beach.

contiguous fluvial and marine coastal systems.

Therefore, six stations representing different situations were chosen accordingly (see Fig. 1). The results of the first year monitoring, based on the determination of hydrocarbon concentrations and microbiological populations, are presented in Figs. 2 and 3 and summarized in Table I.

Hydrocarbon concentrations have been estimated by UV-fluorescence using a middle distillate (Kuwait diesel-oil) as a reference standard. Although this procedure exhibits some limitations as a means of quantitating oil in water, basically due to the arbitrary nature of the selected standard,[14] it was nevertheless adequate in this case, bearing in mind the sensitivity and simplicity of the technique and that only the

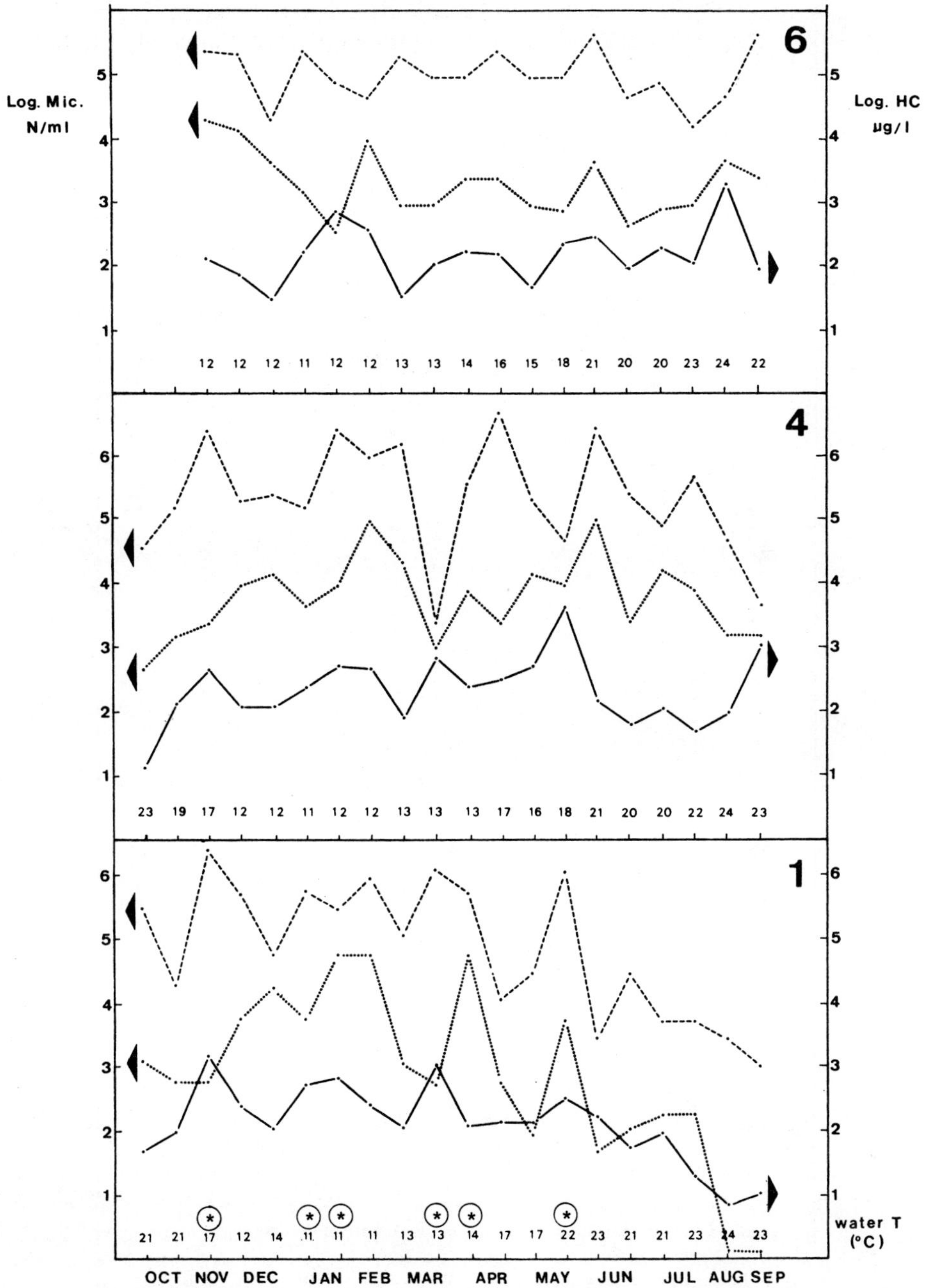

FIGURE 2 Distribution of total heterotrophic (---) and oil-degrading microorganisms (. . .), and hydrocarbon concentrations (———) in marine stations (1, 4, 6).

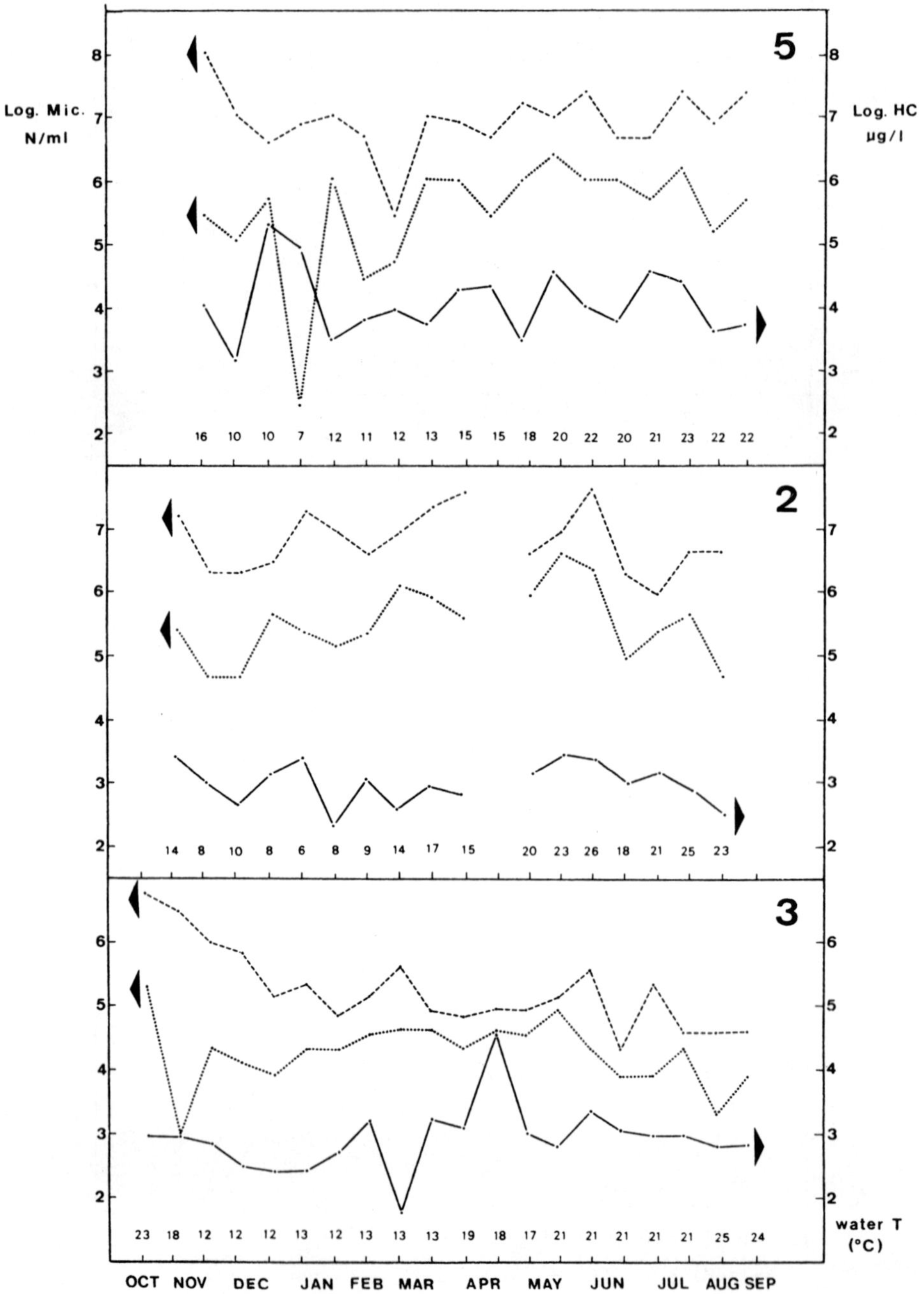

FIGURE 3 As Fig. 3 for river (2, 5) and harbour (3) stations.

TABLE I

Range and mean values of the parameters determined in the monitored stations

Station	Identification	Total hydrocarbons (μg/ml)	Total heterotrophes (N/ml)	Oil-degrading microorganisms (N/ml)	Oil-degrading total heterotrophes (%)
1	coastal	1186–6 (av. 231)	$2.0 \times 10^6 - 9.0 \times 10^2$ (av. 3.4×10^5)	$4.5 \times 10^4 - 1.1 \times 10^0$ (av. 8.3×10^3)	4.29
6	coastal	1922–32 (275)	$4.5 \times 10^5 - 1.5 \times 10^4$ (1.5×10^5)	$2.0 \times 10^4 - 3.5 \times 10^2$ (3.1×10^3)	4.80
4	coastal	4452–13 (491)	$4.5 \times 10^6 - 3.5 \times 10^3$ (8.2×10^5)	$9.5 \times 10^4 - 4.5 \times 10^2$ (1.6×10^4)	8.02
3	harbour	42300–64 (3546)	$6.5 \times 10^6 - 2.5 \times 10^4$ (7.1×10^5)	$1.1 \times 10^5 - 1.1 \times 10^3$ (3.1×10^4)	22.79
2	river	3005–214 (1248)	$4.5 \times 10^7 - 2.0 \times 10^6$ (1.2×10^7)	$4.5 \times 10^6 - 4.5 \times 10^4$ (7.7×10^5)	9.77
5	river	176570–1153 (24357)	$9.5 \times 10^7 - 2.5 \times 10^5$ (1.4×10^7)	$2.5 \times 10^6 - 2.5 \times 10^2$ (7.1×10^5)	8.23

measurement of relative concentrations of hydrocarbons in the samples was intended.

It is apparent from Figs. 2 and 3 that hydrocarbon concentrations have a large variability between stations, depending on the vicinity of industrialized or populated areas. Open marine stations (No. 1, 4 and 6) are the less polluted (av. 231–491 μg/l) because most of the urban waste waters receive appropriate treatment in depuration plants. However, concentrations in station 4 are slightly higher (av. 491 μg/l), probably due to the influence of the Besós river outflow, which follows the main sea current parallel to the coast in a NE-SW direction (Fig. 1).

In any case the observed levels are approximately two orders of magnitude higher than those generally found far from the coast,[15,16] pointing to the definite influence of continental activities on marine coastal pollution. The two rivers (stations 2 and 5), which are small streams (715 and $98 \times 10^6\ m^3$/year, respectively) running through very industrialized areas, are contributing more significantly to the coastal pollution in this area (av. 1.2–24.3 mg/l).

Finally, the harbour, as a typically commercial port, (about 8000 vessels/year) exhibits a rather constant and elevated concentration (av. 3.5 mg/l).

It is interesting to note that the hydrocarbons variability is not only quantitative but qualitative. Figure 4 shows the types of spectra displayed by the samples from the different stations. Type A, which is largely predominant in river samples, is produced by petroleum products containing one or two ring aromatic compounds, such as diesel or lube oils. On the other hand, type B, which is mainly observed in marine samples, corresponds to petroleum products containing three and four rings, as crude oils or heavy residues (tanker washings, fuel-oils, etc.).

As far as microbial populations is concerned, it can be seen from Table I that the numbers and types of microorganisms reflect to some extent the nature and degree of water pollution.

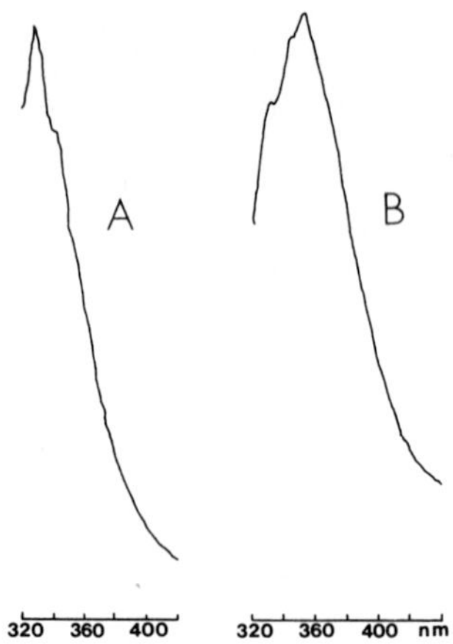

% OCCURRENCE		
STATION	TYPE A	TYPE B
1	56	44
2	96	4
3	95	5
4	65	35
5	72	28
6	50	50

FIGURE 4 Types of emission fluorescence spectra displayed by the analyzed samples. The table indicates the occurrence of each type in the different stations.

Total heterotrophes exhibit a large variability within and between the stations, paralleling that of the hydrocarbon concentrations, as indicative of the organic pollution. The numbers are of the same order of magnitude as those observed in other coastal areas, e.g., Mulkins *et al.*[13] reported values in the range of $3.0 \times 10^5 - 2.2 \times 10^7$ in the Northwestern Atlantic and Gunkel *et al.*[7] of $5.3 \times 10^4 - 4.0 \times 10^7$ in the North Sea.

A rough correlation also exists between the concentrations of hydrocarbons in water and those of oil-degraders, whose occurrence in aquatic environments has been recently reviewed by Atlas.[6] However, it appears that these microorganisms have only a response capacity when hydrocarbons increase up to approximately 500 μg/l as in shoreline stations. The relative importance of oil-degraders within total microorganisms, is also increasing with hydrocarbons concentrations, to the point that for the marine sites (stations 1, 6, 4 and 3) the correlation shows a regression coefficient of 0.9887, so providing a reliable microbiological criteria for evaluating hydrocarbon pollution in these systems.

In fact, Walker and Colwell[17] suggested that counts of petroleum-degrading microorganisms be expressed as a percentage of the total population rather than a total number of petroleum degraders per se. Subsequently, the above mentioned correlation was observed by several authors.[2,18] Le Petit, *et al.*[19] reported that hydrocarbon-utilizing microorganisms accounted for 10% of the total heterotrophes in the area of a refinery effluent, whereas it was only of 4% in areas unrelated with hydrocarbon pollution. In the present study the less polluted marine stations (Nos. 1 and 6) exhibit only 4.2–4.8% of oil-utilizing microorganisms, by contrast with the harbour which has 22%. This feature is not followed by the river stations, where a much higher increment of the number of total heterotrophes against oil-degraders takes place, despite the considerable amount of hydrocarbons present. Unlike the case of the harbour, where hydrocarbons are the major pollutant type, the total organic charge of the rivers is composed of a large variety of pollutants, probably exhibiting different toxic or bacteriostatic effects for the indigenous microorganisms and particularly for the oil-degraders. This will explain the behaviour deviation of the rivers in comparison with the other sites.

The comparison of the chemical and microbiological monitoring profiles obtained for each station permits one to approach another important problem such as the autodepuration capacity of the systems or the response capability of the microbiological populations to an input of hydrocarbon pollutants.

Several studies have shown that within a few days after an oil spill the population of hydrocarbon-utilizing microorganisms rises by several orders of magnitude,[20] thus contributing to the degradation of the oil.

When we observed in Fig. 2 the profiles corresponding to stations No. 1, 6 and 4 it can be realized that an increase in hydrocarbons correspond to an increase in oil-degraders. However, from October–May there is a time-lag of response for the latter which is minimized during June–August. Environmental factors are most probably the responsibles for such a behaviour. Particularly, the temperature has been recognized to have a marked effect on microbial growth and enzymatic activities,[21] being the limiting factor for the degradation of dissolved hydrocarbons. In these stations the response delay of the microorganisms is clearly shortened when the water temperature reaches to about 20°C.

The autodepuration capacity is particularly significant in station 1, where periodical inputs of urban and industrial waste waters were known.

These inputs are indicated by asterisks (*) and they coincide with an increase in hydrocarbon concentration. At the end of the monitoring period the declination of the hydrocarbons produces also a diminution in the microbial populations, clearly pointing to the close correlation between both parameters.

The situation in the river systems (stations 2 and 5, Fig. 3) is more random. However, during summer the correlation between hydrocarbons and microorganisms seems to improve slightly.

In conclusion it can be said that microbiological methods are suitable for monitoring hydrocarbon pollution in ecosystems where continuously small amounts of organic substances or much higher, but predominantly hydrocarbons, are introduced.

These systems also exhibit a higher degradation potential. The microbial response seems to be faster when the water temperature increases.

Acknowledgement

The technical assistance of Angela Gallifa is much appreciated.

References

1. J. Albaigés, in *Biogéochimie de la matiere organique a l'interface eau-sediment marin* (Ed. CNRS, No. 293, 1980) pp. 233–247.
2. J. D. Walker and R. R. Colwell, *Can. J. Microbial.*, **21,** 305 (1975).
3. B. K. Afghan and D. McKay, *Hydrocarbons and halogenated hydrocarbons in the aquatic environment*, Plenum Press (1980).
4. S. C. Brassell and G. Eglinton, in *Analytical Techniques in Environmental Chemistry* (Ed. J. Albaigés, Pergamon Press, 1980) pp. 1–22.
5. D. C. Malins, *Effects of Petroleum on Arctic and Subarctic Marine Environments and Organisms*, Academic Press (1977).
6. R. M. Atlas, *Microbiological Rev.*, **45,** 180 (1981).
7. W. Gunkel, G. Gassmann, C. H. Oppenheimer and I. Dundas, *Proc. International Symposium on Marine Microbiology*, Santiago de Compostela (Spain), (1980), pp. 223–247.
8. S. Galassi and W. Canzonier, *Atti dell'Istituto Veneto di Scienze Lettere ed. Arti*, **135,** 157 (1977).
9. *Marine Pollution Monitoring*, N.B.S. Special Pub. 409, pp. 281–284 (1974).
10. A. L. Mills, C. Breuil and R. R. Colwell, *Can. J. Microbiol.*, **24,** 552 (1978).
11. R. R. Colwell, in *Native aquatic bacteria: enumeration, activity and ecology* (J. W. Costerton and R. R. Colwell, ASTM STP-695, 1978) pp. 56–61.
12. L. D. Bushnell and H. F. Hass, *J. Bacteriol.*, **41,** 653 (1941).
13. G. J. Mulkins-Phillips and J. E. Stewart, *Can. J. Microb.*, **20,** 955 (1974).
14. D. C. Gordon, P. D. Keizer and J. Dale, *Marine Chem.*, **2,** 251 (1974).
15. F. Faraco and J. Ros, *IV Journées Etud. Pollutions*, (Antalya, C.I.E.S.M., 1978) pp. 117–121.

16. R. A. Brown and H. L. Huffman, *Marine Poll. Bull.*, **10,** 291 (1979).
17. J. D. Walker and R. R. Colwell, *Appl. Environ. Microbiol.*, **31,** 198 (1976).
18. G. Roubal and R. M. Atlas, *Appl. Environm. Microbiol.*, **35,** 897 (1978).
19. J. Le Petit, M. H. N'Guyen and S. Tagger, *Environ. Pollut.*, **13,** 41 (1977).
20. R. Atlas and A. Bronner, *Proc. of the International Symposium on the Amoco Cadiz: Fates and Effects of the Oil Spill*, Brest (France), (1980).
21. R. M. Atlas and R. Bartha, *Can. J. Microbiol*, **18,** 1851 (1972).

Comparaison de Differentes Methodes Analytiques Appliquees au Dosage des Hydrocarbures Dissous dans L'Eau a L'etat de Traces†

J. DUCREUX, R. BOULET, N. PETROFF and J. C. ROUSSEL
Institut Français du Pétrole, 92596 Rueil-Malmaison, France

(*Received January 26, 1982*)

Many physical-chemical methods are employed for the evaluation of trace hydrocarbons dissolved, in underground or surface waters.

Comparison is done of three techniques chosen amongst the most commonly used in the studies conducted on water pollution caused by oil products:

—infrared spectroscopy;
—ultraviolet spectrofluorimetry;
—gas chromatography.

To enable the application of these techniques to very low concentrations of hydrocarbon all these techniques require a concentration step.

This can be carried out, either by the liquid–liquid extraction of the hydrocarbons with the assistance of an organic solvent and in this case, the concentration ratios are poor, or by the trapping technique, where the hydrocarbon concentration is effected by adsorption on a resin, followed by desorption by a relatively small amount of an appropriate organic solvent. The water amounts so treated can be 10 to 100 larger than those treated by a liquid–liquid extraction.

A trapping system is described which has been used in an initial phase for the detection and quantitation of aromatic hydrocarbons up to C_9. The operational conditions are described for the measurement of heavier hydrocarbons (sampled water quantity and flow, desorption solvent, amount of resin).

The importance of the concentration method and of the sensitivity of the analytical techniques chosen for the definition of the detection threshold of the dissolved hydrocarbons is pointed out.

In addition to the detection level problems for certain types of hydrocarbons, the selectivity of these techniques is considered.

---

†Presented at the Workshop on the Chemistry and Analysis of Hydrocarbons in the Environment, Barcelona, November 1981.

In conclusion, the various methods are compared amongst themselves from the point of view of their application, of the quality of the results which are obtained and of their applicability to the different real water samples polluted by hydrocarbons.

KEY WORDS: Water pollution; hydrocarbons; trapping; extraction.

INTRODUCTION

La mesure de la concentration des hydrocarbures dissous dans l'eau s'est longtemps effectuée par gravimétrie (méthode du Dr Coin)[1] puis par spectrometrie infra-rouge (norme Afnor 90203),[2] la recherche de concentrations de plus en plus faibles (de l'ordre de 10 μg/l) et également de produits plus nocifs tels que les hydrocarbures aromatiques polynucléaires, a entraîné l'utilisation de la spectrofluorimétrie ultraviolette.[3,4,5,6] Aucune de ces deux méthodes spectrales ne peut prétendre analyser la totalité des hydrocarbures dissous. L'emploi de la chromatographie en phase gazeuse, avec un détecteur approprié, permet de s'affranchir d'un étalonnage, à condition que tout le produit injecté soit élué de la colonne. Nous avons comparé ces différentes méthodes dans le cas d'une pollution pétrolière artificielle.

Nous avons utilisé deux méthodes de concentration:
—l'extraction liquide–liquide;
—l'adsorption sur résine macromoléculaire, suivie d'une désorption par un solvant approprié.

Etant donné les faibles quantités d'eau traitées par extraction liquide–liquide cette méthode fournit des rapports de concentration faibles (environ 20 fois) et nécessite l'utilisation de méthodes de détection sensibles.

Depuis plusieurs années, de nombreuses publications, relatives à l'analyse des polluants, rapportent l'utilisation de résines poreuses de polymères comme adsorbants de composés organiques solubles dans l'eau à l'état de traces. Ainsi plusieurs auteurs ont mis en évidence la recupération des hydrocarbures aromatiques en utilisant la résine XAD-2.[7,8,9,10] Parmi ceux-ci Petroff *et coll.*[11,12] se sont attachés à doser les hydrocarbures aromatiques légers dans des nappes phréatiques polluées, d'autres auteurs tels que Strup *et coll.*[13] ont étudié la séparation d'hydrocarbures aromatiques plus lourds, dans des eaux polluées principalement par des hydrocarbures aromatiques polynucléaires.

C'est en s'appuyant sur ces travaux, que nous avons choisi ce type de résine pour piéger les hydrocarbures totaux dissous. Cette méthode de concentration, par l'obtention de taux de concentration plus élevés que l'extraction liquide–liquide, devrait permettre d'abaisser le seuil de détection de certaines techniques de dosages. Cette méthode nécessite une étude du débit de filtration de l'eau et de la technique de désorption afin d'optimiser les taux de piégeage et de récupération.

PARTIE EXPERIMENTALE

Preparation des echantillons

Le Système de Simulation de la Pollution: La pollution de l'eau douce a été effectuée dans un bac en acier galvanisé d'une contenance de 6 m3 et d'un diamètre de 2,20 m. L'eau utilisée est celle distribuée par la ville de Rueil–Malmaison. Le polluant est versé à la surface. L'homogénéisation des hydrocarbures dissous dans la colonne d'eau est effectuée par une pompe "Japsco" d'une puissance de 2 CV. L'eau prévelée à une vingtaine de centimètres sous la surface libre est refoulée dans le fond du bac. Ce dispositif permet une homogénéisation sans mettre en suspension de particules d'hydrocarbures. A la sortie de la pompe un "by-pass" muni d'une vanne à pointeau permet de prélever l'eau du bac à différents débits. Les pièges de concentration sont fixés à ce dispositif. Un compteur Micro-oval II (Emerson Electrique-France SA) est utilisé afin de connaître le volume d'eau prélevé lors de l'utilisation des pièges de concentration.

Caractéristiques du Polluant: Le polluant est constitué d'une coupe d'un pétrole brut Nigerian Light isolée par distillation et dont l'intervalle de point d'ébullition est compris entre 240 et 480°C.

L'élimination de la fraction supérieure à 480°C permet de supprimer les composés polaires et les asphaltènes et d'obtenir ainsi une coupe pétrolière constituée d'hydrocarbures dosables par chromatographie en phase gazeuse. L'éliminsyion de la fraction légère déjà étudiée, correspond au soin d'éviter les pertes par évaporation.

Un fractionnement de cette coupe par grandes familles chimiques a été effectuée par chromatographie liquide basse pression. La répartition est la suivante:

—hydrocarbures saturés:	67% pds
—hydrocarbures aromatiques:	27% pds
—résines:	6% pds

L'analyse, par spectrométrie de masse, de la fraction: hydrocarbures saturés, nous a permis d'obtenir la répartition suivante (en % vol.):

—(n + iso) paraffines:	35,7
—naphtènes à 1 noyaux:	7,7
—naphtènes à 2 noyaux:	8,5

—naphtènes à 3 noyaux:	6, 9
—naphtènes à 4 noyaux:	5, 2
—naphtènes à 5 noyaux:	2, 1
—naphtènes à 6 noyaux:	0, 9

Par cette méthode, nous avons également déterminé, à partir de la fraction hydrocarbures aromatiques, les teneurs en mono, di, tri + polyaromatiques respectivement égales à 6, 5; 11, 3; 9, 2 % pds.

Methodes de concentration

La Fig. 1 rassemble les modes opératoires suivis selon la technique d'analyse choisie.

Extraction Liquide–Liquide: L'extraction s'effectue dans une ampoule à décanter en agitant l'eau et le solvant pendant dix minutes.

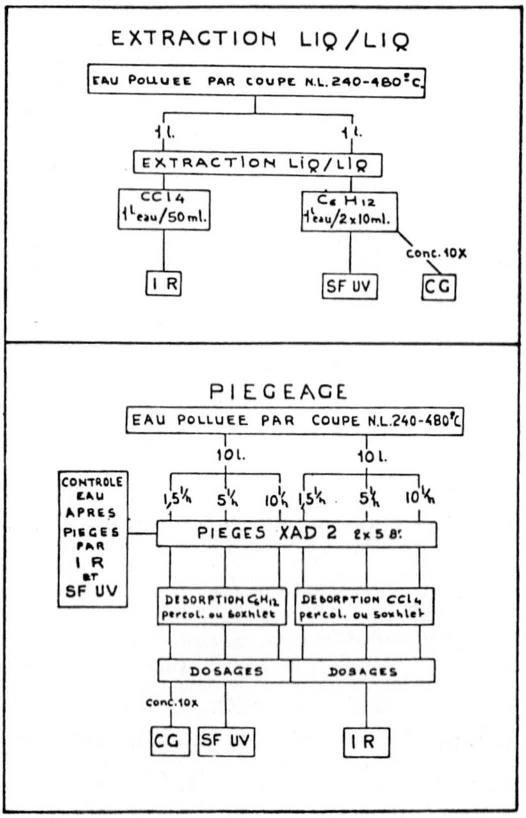

FIGURE 1 Schéma d'expérimentation.

Deux types de solvants sont utilisés pour extraire les hydrocarbures:

—le tétrachlorure de carbone (Merck pour spectrométrie—Art. 2209), lorsque le dosage est effectué par spectrométrie infra-rouge. Dans ce cas le rapport eau/solvant est de 1 1/50 ml, soit un rapport de concentration de 20.

—le cyclohexane (Pestipur de SDS—13124 Peypin) pour les dosages effectués par spectrofluorimétrie UV (SFUV) et par chromatographie en phase gazeuse (CPG) en traitant 1 litre d'eau par deux fois 10 ml de solvant, soit un rapport de concentration de 50. La pureté du solvant est testée par chromatographie en phase gazeuse dans les mêmes conditions opératoires que celles utilisées pour le dosage des extraits (solvant concentré 10 fois avant injection) (Fig. 2(a)).

Piégeage sur Résine Adsorbante

● *Adsorption* Le système est constitué d'un piège en acier de 38 cm de longueur, d'un diamètre intérieur de 10 mm, et rempli de 10 g de résine Amberlite-XAD-2 (Rohm et Haas-France SA). Avant utilisation la résine est mise dans une colonne (diamètre intérieur: 3 cm) et nettoyée par percolations successives de méthanol, de dichlorométhane, de tétrachlorure de carbone et de cyclohexane. Après ce traitement, on effectue une désorption comme pour un piège réel et on analyse l'extrait par chromatographie en phase gazeuse dans les mêmes conditions analytiques que pour les extraits réels (Fig. 2(b)).

Le piége contenant la résine est obturé aux deux extrémités par un bouchon de laine de verre (préalablement nettoyée par passage au soxhlet avec du tétrachlorure de carbone) maintenu par une grille métallique en acier inoxydable et un raccord union en acier.

Ce piège est monté sur le "by-pass" du système de pompage de l'eau du bac. Une vanne pointeau permet de régler les débits qui seront de 1, 5; 5 et 10 l/h.

Le volume d'eau filtrée est de 10 litres quel que soit le débit choisi.

● *Désorptions* Pour désorber les hydrocarbures fixés sur la résine, deux méthodes ont été employées.

—Désorption par simple percolation d'un solvant organique

Elle se fait en sens inverse du prélèvement d'eau. L'élution des hydrocarbures adsorbés sur la résine s'effectue par 50 ml de tétrachlorure de carbone pour les dosages par spectrométrie infra-rouge ou par 30 ml de cyclohexane pour les dosages par spectrofluorimétrie UV et chromatographie en phase gazeuse.

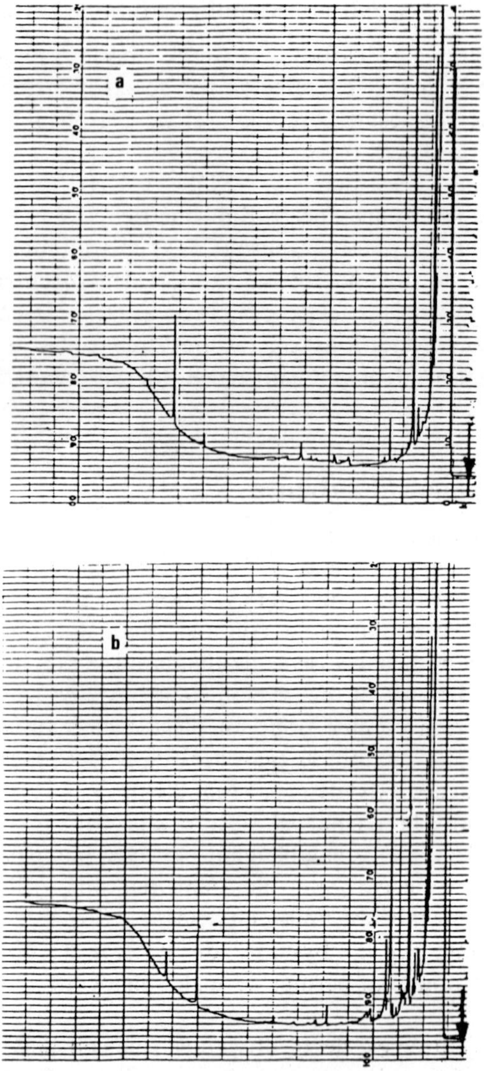

FIGURE 2 Chromatogrammes tests: (a) du cyclohexane "PESTIPUR"; (b) du cyclohexane après extraction de la résine XAD-2 avant les essais de piégeage.

—Désorption en continu à l'aide d'un soxhlet modifié

Nous utilisons un soxhlet modifié (Fig. 3). La résine contenue dans le piège est transvasée dans le soxhlet. Le volume de solvant utilisé est de 50 ml, le temps d'extraction est de 8 h.

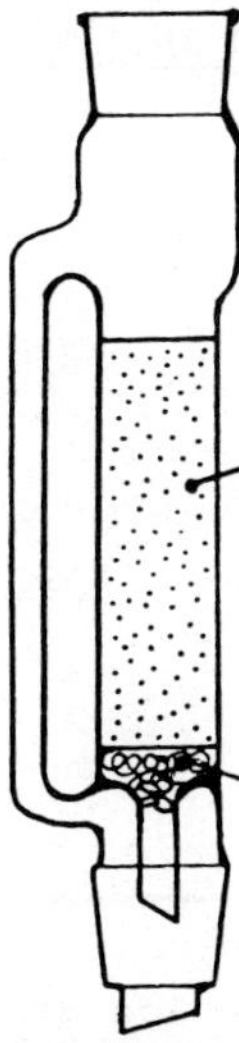

FIGURE 3 Appareil soxhlet modifié (extraction en continu de la résine XAD-2).

Techniques de dosage

Chromatographie en Phase Gazeuse (CPG): Le chromatographie utilisé est un appareil Carlo-Erba-Fractovap 4160 équipé d'un injecteur "on-column" et d'un détecteur à ionisation de flamme. Le volume injecté est de 1 μl. La séparation se fait sur une colonne capillaire de verre Cp SIL 5 (Chrompack France SA) dont la longueur est de 27 mètres, le diamètre intérieur de 0, 3 mm et l'épaisseur du film de 0, 29 micron. La programmation de température s'effectue de 60°C à 310°C à une vitesse de programmation de 10°C/mn.

L'intégration des pics est faite par un ordinateur Hewlett Packard 3354.

Le dosage quantitatif est effectué par ajout d'un standard interne (le n-décane) dans l'extrait au cyclohexane; celui-ci est concentré 10 fois avant analyse.

Spectrofluorimétrie UV (SFUV): L'appareil utilisé est un spectrofluorimètre Fica 55 MK II. Les conditions expérimentales sont fixées dans la méthode Igoss.[14] Pour la longueur d'onde d'excitation $\lambda_{exc} = 310$ nm, on enregistre la spectre d'émission et on mesure le signal à $\lambda_{ém} = 360$ nm. Les mesures se font à température ambiante.

L'étalonnage est effectué soit à partir de la coupe pétrolière Nigerian Light 240–480°C dans le cyclohexane, soit à partir de la fraction aromatique de cette même coupe diluée dans le même solvant (Fig. 4).

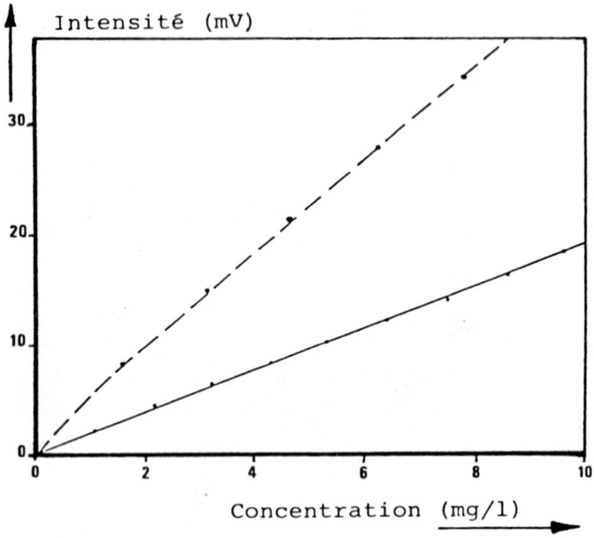

FIGURE 4 Courbe d'étalonnage de la spectrofluorimétrie ultra-violette.——: à partir de la coupe Nigerian Light 240–480°C. - - -: à partir de la fraction aromatique de cette coupe.

Spectrométrie Infra-rouge (IR): Le spectrométre employé est un Beckman 4240 avec possibilité d'expansion d'échelle en ordonnée.

La méthode de dosage utilisée est voisine de la méthode Afnor 90 203,[2] seule la purification sur florisil en a été supprimée. Les cellules sont en quartz de 1 et 5 cm de trajet optique. Le dosage se fait en différentiel, l'absorbance est mesurée à 2925 cm^{-1}. La concentration en hydrocarbures dans l'extrait est déterminée par comparaison avec les courbes d'étalonnages établies soit avec la coupe polluante, soit avec sa fraction aromatique (Fig. 5).

RESULTATS

Les résultats obtenus figurent dans les tableaux 1 et 2. Le tableau 1 regroupe les concentrations en hydrocarbures obtenues après les extractions liquide–liquide pour trois essais indépendants.

Chacun de ces essais a été fait en même temps qu'un essai de piégeage, de manière à contrôler la variation possible du degré de pollution du bac d'expérience.

Le tableau 2 rassemble, d'une part les concentrations en polluant retrouvées après piégeage à diffèrents débits et d'autre part les taux d'adsorption et de désorption. Ceux-ci ont été mesurés après extraction

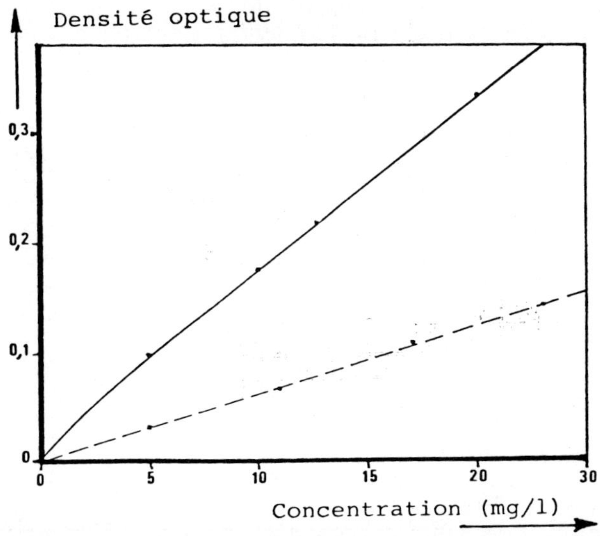

FIGURE 5 Courbe d'étalonnage de la spectrométrie infra-rouge. ——: à partir de la coupe Nigerian Light 240–480°C. - - -: à partir de la fraction aromatique de cette coupe.

TABLEAU I

Extraction liquide–liquide sur 1 litre d'eau, teneurs en hydrocarbures dissous dans l'eau (μg/l)

CP G	SF UV		IR	
Etalon interne nC10	Etalon charge polluante	Etalon fract. arom.	Etalon charge polluante	Etalon fract. arom.
300	237	101	77	252
280	247	106	54	176
300	245	105	53	173

liquide–liquide par la spectrofluorimétrie UV, en raison de sa plus grande sensibilité.

On définit:

—*le taux de piégeage*: pourcentage des teneurs en hydrocarbures dans l'eau en amont et en aval du piège;

—*le taux de désorption*: pourcentage entre la teneur en hydrocarbures désorbés et celle des hydrocarbures réellement adsorbés.

TABLEAU II
CONCENTRATION PAR PIEGEAGE XAD-2
Volume d'eau filtrée: 10 litres
Influence du débit

	Contrôle par SF UV		Teneurs en hydrocarbures (μg/l eau)				
			CPG	SF UV		IR	
Débits l/h	Taux de piégeage (%)	Taux de désorption (%)	Etalon interne nC10	Etalon charge polluante	Etalon fract. arom.	Etalon charge polluante	Etalon fract. arom.
1,5	93	58 (soxhlet)	133	127	50	128	398
5	79	39 (percolation)	87	76	30	152	472
10	79	79 (soxhlet)	152	154	61	112	348

DISCUSSION

Extraction liquide-liquide

Les teneurs mesurées par spectrométrie infra-rouge avec la coupe entière comme étalon sont trés inférieures à celles trouvées par les deux autres méthodes. Ce résultat est explicable par le principe même du dosage puisque la méthode infra-rouge ne détecte, à la longueur d'onde de mesure ($2925\,cm^{-1}$) que les CH, CH_2, CH_3 des groupes saturés. Dans ces coupes pétrolières, les hydrocarbures aromatiques, qui sont en quantité importante, sont mal pris en compte, car ils ne comportent que des chaînes alkyles courtes de quelques atomes de carbone. Par ailleurs les hydrocarbures dissous ne forment pas une partie aliquote du mélange employé pour la pollution et les hydrocarbures aromatiques sont plus solubles que les saturés, ce qui diminue encore la concentration que l'on peut calculer par infra-rouge.

Les résultats trouvés par chromatographie et spectrofluorimétrie ultra-violette semblent proches les uns des autres, il faut cependant se garder de conclure à leur quasi identité: en effet si la chromatographie, dans les conditions où nous avons travaillé, semble doser la totalité du polluant, la spectrofluorimétrie, elle, est encore plus sélective que l'infra-rouge mais dans un sens opposé: elle ne répond qu'aux hydrocarbures aromatiques polynucléaires et de plus avec un coefficient de sensibilité très différent selon les espèces d'hydrocarbures.

Tels qu'ils apparaissent ces résultats indiquent clairement une solubilité dans l'eau nettement plus élevée pour les hydrocarbures aromatiques. En spectrométrie I.R., le dosage à partir de la fraction aromatique, donne une concentration plus élevée (elle est en fait multipliée par 3), chiffre optimiste bien entendu puisqu'une partie des paraffines doit être solubilisée. En spectrofluorimétrie UV, l'emploi des deux étalons conduit à un effet opposé puisque cette fois un même poids de référence (fraction aromatique) donne un signal plus intense que pour la coupe intégrale et de fait, le résultat est divisé par deux (facteur 2 et non facteur 3 comme l'IR car les hydrocarbures mono-aromatiques montrent peu ou pas de fluorescence).

Ceci nous ramène à des chiffres plus cohérents en ce qui concerne les méthodes spectrales, mais qui restent bien inférieurs à ceux trouvés par CPG. Ces différences tiennent évidemment au fait que chaque méthode ne dose pas les mêmes produits et de ce point de vue, toujours dans notre étude, la CPG paraît plus appropiée puisque plus générale. Actuellement, elle apparaît moins sensible: dans nos conditions opératoires le chromatogramme est très étalé, ce qui permet d'avoir une information descriptive sur la qualité du polluant, mais diminue la détectabilité en hydrocarbures totaux. A titre indicatif, sur une extraction liquide–liquide, compte tenu des rapports de concentration employés, on peut évaluer les limites de sensibilité (en μg.hydrocarbures/l.eau) à 1 pour la SFUV, à 20 pour l'IR, à 100 pour la CPG. C'est ainsi que nous avons été amenés à concentrer davantage les hydrocarbures avant leur quantification par CPG.

Piegeages

En passant une eau contaminée sur une colonne de résine macromoléculaire, lipophile et hydrophobe, telle que l'amberlite XAD-2, les composés organiques sont adsorbés et concentrés dans le rapport volume d'eau filtrée sur volume du solvant d'extraction.

Les différents bilans ont été calculés à partir des teneurs déterminées sur l'eau, par extraction liquide–liquide, avant et après le piège. La mesure finale a été faite par spectrofluorimétrie UV, seule capable de détecter les hydrocarbures dans l'eau filtrée. Les résultats sont consignés dans le Tableau II.

Pour l'adsorption le meilleur taux de piégeage est obtenu pour le débit le plus lent (1,5 l/h) au-delá, il décroît pour se stabiliser aux environs de 79%. Le temps de contact eau-résine joue donc un rôle important dans l'efficacité du piégeage et un débit inférieur à 5 l/h semble recommandé.

Il nous faut souligner que le nombre d'expériences est limité et que les conclusions concernant la désorption, surtout par la méthode d'extraction en continu, doivent être considérées avec prudence. En tout état de cause cette technique donne de meilleurs rendements que la percolation simple mais quelques artéfacts nous ont montré que nous ne la maîtrisons pas encore complètement. En particulier, nous avons constaté qu'il est très important de veiller à un réglage constant et suffisamment lent du reflux car la présence d'air dans la résine crée dans celle-ci des chemins préférentiels qui nuisent au bon rendement de la désorption.

Dans les calculs de rendement par spectrofluorimétrie on est amené à supposer que les hydrocarbures piégés (ou non piégés) sont partie aliquote des hydrocarbures dissous ce qui n'est pas prouvé dans l'état actuel de l'étude. Cette restriction paraît cependant mineure puisque les teneurs obtenues par CPG ou par SFUV restent proches.

Pour ce qui est des quantités d'hydrocarbures mesurées, ici encore les quantités trouvées par CPG et SFUV sont trés proches l'une de l'autre, mais il nous faut renouveler la mise en garde au sujet de la similitude de ces résultats, nous en voulons pour preuve la différence entre les chromatogrammes de la charge, de l'extrait séparé par extraction liquide–liquide et de l'extrait du piège (Fig. 6a, b, c). On y voit nettement que la nature des hydrocarbures a changé (en particulier la plus grande partie des paraffines normales ne s'est pas dissoute) et qu'il est illusoire de doser l'extrait en prenant la charge pour référence, dans le cas du dosage spectrofluorimétrique. Pour la chromatographie en phase gazeuse ceci a peu d'importance, car les résponses des divers types d'hydrocarbures sont peu différentes les unes des autres.

Si l'on s'attache aux valeurs absolues mesurées, on s'aperçoit que le piégeage donne des valeurs à peu près deux fois plus faibles que l'extraction liquide–liquide. Les valeurs trouvées par spectrométrie IR (Tableau II) sont beaucoup trop élevées. Cet excès pourrait être causé par la présence dans l'eau utilisée de composés qui ne sont pas des hydrocarbures et concentrés sur la résine au cours du piégeage: c'est ainsi que dans des essais antérieurs des composés chlorés légers à faible nombre de carbones avaient été mis en évidence dans l'eau.

CONCLUSION

Pour le dosage des hydrocarbures dissous dans l'eau, l'extraction liquide–liquide apparaît comme une bonne méthode de concentration des hydrocarbures et même si le rapport eau/solvant reste relativement faible (=20) il n'est pas indispensable, comme le montre le Tableau I, de recourir aux méthodes globales plus sensibles dans le cas de mélanges

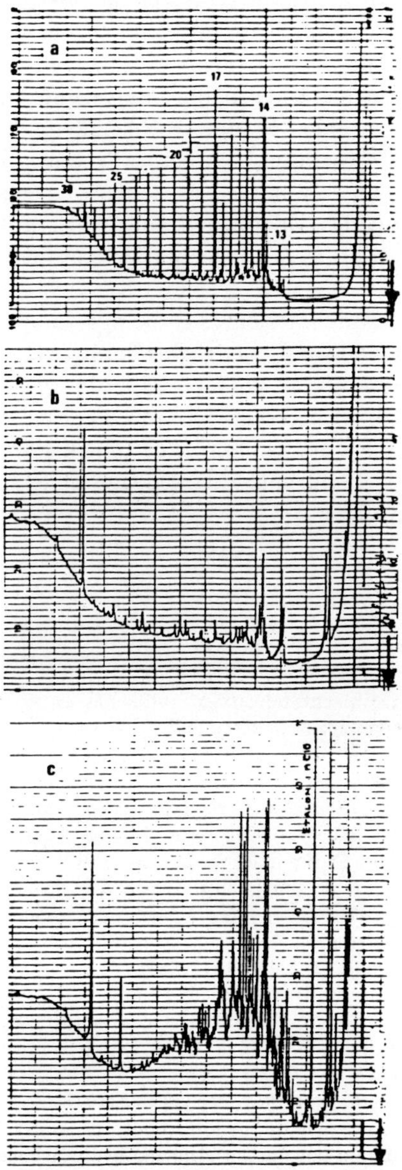

FIGURE 6 Chromatogrammes: (a) du polluant (coupe Nigerian Light 240–480°C). (b) des hydrocarbures dissous dans l'eau, séparés par extraction liquide–liquide. (c) des hydrocarbures dissous dans l'eau, concentrés sur piège de résine XAD-2.

complexes: on voit que la chromatographie en phase gazeuse convient parfaitement à des concentrations de 200 à 100 ppb. En fait, le choix d'une des méthodes dépendra du problème posé, des considérations de disponibilité d'appareillage et de délai d'obtention des résultats, mais la chromatographie est la méthode la plus universelle dans la mesure où la totalité du polluant est éluée ce qui semble devoir être le cas si l'on ne considére que les hydrocarbures dissous, à l'exclusion de toute émulsion. En tout état de cause, l'examen des teneurs mesurées par extraction liquide–liquide montre des valeurs extrêmes différant d'un facteur 2 au plus, ce qui paraît tout à fait acceptable compte tenu du niveau de pollution étudié.

Pour des concentrations inférieures à 50 ppb la spectrofluorimétrie UV convient, avec les restrictions citées plus haut, qui concernent la haute sélectivité de la méthode. Pour ces teneurs le dosage par chromatographie gazeuse devient difficile et nécessite un taux d'enrichissement plus élevé; le piégeage sur résines est alors impératif, ce qui justifie la poursuite de nos études pour améliorer l'efficacité du piégeage et le taux de récupération des hydrocarbures adsorbés.

References

1. Norme T Afnor 90202, Effluents aqueux des raffineries de pétrole—Dosage des matières organiques en suspension dans l'eau extractibles à l'hexane.
2. Norme T Afnor 90203, Fév. 1979, Effluents aqueux des raffineries de pétrole—Dosage des hydrocarbures totaux.
3. D. C. Gordon and P. D. Keizer, Technical report No. 448, Department of the Environment, Fisheries and Marine Service, Marine Ecology Laboratory, Bedford Institute of Oceanography, Dartmouth, Nova Scotia, (1974).
4. P. D. Keizer, D. C. Gordon Jr. and J. Dale, *J. Fish. Res. Board Can.*, **34,** 347–353 (1977).
5. U. Franck, *Analyt. Quality Contr. Newsl.,* U.S. Environmental Protection Agency, Cincinnati, Ohio, **18,** 9–10 (1973).
6. R. J. Law, *Mar. Pollut. Bull.*, **12,** 5, 153–157 (1981).
7. G. A. Junk, J. J. Richard, M. D. Griesen, D. Witiak, J. L. Witiak, M. D. Argnello, R. Vick, H. J. Suee, J. S. Fritz and G. U. Galder, *J. Chromatogr.*, **99,** 745–762 (1974).
8. C. D. Chriswell, R. L. Ericson, G. A. Junk, K. W. Lee, J. S. Fritz and H. J. Svec, *J. Amer. Water Works Assoc.*, **69,** 674–699 (1977).
9. S. F. Stefan and J. F. Smith, *Water Res.*, **11,** 339–342 (1977).
10. B. Olufsen, Polynuclear aromatic hydrocarbons: chemistry and biological effects. International Symposium on polynuclear aromatic hydrocarbons, 4th Battelle Columbus Laboratories, 333–343 (1979).
11. N. Petroff and C. Gatellier, Concawe—UCSIP project, IFP Report No. 25 346, (1977).
12. N. Petroff, Rapport IFP No. 25 502, (1977).
13. P. E. Strup, J. E. Wilkinson and P. W. Jones, Polynuclear aromatic hydrocarbons, Carcinogenesis, Raven Press, New York, **3,** 131–138 (1978).

14. International Oceanographic Commission/World Meteorological Office. Guide to the Operational procedures for the IGOSS project on marine pollution (petroleum) monitoring—UNESCO Paris, (1976).

Résumé

De nombreuses méthodes physico-chimiques sont employées pour doser les hydrocarbures dissous, à l'état de traces, dans les eaux souterraines ou de surface.

Trois techniques, choisies parmi les plus couramment utilisées dans les études de pollution des eaux par des produits pétroliers, sont comparées: la spectrométrie infra-rouge, la spectrofluorimétrie ultra-violette, la chromatographie en phase gazeuse. Pour être appliquées à des dosages d'hydrocarbures à faible concentration ces techniques nécessitent toutes une phase de concentration. Celle-ci peut s'effectuer, soit par extraction liquide–liquide, soit par piégeage sur résine. Cette dernière technique permet d'envisager des rapports de concentrations supérieurs à ceux obtenus par extraction liquide–liquide. On décrit un système de piégeage dont la mise au point a été faite pour détecter les hydrocarbures légers jusqu'en C9 dans une nappe phréatique polluée. Les conditions opératoires requises pour doser les hydrocarbures plus lourds (volume et débit d'eau échantillonée, nature du solvant de désorption) sont décrites.

L'importance simultanée de la méthode de concentration, de la sensibilité et de la sélectivité des techniques analytiques choisies a été mise en évidence pour la définition des seuils de détection des hydrocarbures dissous. Dans le cas d'une pollution artificielle provoquée par une coupe pétrolière 240–280°C les résultats obtenus par les différentes techniques proposées sont comparés, leur représentativité pour le dosage des hydrocarbures dissous est discutée.

Application des Capteurs Automatises DTO, COT et DCO au Dosage des Hydrocarbures dans L'Eau

C. LEGRAND,† B. CAPDEVILLE,† H. ROQUES,† M. ANGLES‡

†*Laboratoire de Chimie et Génie de l'Environnement, INSA. Av; de Rangueil; 31077 Toulouse Cedex, France*

‡*Centre de Recherche Elf-Solaize; St. Symphorien D'Ozon 69360, France*

(*Received November 27, 1981*)

Among the objectives of this study was the determination of the oxidation yields for pure hydrocarbons in water (alkanes, alkenes, aromatics), following the TOD, TOC and COD methods, and the extraction efficiency of carbon tetrachloride followed by infra-red dosage (NF T 90 203).

For the global oxidative pollution tests, the best oxidation yields were obtained with TOC, followed by the TOD; COD results were clearly poor. The solvent extraction procedure afforded reliable results.

Correlation studies with the methods proposed enable the grouping of hydrocarbons in two main categories, according to the oxidation criteria: aliphatics and aromatics. Generally, it can be said that the correlations found were satisfactory.

KEY WORDS: Hydrocarbons in water, Total oxygen demand (TOD), carbon organic total (COT), chemical oxygen demand (COD), infra-red method (IR).

INTRODUCTION

Le dosage des hydrocarbures totaux en phase aqueuse fait l'objet actuellement en France de protocoles expérimentaux normalisés: normes API 73-358[1] et AFNOR T 90 203[2] (extraction par le tétrachlorure de carbone, puis dosage à l'infra-rouge (IR) à une longueur d'onde

---

†Presented at the Workshop on the Chemistry and Analysis of Hydrocarbons in the Environment, Barcelona, November 1981.

d'absorption donnée), norme AFNOR T 90 202[3] (dosage par gravimétrie des matières organiques d'effluents de raffinerie extractibles à l' hexane) et projet de norme AFNOR T 90 114[4] (extraction par le CCl_4, puis dosage à l'infra-rouge à quatre longueurs d'onde, la gamme étalon étant réalisée à partir d'un mélange connu d'hydrocarbures).

D'autre part, des critères globaux de pollution normalisés (DCO, DBO_5) sont utilisés pour déterminer les teneurs en matières organiques dans des rejets aqueux, quelles que soient leurs origines. Depuis quelques années d'autres techniques d'oxydation (DTO, COT), présentant l'avantage d'être automatisables, ont été proposées.

Le but de notre étude est de comparer ces diverses méthodes pour évaluer de manière globale des teneurs en hydrocarbures de solutions aqueuses.

Les objectifs de notre travail concernent:

— les déterminations préalables pour plusieurs hydrocarbures purs (alcanes, alcènes, aromatiques), des différents rendements d'oxydation obtenus avec les méthodes DTO, COT et DCO et d'extraction au CCl_4 selon les différentes normes AFNOR.

— la recherche de corrélations entre ces quatre méthodes et plus particulièrement entre les tests d'oxydation et l'infra-rouge.

PROTOCOLE EXPERIMENTAL

Nous rappellerons brièvement les principes des différentes techniques analytiques utilisées au cours de notre étude.

Demande totale en oxygène[5–6] Mesure globale de la pollution hydrocarbonée contenue dans une eau. La DTO représente la quantité d'oxygène consommée au cours de l'oxydation catalytique, sur du platine et à 900°C, de la matière organique dissoute ou émulsionnée. Les expériences ont été faites avec un appareil IONICS 225 de la DOW Chemical. Cette méthode présente notamment l'avantage de donner des réponses très rapides (quelques minutes) et il existe d'autre part des appareils process, entièrement automatiques.

Carbone organique total[5,6,7] Mesure la teneur globale en carbone organique, par oxydation thermique à 900°C et sur catalyseur céramique des matières organiques. L'appareil utilisé est un COT mètre IONICS type 1258, qui présente les mêmes avantages que le DTO mètre précédent.

Quelques remarques sont toutefois à faire quant à l'utilisation de cette technique pour le dosage de produits très volatils. En effet, la mesure directe du carbone organique total nécessite une acidification préalable de l'échantillon, pour stripper le carbone minéral de la phase liquide par bullage d'azote, mais au cours de cette étape, une grande partie des hydrocarbures volatils est aussi éliminée. Dans ce cas, il est alors nécessaire de mesurer le carbone organique de manière indirecte, c'est-à-dire à partir du carbone total (CT) auquel on soustrait la contribution du carbone minéral (CM) de la phase aqueuse.

Demande chimique en oxygène[8] Selon certains auteurs,[9] il apparait qu'une partie des composés volatils sont éliminés lors de l'introduction de l'acide sulfurique concentré dans le milieu réactionnel. Ces auteurs proposent alors une modification du protocole de la norme, en introduisant ce réactif par le tube de réfrigération. Dans le cadre de notre travail, nous avons tout de même suivi scrupuleusement les indications de cette norme, afin de se placer dans les mêmes conditions opératoires que celles utilisées dans les laboratoires agréés d' analyses.

Dosage à l'Infra-rouge D'une manière générale, les hydrocarbures en phase aqueuse sont extraits par du CCl_4, puis dosés en infra-rouge. Ces méthodes sont normalisées: norme AFNOR T 90 203,[2] projet T 90 114,[4], les principales différences résidant dans le choix des produits d'étalonnage et des longueurs d'onde d'absorption.

— NF T 90 203: étalonnage par rapport à l'hydrocarbure utilisé et mesure à 2920 cm^{-1}.

— projet T 90 114: étalonnage par rapport à un mélange étalon (37, 5% isooctane, 37, 5% héxadécane, 25% benzène) et mesure de la somme des absorbances: 2960 cm^{-1} (CH_3), 2920–2850 cm^{-1} (CH_2) pour les hydrocarbures aliphatiques et 3040 cm^{-1} (CH) pour les hydrocarbures aromatiques.

L'analyseur IR utilisé est l'Acculab 4 de Beckman.

Compte tenu des propriétés physico-chimiques des hydrocarbures (volatilisation et solubilisation), il est nécessaire d'adopter un protocole particulier, tout en tenant compte de certains impératifs technologiques des appareillages de DTO et COT.

Les solutions étalons ont été préparées selon la méthode suivante: injection de 5, 10, 15, 20, 25 μl d'hydrocarbure pur dans 250 cc d'eau permutée (ces concentrations correspondent approximativement à des DTO théoriques comprises entre 50 et 250 mgO_2/l, ce qui permet de rester dans la gamme de linéarité de l'appareil). L'homogénéité des échantillons étant un préalable aux mesures de DTO et COT (injection automatique d'une prise d'échantillon de 20 μl), les solutions étalons sont passées aux ultrasons afin de disperser les hydrocarbures en microgouttelettes. D'autre part, pour avoir une meilleure précision lors de l'étude des corrélations, les mesures relatives aux tests d'oxydation (DTO, COT, DCO) sont réalisées sur les mêmes solutions standard et immédiatement après l'étape d'homogénéisation. Pour les mesures en infra-rouge, les solutions étalons sont préparées in situ et indépendamment des autres car les normes conseillent d'effectuer l'extraction des hydrocarbures directement dans le flacon d'origine.

Les expériences ont été effectuées sur les hydrocarbures suivants: alcanes (du C_6 au C_{15}), alcènes (du C_6 au C_{10}), aromatiques (benzène, *m*-xylène, toluène, ethylbenzène, propylbenzène et cumène).

RESULTATS EXPERIMENTAUX

Détermination des rendements d'oxydation

Pour évaluer précisément les rendements d'oxydation de chaque hydrocarbure pour les trois méthodes (DTO, COT, DCO), nous déterminons au préalable, à partir des mesures correspondant aux solutions standard l'équation de la droite d'oxydation expérimentale. Les rendements d'oxydation correspondent alors aux rapports des pentes entre les droites expérimentale et théorique (Tableau I).

TABLEAU I
Rendements d'oxydation, pour différents hydrocarbures et selon les techniques DTO, COT, DCO

Composés	DCO (% d'oxydation)	DTO (% d'oxydation)	COT (% d'oxydation) CT-CM	COTexp.
Décane	31	42	46	28
Undécane	41	43	36	26
Dodécane	25	40	53	22
Tridécane	44	42	47	11
Tétradécane	19	40	33	19
Pentadécane	25	41	39	17
Nonène	41	47	48	10
Décène	47	51	56	18
Benzène	61	78	89	11
Toluène	58	64	96	5
m-Xylène	53	61	100	6
Ethylbenzène	72	78	94	6
Propylbenzène	65	73	91	5
Cumène	66	77	100	9

Ces résultats montrent d'une part que les oxydations sont différentes selon le type d'hydrocarbure, mais restent comparables à l'intérieur d'une même famille. Ainsi par exemple pour les aromatiques, le nombre d'atomes de carbone ou la position du groupement fonctionnel ne modifient pas le degré moyen d'oxydation. Pour les hydrocarbures aliphatiques dont le nombre d'atomes de carbone est inférieur à 10, la détermination de ces rendements est rendue particulièrement difficile, car ces hydrocarbures sont d'une part insolubles dans l'eau et d'autre part très volatils.

Les rendements que l'on peut être amené à déterminer, malgré de nombreuses précautions lors de la préparation des échantillons, sont alors incorrects, et ces résultats ne peuvent pas être pris en compte.

Si l'on compare, pour un même type d'hydrocarbure, les rendements obtenus selon les différentes méthodes étudiées, on remarque que la méthode d'oxydation la moins performante est la demande chimique en oxygène. De plus, on doit noter que les mesures de DTO sont toujours inférieures aux mesures de COT obtenues selon la méthode indirecte. Ceci permet en outre d'introduire une remarque très importante quant à l'utilisation de cette dernière technique pour le dosage en général des matières organiques volatiles. Si ces appareillages ne disposent pas d'un circuit de mesure spécifique du carbone organique volatil (COV), il est indispensable d'utiliser des appareils donnant le carbone total plutôt que le carbone organique total qui necéssite toujours la phase d'acidification des échantillons.

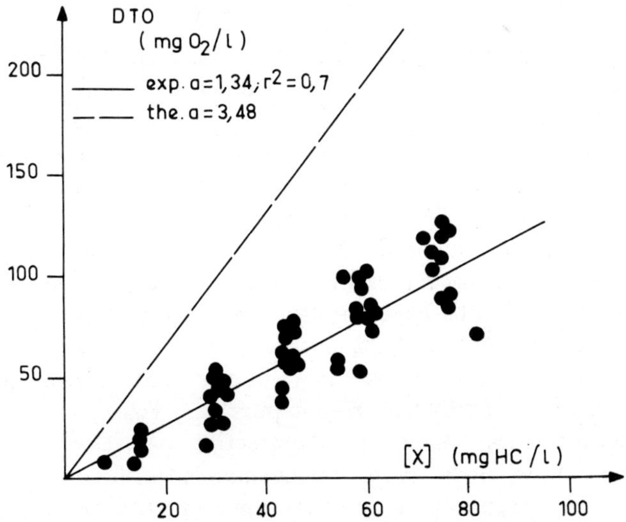

FIGURE 1 Oxydation des alcanes DTO mètre.

La principale conclusion de ces expériences est que le mesure du carbone total permet d'obtenir dans tous les cas de meilleurs rendements d'oxydation et pour les hydrocarbures aromatiques ces taux d'oxydation sont proches de 100%.

Si l'on regroupe l'ensemble des résultats expérimentaux obtenus pour chaque produit étudié (alcanes, alcènes, aromatiques) et en ne considérant que les composés dont le nombre d'atomes de carbone est supérieur à 10, on peut alors calculer des rendements moyens d'oxydation. Un exemple de corrélations est donné dans la Figure 1 pour la famille des alcanes.

Les résultats permettent alors de classer les hydrocarbures, du point de vue de leur oxydation par les différentes méthodes, en deux catégories: aliphatiques-aromatiques (Tableau II).

TABLEAU II
Rendements moyens d'oxydation, selon les méthodes DTO, COT, DCO, des hydrocarbures aromatiques et aliphatiques

Composés	DCO (%)	DTO (%)	COT (%) CT-CM	COTexp.
Aliphatiques	34	43	45	19
Aromatiques	62	72	95	7

Détermination des rendements d'extraction

Nous avons tout d'abord réalisé quelques essais préliminaires quant à l'utilisation de l'adsorbant Florisil prévu dans le protocole de la norme T 90 203, car les données de la littérature sont partagées sur ce point. Ford *et al.*[10] ont en effet constaté lors de cette étape, des pertes importantes d'hydrocarbures alors que Ranchet et Clement[11,12] montrent que l'on peut réduire l'adsorption de ces produits de 15 à 2% si certaines précautions sont prises: humidification préalable au CCl_4 de la colonne de Florisil, volume de rinçage représentant deux fois et demie le volume de la colonne. Au cours de nos expériences nous n'avons d'une part constaté aucune perte de produit lors du passage des solutions étalons sur du Florisil et d'autre part cette étape s'avère inutile puisque ces solutions sont constituées uniquement d'hydrocarbures purs dans l'eau.

Pour déterminer les rendements d'extraction par le CCl_4 de chaque hydrocarbure étudié, nous avons donc suivi le protocole de la norme NF T 90 203 à l'exception de la phase préliminaire correspondant au Florisil. L'étalonnage est fait à partir des solutions standard du produit étudié dissous dans du CCl_4 pur, les mesures en IR sont effectuées à la longueur d'onde de 2920 cm^{-1} qui correspond à la vibration de valence la plus intense des liaisons CH_2. Les rendements d'extraction (Tableau III) sont obtenus en faisant le rapport entre les valeurs expérimentales déduites de la courbe d'étalonnage et les valeurs théoriques.

TABLEAU III
Rendements d'extraction de différents hydrocarbures purs par le tétrachlorure de carbone

Composés	% d'Extraction
Décane	90
Undécane	94
Dodécane	96
Tridécane	98
Tétradécane	100
Pentadécane	98
Nonène	60
Décène	66
Benzéne	0
Toluène	100
m-Xylène	100
Ethylbenzène	96
Propylbenzène	98
Cumène	98

Ces résultats montrent que les extractions sont en général totales quel que soit l' hydrocarbure étudié, à l' exception du benzène. En effet, dans le cas de ce produit, si on suit scrupuleusement la norme T 90 203, le rendement déterminé est nul. Ceci est dû au fait que le benzène n'absorbe pas à cette longueur d'onde, mais si l'on effectue la mesure à 3040 cm^{-1}, le rendement est alors similaire à celui des autres composés (94%). Cette norme présente donc sur ce point, une lacune importante qui n'existe pas par contre pour les hydrocarbures aromatiques dont le noyau benzénique est substitué par des chaines aliphatiques.

En conclusion préliminaire de cette première partie de l'étude, on doit noter que le dosage en IR reste la meilleure méthode d'évaluation des hydrocarbures totaux en phase aqueuse, comparée aux tests globaux d'oxydation (DTO, COT, DCO) utilisés pour évaluer en général des pollutions organiques.

ETUDE DE CORRELATIONS ENTRE LES DIFFERENTES METHODES

L'utilisation possible des capteurs automatisés (DTO, COT) comme indicateurs de pollution rend necessaire l'étude de corrélations avec les méthodes normalisées d'analyse (DCO, IR). Des corrélations linéaires peuvent être établies pour chaque hydrocarbure, mais il est plus interessant de regrouper les données par type de produit (alcanes, alcènes, aromatiques) afin d'évaluer globalement les performances d'une méthode par rapport à une autre. Celles-ci correspondent aux pentes ($a_{exp.}$) des droites expérimentales qui peuvent être d'autre part comparées aux coefficients théoriques (a_{th}) (Tableau IV).

Les corrélations entre les techniques d'oxydation automatisées (DTO, COT) et la méthode normalisée DCO montre que cette dernière méthode est dans tous les cas nettement moins performante. D'autre part, les corrélations DTO/COT indiquent une meilleure oxydation en général des matières hydrocarburées au COTmètre. On doit remarquer que ces résultats confirment des données de la littérature bien que celles-ci soient obtenues en général sur d'autres composés organiques. Ford *et al.*[10,13] proposent dans le cas d'effluents de raffinerie des valeurs comprises entre 0,3 et 0,5 pour la corrélation COT/DCO et Bories[14] confirme ces résultats pour diverses eaux résiduaires. Pour les corrélations du type DTO/DCO, les rapports donnés dans la littérature[6,15,16] varient entre 1 et 1,5. Des exemples de corrélations obtenues à partir de ces différentes méthodes sont représentées Figures 2, 3 et 4.

Pour l'étude des corrélations entre l'IR et les techniques globales d'oxydation des matières organiques, il est nécessaire, pour faciliter les comparaisons, de convertir les réponses des différents capteurs en mg HC/1. Les résultats confirment que la méthode infra-rouge reste la seule technique susceptible de déterminer convenablement la teneur en hydrocarbures totaux en phase aqueuse. On peut toutefois remarquer que les corrélations IR/COT conduisent à un rapport voisin de l'unité, dans le cas particulier des hydrocarbures aromatiques étudiés. Par contre, pour les hydrocarbures aliphatiques saturés, ce rapport est égal à 2,33, ce qui confirme la mauvaise oxydation de ces produits au COTmètre.

CONCLUSIONS

La recherche systématique des rendements d'oxydation et d'extraction d'hydrocarbures purs en solution aqueuse, selon les méthodes DTO, DCO, COT et IR permet de dégager les conclusions suivantes:

— La méthode normalisée de DCO manque de reproductibilité et donne des oxydations très incomplètes.

TABLEAU IV

Etude des corrélations entre les différentes méthodes

Produits	DTO/DCO mgO_2/mgO_2		COT/DCO mgC/mgO_2		DTO/COT mgO_2/mgC		IR/DTO mgHC/mgHC		IR/DCO mgHC/mgHC		IR/COT mgHC/mgHC	
	a_{th}	$a_{exp.}$	a_{th}	$a_{exp.}$	a_{th}	$a_{exp.}$	a_{th}	$a_{exp.}$	a_{th}	$a_{exp.}$	a_{th}	$a_{exp.}$
Alcanes	1	1,39	0,24	0,35	4,12	3,96	1	2,32	1	3,2	1	2,33
Alcenes	1	1,01	0,25	0,30	4	3,50	1	1,43	1	1,53	1	1,21
Aromatiques	1	1,20	0,29	0,43	3,46	2,79	1	1,41	1	1,65	1	1,02

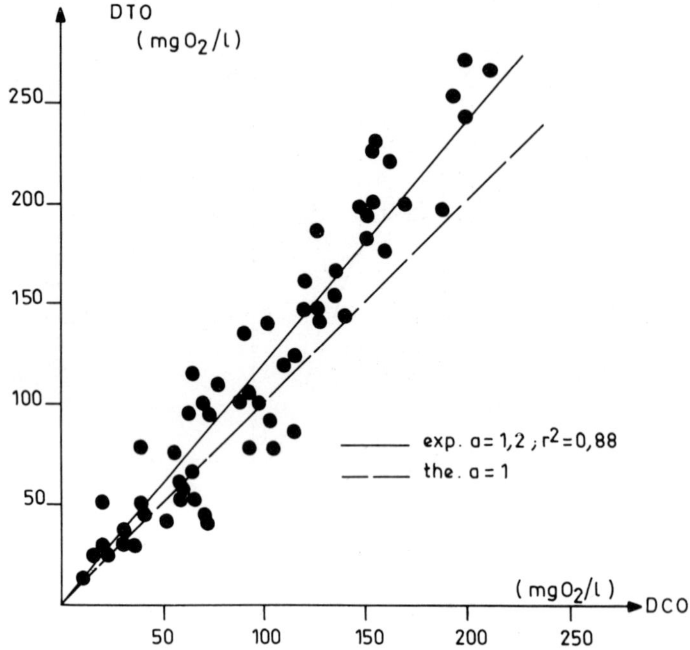

FIGURE 2 Corrélation DTO/DCO pour les hydrocarbures aromatiques étudiés.

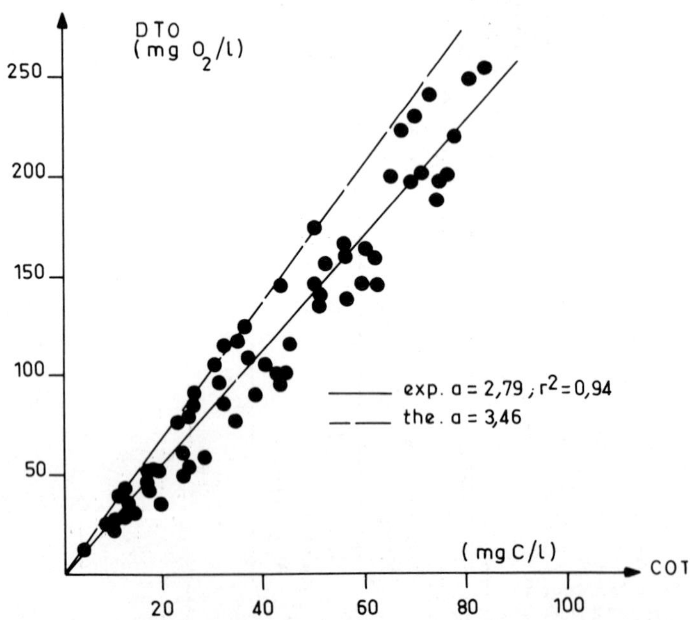

FIGURE 3 Corrélation DTO/COT pour les hydrocarbures aromatiques étudiés.

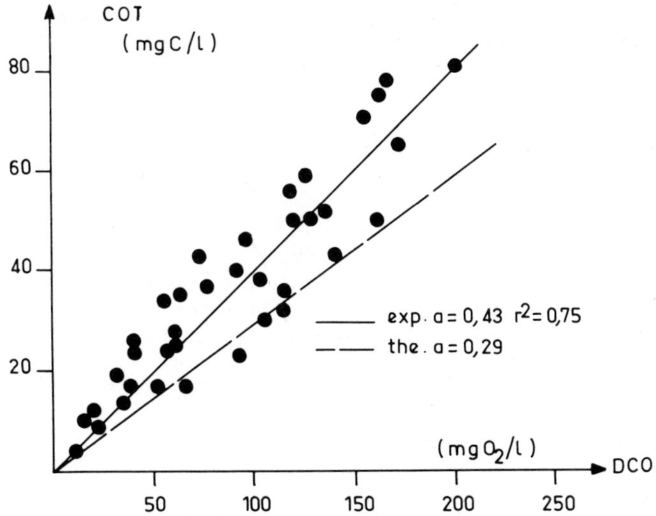

FIGURE 4 Corrélation COT/DCO pour les hydrocarbures insaturés.

— Les capteurs automatisés DTO et COT donnent aussi en général des oxydations inférieures aux valeurs théoriques, sauf pour les hydrocarbures aromatiques, le COTmètre ayant des rendements plus satisfaisants. Il faut toutefois noter que l'utilisation de ce type d'appareillage (COT)nécessite des mesures indirectes du carbone organique à partir du carbone total.

— La technique IR (selon la norme NF T 90 203) permet d'obtenir pour tous les types d'hydrocarbures des résultats satisfaisants, sauf dans le cas du benzène qui n'absorbe pas à la longueur d'onde de 2920 cm^{-1}. Le projet de norme NF T 90 114 permet d'éliminer cette lacune.

— Les corrélations obtenues entre les différentes méthodes sont dans l'ensemble correctes. Elles montrent que les performances les meilleures par rapport à l'infra-rouge sont obtenues avec le COTmètre. Ces comparaisons permettent de regrouper les hydrocarbures en deux grandes catégories (aliphatiques et aromatiques) quant á leurs déterminations par ces techniques.

REFERENCES

1. Norme API 73-358, *Determination of volatile and non volatile oily material. Infra-red spectrometric method.* (1973).
2. Norme AFNOR T 90 203, *Effluents aqueux des raffineries de pétrole. Dosage des hydrocarbures totaux.* (Août 1973).

3. Norme AFNOR T 90 202. *Dosage par gravimétrie des matières organiques en suspension dans les effluents acqueux de raffinerie le pétrole et extractibles à l'hexane.* (Août 1973).
4. Projet de norme AFNOR T 90 114, *Dosage des hydrocarbures totaux (méthode par spectrophotométrie infra-rouge).* (Fev. 1979).
5. J. L. Marty, *Demande totale en oxygène; étude critique d'un appareillage et application à la mesure de la pollution des eaux.* DEA; Université des Sciences et Techniques du Languedoc. Montpellier (1974).
6. C. Laffont, *Contribution à l'étude des interférences et des correlations avec les méthodes DTO, COT, N.-total.* Thèse de 3° cycle No. 8 INSA Toulouse (1980).
7. J. L. Marty and J. Vezinhet, *Etude bibliographique: signification des résultats obtenus par les méthodes automatiques de mesure de pollution organique: COT, N-total, DTO, absorption UV.* Rapport interne. Agence financière de Bassin Adour Garonne. (fev. 1978).
8. Norme AFNOR T 90 101, *Détermination de la DCO (méthode par $K_2Cr_2O_7$).* (Sept. 1971).
9. R. Beauchamp and M. Bonnefille, *T.S.M.* **4,** 247–252 (1978).
10. D. L. Ford, J. M. Eller, E. F. Gloyna, *J.W.P.C.F.* **43,** 1712–1723 (1971).
11. J. Ranchet and P. Clement, *T.S.M.* **72,** 23–27 (1977).
12. J. Ranchet and P. Clement, *Bull. Lab. Ponts et Chaussées* **91,** 67–71 (1977).
13. D. L. Ford, *Proc. 23rd Ind. Waste Conf.* **132,** 989–999 (1968).
14. A. Bories, *Ann. Tech. agric.* **25,** 273–289 (1976).
15. G. Voorn, J. Lee-Frampton and M. Roques, *Eau et Industrie* **14** (1977).
16. J. L. Marty and R. B. Ben Aim, *T.S.M.* **71,** 531–540 (1976).

Un des objectifs de cette étude est de déterminer les rendements d'oxydation, selon les méthodes de DTO, COT et DCO, et d'extraction au tétrachlorure de carbone puis dosage en Infra-rouge (norme NF T 90 203) sur des hydrocarbures purs (alcanes, alcènes, aromatiques). Les résultats montrent que la méthode d'extraction et dosage à l'IR (méthode spécifique) donne des rendements proches de 100%. Pour les tests globaux de pollution organique, les rendements d'oxydation les meilleurs sont obtenus en COT, puis DTO; la DCO (méthode normalisée) donne des résultats nettement plus faibles.

L'étude des corrélations entre toutes ces méthodes permet de regrouper, par rapport au critère d'oxydation, les hydrocarbures en deux grandes catégories: aliphatiques et aromatiques. Ces corrélations sont, dans l'ensemble, satisfaisantes.

Fingerprinting of Crude Oils for Pollution Control†

SALEH K. HAJIBRAHIM

Department of Chemistry, Faculty of Science, University of Riyadh, P.O. Box 2455, Riyadh, Saudi Arabia

(*Received November 26, 1981*)

Total, demetallated porphyrins from selected "export" and "weathered" crude oils were analyzed by high performance liquid chromatography (HPLC). The aim was to study the effect, if any, of oil weathering processes on the type and distribution of these compounds and hence, assess the applicability of their HPLC fingerprints in the area of pollution control. The results revealed that petroporphyrins are not readily susceptible to natural weathering processes and that their HPLC fingerprints may be applied to crude oil classification and could supplement other techniques of identifying oil pollutants.

KEY WORDS: Petroporphyrins; HPLC; fingerprinting of oil pollutants; pollution control.

INTRODUCTION

Nickel and vanadyl porphyrins are common trace components of crude oils, ranging in concentration from a few parts per million (ppm) in light crudes to hundreds ppm in heavy oils. Lab simulation experiments‡ and field observations§ have shown that porphyrins are less volatile, less soluble in sea water and more resistant to weathering processes (bacterial degradation, water washing and oxidation) than the hydrocarbon fraction of the oil. Thus, they should prove more useful as "passive tags" for the recognition of oil pollutants. The HPLC methods developed for petroporphyrins analysis[1] provided a new analytical technique for oil

---

†Presented at the Workshop on the Chemistry and Analysis of Hydrocarbons in the Environment, Barcelona, November 1981.

‡Artificial weathering of a Kuwaiti crude oil was simulated in the laboratory (Warren Spring Laboratory, England) using sea water in a constructed tank for a duration of 100 days.

§Tar residue believed to be due to oil spills over the past five years was collected from the shore of an uninhabited island in the Arabian Gulf. Some of the specimens collected were offshore under a few feet of water, others were exposed on land.

fingerprinting. This work discusses, through representative examples, the applicability of these methods to crude oil classification and recognition for pollution control.

EXPERIMENTAL

Isolation of petroporphyrins

A known amount (*ca.* 5 g) of the crude oil, its weathered form (tar residue) or the tar sand was dissolved in minimum amount of toluene. Cellulose powder (*ca.* 5 g) was added to the oil solution (to prevent formation of a stable emulsion) followed by anhydrous methanol (100 ml). Sonication (3 min) and centrifugation (3000 rpm/5 min) afforded an upper layer containing the metalloporphyrins concentrate. The supernatant was filtered and the residue was repeatedly extracted in a similar way until the supernatant contained no porphyrins as indicated by uv/visible spectrometry. The combined supernatants were concentrated using a rotary evaporator. The residue was purified by alumina column chromatography (Woelm, Neutral Act. I). Elution with chloroform removed the purified metalloporphyrins which were assayed by uv/visible spectrometry to get their type (nickel and/or vanadyl complexes) and concentration. Methane sulphonic acid (MSA) was added to the toluene solution of the purified metalloporphyrins (acid to porphyrin ratio, 1 ml:2 μmole). The mixture was refluxed (110°C) for one hr. After cooling, the reaction mixture was diluted with equal volume of hexane/toluene (1:1 v/v) and distilled water (2 volumes). The aqueous layer (purple) contained the porphyrin dictations. After separating the layers, the organic one was repeatedly extracted with aq. MSA (25%) until the acidic layer was colourless. The combined acidic layers were extracted with methylene chloride (5 times). The organic extracts were neutralized with sodium bicarbonate solution (5%), washed with water, concentrated (rotary evaporator) and chromatographed on alumina (Act. II) column. Elution with toluene/methylene chloride (1:1 v/v) removed the purified free-base petroporphyrins used for LC fingerprinting.

HPLC fingerprinting of free-base petroporphyrin mixtures

The development of HPLC for petroporphyrin analysis has been described elsewhere.[1]

Equipment: A Waters Associates ALC/GPC-244 liquid chromatograph including second M6000A solvent delivery system, M660 solvent programmer, M450 variable wavelength detector and U6K universal injector.

Samples: 2 μl of petroporphyrins solution in methylene chloride (*ca.* 1 mg/ml) were introduced from a 10 μl syringe.

Column: Stainless steel 24 cm × 4.6 mm i.d. packed with 5 μ silica (Partisil-5, Whatman) using a balanced density method.[2]

Mobile phase: The solvent system, using gradient elution, was toluene-hexane (1:1 v/v) as solvent B, programmed (40 min) from 20% to 100% B (curve 9) at a flow rate of 1.5 ml/min. The methylene chloride contained 1% ethanol as a modifier. Solvents were HPLC grade or glass-distilled

Detection: uv/vis at 400 nm and 0.1 AUFS.

Recorder: Bryans 28000. Response 0.1 mV, chart speed 5 min/cm.

RESULTS AND DISCUSSION

Fourteen samples were examined by the method described in the Experimental section. They came from Alberta, Canada; the North Sea; East Venezuela and the Arabian Gulf. The Alberta samples were comprised of an Athabasca tar sand and three related oils of different stages of biodegradation.[3] The Arabian Gulf samples (from Kuwait, Iran and Saudi Arabia) consisted of five production crudes, one artificially weathered oil and a tar residue from oil spills (Fig. 1). For peak

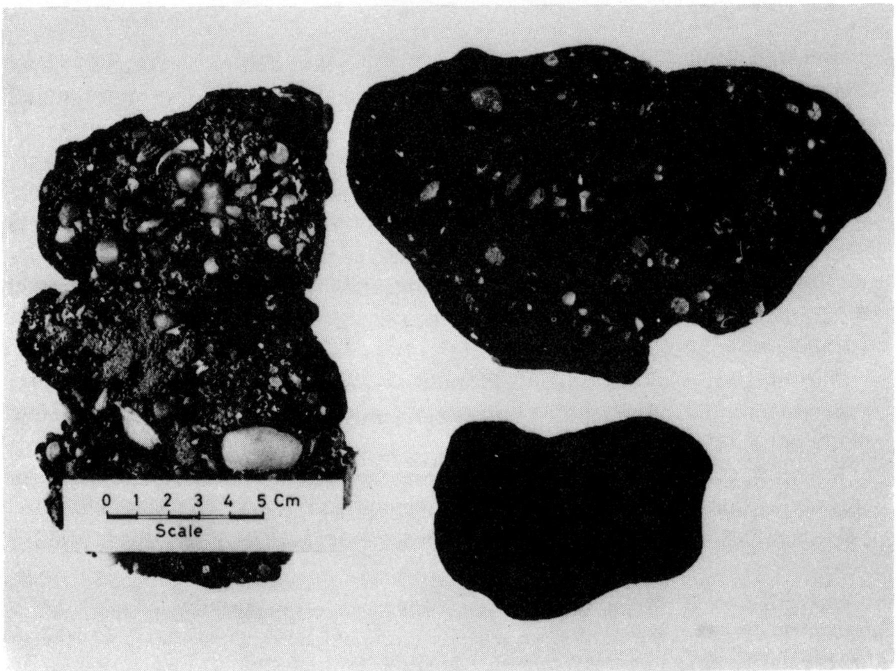

FIGURE 1 Specimens of the tar residue of oil spills in the Arabian Gulf.

TABLE I

Mass spectral characteristics of the major peaks in the liquid chromatogram of demetallated porphyrin mixture (*ex.* boscan oil)

		Petroporphyrin molecular ions[b]		
Peak no.	t_R^a (min)	Major component	Minor component(s)	(%)[c]
4	11.00	$C_{30}E$	$C_{28}E$(26)	
5	12.5	$C_{32}E$	—	
6	14.25	$C_{29}E$	—	
7	16.00	$C_{31}E$	—	
8	18.00	$C_{28}E$	—	
9	20.75	$C_{30}E$	—	
11	26.25	$C_{29}E$	—	
14	33.00	$C_{32}D$	—	
16	37.00	$C_{31}D$	—	
17	39.00	$C_{30}D$	—	

[a]Absolute retention time on Partisil-5 silica column using gradient elution. Reproducibility, ±1 min.

[b]From the EI mass spectra of the trapped peaks expressed as carbon number; E = etio-type; D = DPEP-type.

[c]Relative to the major ion = 100. The presence of the same porphyrins in the mass spectra of adjacent peaks is likely due to overlapping of the peaks rather than different isomers.

characterization see Table I. The North Sea (well 2/1–1) and East Venezuela (wells Ostra and Merey, Muturin Basin) crudes represented single-oil wells.

Fig. 2 depicts the petroporphyrin fingerprints for the Kuwaiti export oil and its artificially weathered form, whereas Fig. 3 presents the fingerprints for the Athabasca tar sand and a conventional crude oil (Bellshill Lake) from the same region. Inspection of Figs. 2 and 3 shows that artificial weathering and in-reservoir water washing and biodegradation[4]† had little effect, if any, on the distribution of petroporphyrin compounds. Furthermore, petroporphyrins were not readily affected by natural weathering processes acting on oil spills as indicated by the close pattern similarity in the fingerprints (Figures not shown) of the oil spills and few oils from fields in the vicinity.

On the other hand, the comparison of Figs. 2 with 3 reveals, as expected, that oils from different geographical locations have different petroporphyrin fingerprints, whereas those from one location show similar

†According to G. T. Demaison,[4] water washing and biodegradation of crude oil in reservoirs have taken place in most supergiant tar sand or heavy oil deposits from Western Canada (Athabasca) and Eastern Venezuela (Orinoco heavy oil belt).

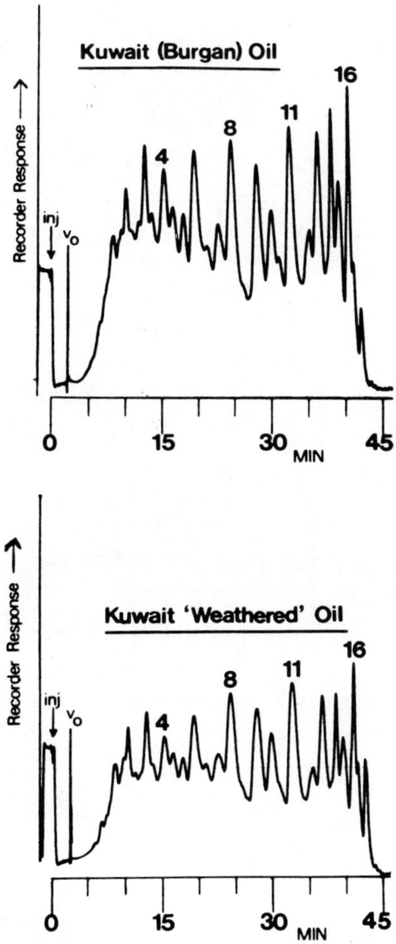

FIGURE 2 HPLC fingerptints of the petroporphyrins obtained from a Kuwaiti export oil and its weathered form. (The numbering of certain peaks in Figs. 2–5 is meant to facilitate correlation between fingerprints. For the identity of numbered peaks see Table I, Reference 1.)

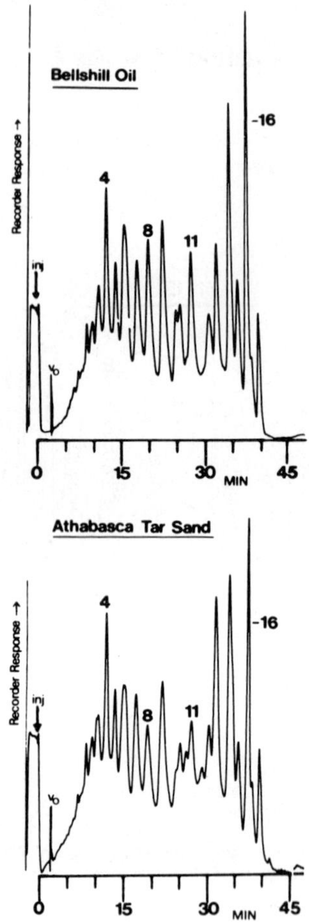

FIGURE 3 HPLC fingerprints of the petroporphyrins obtained from Athabasca tar sand and Bellshill Lake oil, Alberta.

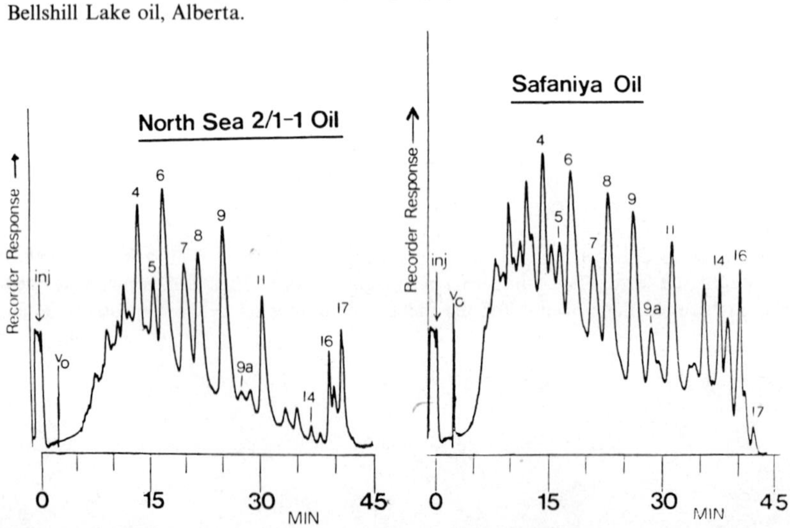

FIGURE 4 HPLC fingerprints of the petroporphyrins obtained from a North Sea and a Saudi Arabian (Safaniya) oil.

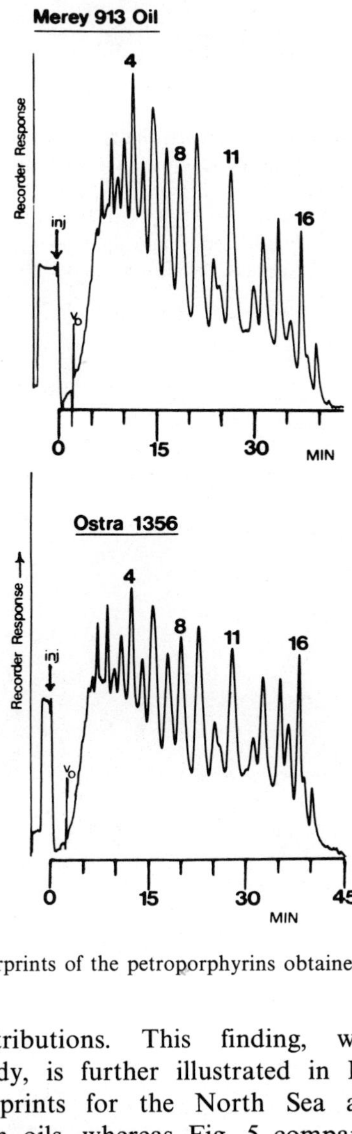

FIGURE 5 HPLC fingerprints of the petroporphyrins obtained from two East Venezuelan crudes.

petroporphyrin distributions. This finding, which was consistent throughout this study, is further illustrated in Figs. 4 and 5. Fig. 4 contrasts the fingerprints for the North Sea and a Saudi Arabian (Safaniya) production oils, whereas Fig. 5 compares the petroporphyrin distributions in the two East Venezuelan oils.

The results of this study ascertain the usefulness of petroporphyrins fingerprinting by HPLC as an analytical technique for oil classification. However, the real potential of this technique in pollution control will be established when fingerprints of a much wider collection of crude oils have been catalogued.

Acknowledgement

I wish to thank G. Guiochon and H. Colin (Ecole Polytechnique, Paliseau, France) for useful discussions on the development of HPLC and for use of laboratory facilities to pack the silica columns.

The following companies and/or individuals kindly supplied the samples indicated: the weathered Kuwaiti oil by Warren Spring Laboratories, Stevenage, England; the North Sea oil by H. V. Dunnington, Cluff Oil Ltd., London: Esso Researcher Centre, Abingdon England supplied part of the Arabian oils; Shell Netherland provided the East Venezuelan oil; the Institut Francais du Petrole (G. Deroo) and the Geological Survey of Canada (R. G. McCrossan) supplied the Alberta samples. Separations and HPLC analyses were carried out by Z. A. Isa.

References

1. S. K. HajIbrahim, *J. Liq. Chromatog.* **4,** 749 (1981).
2. M. Martin, C. Eon and G. Guiochon, *Res/Dev.* **26,** 24 (1975).
3. G. Deroo, B. Tissot, R. G. McCrossan and F. Der, *Can. Soc. Petroleum Geologists Mem.* **3,** 148 & 186 (1974).
4. G. T. Demaison, *AAPG Bull.* **61,** 1950 (1977).

Dibenzothiophene Derivatives as Organic Markers of Oil Pollution†

M. P. FRIOCOURT, F. BERTHOU and D. PICART

Laboratoire de Chromatographie, Service de Biochimie, (Pr. H. H. Floch), Faculté de Médecine BP. 815, 29279 Brest-Cédex, France

(*Received November 27, 1981*)

Sulfur compounds as dibenzothiophene derivatives were used as pollution markers in oyster samples after oil spills. By means of high performance glass capillary gas chromatography (GC) fitted with flame photometric detector and mass fragmentography, specific fingerprints of DBT derivatives were obtained in oils and oyster samples. The retention indices of some alkylated DBT (C_1 to C_4) were measured.

KEY WORDS: Dibenzothiophene, oil pollution markers, GC, FPD and MF fingerprintings.

INTRODUCTION

Accurate quantitation of hydrocarbons (Hcs) in the marine environment is estimated as a difficult task, because it is not easy to differentiate fossil from biogenic Hcs recently biosynthesized. Nevertheless, such a differentiation is necessary to avoid quantitation of fossil and biogenic Hcs in the same pot. Many criteria were suggested in order to solve this problem.[1]

Thus, a series of petroleum Hcs of geochemical importance were proposed as pollution markers in geological samples: pentacyclic triterpanes, C_{20}–C_{40} acyclic isoprenoids, C_{27} steranes[2–4] and isoprenoids.[19] In a previous work, we demonstrated the ability of using isoprenoids (nor-pristane, pristane and phytane) as pollution markers in oyster samples.[5] Each of them, by means of high efficiency gas chromatography (GC) was resolved

†Presented at the Workshop on the Chemistry and Analysis of Hydrocarbons in the Environment, Barcelona, November, 1981.

into doublets in polluted samples, whereas only the rear peak of the doublet corresponding the biogenic isoprenoid was detected in unpolluted samples. So aliphatic Hcs are concerned with most of the markers named in literature. In the same purpose, the idea of using sulfur polycyclic aromatic hydrocarbons as markers of pollution in aromatic fraction came to several authors.[6–10] Dibenzothiophene (DBT) and its derivatives are common to most of oils and oil products, but are not synthesized by biota. It is well-known that among the most resistent components to physical, chemical and biological processes are the alkylated DBT.[9–13] From the *Amoco Cadiz* oil spill monitoring, it was clearly asserted that the DBT derivatives were persistent in sediments[9,10] and in oyster tissues[11–13] during at least three years.

For the analyst, the use of pollution markers relies on the ability to identify them selectively from other components in a mixture. High efficiency glass capillary gas chromatography is known as a very powerful tool for such a purpose. This method can be improved by using different types of detectors especially the flame photometric fitted with a sulfur-selective filter. Coupling with mass fragmentography (MF) and with high performance liquid chromatography (HPLC) gives more information.

The aim of this paper is to show that DBT derivatives can effectively be used as pollution markers in biota samples by means of high performance chromatographic system coupled to FPD detection and computerized GC-MS.

EXPERIMENTAL

Oyster and oil samples

Flat oysters (*Ostrea edulis*) were collected at different commercial sites along Brittany coasts (Figure 1). The first site Binic was non polluted by *Amoco Cadiz* oil spill, and so was Auray in the south. But Aber-Benoît and Carantec (near Morlaix) were heavily spilled. Oyster tissues (10 g dry weight) were extracted, cleaned-up on Florisil, then fractionated into aliphatics and aromatics on silicagel columns as previously described.[12,13]

Two oil samples were studied. They came from *Amoco Cadiz* and *Tanio* oil spills and consisted of crude oil for *Amoco Cadiz* and fuel no. 2 for *Tanio* (Figure 1). They were purified and fractionated into aliphatics, aromatics and polar compounds as previously described.[13]

Analytical conditions

GC with flame ionization detection (FID) was performed as previously described.[12,13] For sulfur detection, samples were analysed on a Carlo-

FIGURE 1 Map of oysters area and oil spills sites along Brittany coasts. Numbers indicate the different grounding sites.

Erba (Milan, Italy) Model 4160 instrument fitted with a flame photometric detector (FPD) in sulfur mode (Carlo-Erba) Model SSD 250 and a temperature programmer (Carlo-Erba) LT Model 430. Injections were made via a split-splitless injector, with hydrogen as carrier. The temperature was quickly raised from 40°C to 60°C then programmed at a rate of 5°C/min till the final temperature 275°C. The column used was 50 m × 0.25 mm ID and was persilylated at high temperature according to the Grob procedure[14] and coated with an SE-52 liquid film.

Mass spectral analysis and high performance liquid chromatography analysis were performed as previously described.[13,15]

Retention indices

In this work, the retention indices of some DBT derivatives were measured. They are not reported in the system introduced by Kovats in which the retention is determined relative to a series of *n*-alkanes, but to a series of aromatic hydrocarbons. In temperature-programmed GC, the Kovats index system can be replaced by the approximately linear relationship:[16]

$$\mathrm{RI} = 100\frac{\mathrm{Tr}(\mathrm{X}) - \mathrm{Tr}(\mathrm{C}_z)}{\mathrm{Tr}(\mathrm{C}_{z+1}) - \mathrm{Tr}(\mathrm{C}_z)} + 100z$$

where: RI: retention index of the substance X. Tr(X): retention time of the substance X. $\mathrm{Tr}(\mathrm{C}_z)$ and $\mathrm{Tr}(\mathrm{C}_{z+1})$: retention times for *n*-alkane standards.

z: number of carbon atoms of the *n*-alkane standard eluted prior to the substance X. In case of PAHs (Poly Aromatic Hydrocarbons), Lee *et al.*[16] substituted *n*-alkanes with naphthalene, phenanthrene, chrysene, and picene. In the equation given z is the number of rings of the PAH standard eluted prior to X.

RESULTS

1. FPD fingerprinting

In Figure 2 different FPD fingerprints of oyster samples collected on several places are shown. Each sample was previously fractionated into aliphatics and aromatics. The first profile (A) was obtained from oysters collected on Aber-Benoît site on May 1978, two months after the *Amoco Cadiz* oil spill. This area was heavily polluted. The main features were the presence of an envelope of an unresolved complex mixture (UCM) and of numerous peaks. They were identified by mass spectrometry as C_1 to C_4 DBT isomers.

The second fingerprint presented (B) was from oysters collected on the same site three years later. The persistence of the UCM was noticeable. The peaks superimposed on UCM were less numerous than previously and were identified as C_3 and C_4 DBT isomers. This result was in agreement with those previously published.[11–13] DBT and C_1 DBT were degraded at a much greater rate than C_2 to C_4 alkylated derivatives.

The third profile (C) was from an oyster sample collected at Carantec on March 1981. This area was less polluted than Aber-Benoît by *Amoco Cadiz* oil spill. Only a weak hump could be detected.

The control sample (D) was collected in Auray in December 1977. No hump nor peaks were detected by GC-FPD.

The comparison of FID and FPD fingerprints of the same oyster samples (Figure 3) exhibited the advantages of FPD-GC which allowed the detection of minor components of complex mixtures not easily noticeable on FID mode. The persistence of a part of the hump on FPD chromatogram was to be noticed.

2. Retention indices

In order to make easier the comparison of results between laboratories and as most of the sulfur standards are not available, the determination of retention indices (RI) was estimated to be a part of combined data necessary for peaks identification. RI of standard DBT and C_1 to C_4 DBT derivatives were determined by GC-FPD in one oyster sample. To improve the measurement, this aromatic fraction was previously

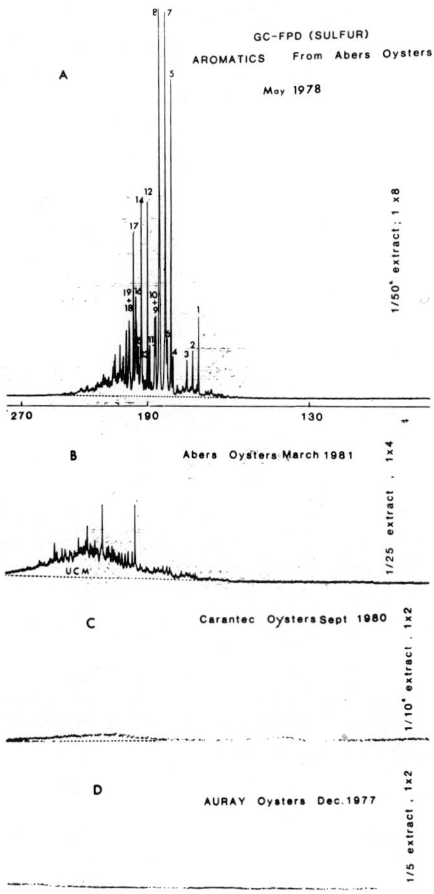

FIGURE 2 GC-FPD fingerprints of flat oysters collected in several places
A: Aber-Benoît in May 1978 (1/50 extract, $S=1\times 8$)
B: Aber-Benoît in March 1981 (1/25 extract, $S=1\times 4$)
C: Carantec in September 1980 (1/10 extract, $S=1\times 2$)
D: control oysters from an unpolluted place Auray (1/5 extract, $S=1\times 2$).
The numbers indicated are those of Table I.

fractionated on Lichrosorb-NH_2 according to the number of condensed aromatic rings.[15,17] Due to the currently available resolving power of glass capillary column, high-precision RI of the three ring sulfur compounds could be obtained at the level of reproducibility of about one retention index unit (Table I). The data obtained were in good agreement with those previously published by Willey *et al.*[18] The slight differences were due to the fact that it was not possible to determine RI on

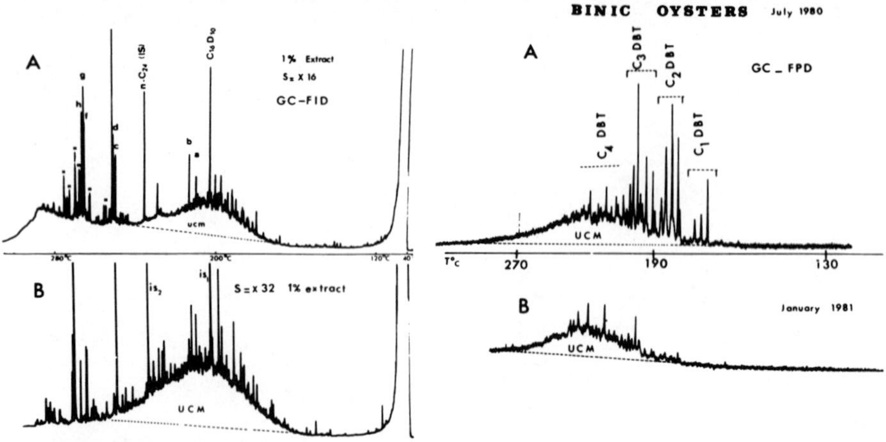

FIGURE 3 GC-FID and GC-FPD fingerprints of Binic oysters (aromatic fraction) collected in July 1980 (A) and January 1981(B).

TABLE I

Retention indices of DBT derivatives contained in oyster samples. Conditions: FPD detection, SE-52 glass capillary column (50 m × 0.30 mm ID).[a]

Peak number	Molecular weight	Formulae	Name	RI measured	Literature results (18)
	128	$C_{10}H_8$	naphthalene[a]	200	
	184	$C_{12}H_8S$	DBT[a]	295	296
	178	$C_{14}H_{10}$	phenanthrene[a]	300	
1	198	$C_{13}H_{10}S$	C_1-DBT	313.9	312.4
2	198			318.1	316.0
3	198			322.2	319.4
4	212	$C_{14}H_{12}S$	C_2-DBT	332.3	327.7
5	212			337.7	332.4
6	212			336.7	335.9
7	212			338	338
8	212			342.3	
9	212			344.9	
10	212			345.7	
11	226	$C_{15}H_{14}S$	C_3-DBT	349.3	352.4
12	226			350.9	
13	226			352.8	
14	226			355.5	
15	226			356.1	
16	226			357.7	
17	226			359.4	
18	226			361.2	
19	226			364.4	
20	240	$C_{16}H_{16}S$	C_4-DBT	366.5	
21	240			368.7	
	228	$C_{18}H_{12}$	chrysene[a]	400	

[a]For analytical conditions, see the text. Standards added to the extract.

isothermal conditions; data were dependent on temperature and on the rate of programming.

3. MF fingerprinting

Mass fragmentograms of the oyster sample collected in Aber-Benoît were compared to those of *Amoco Cadiz* and *Tanio* oil samples. The ions displayed were representative of DBT and C_1 to C_4 alkylated DBT. As the oil fragmentograms were very similar, the *Amoco Cadiz* oil fingerprint only is exhibited on Figure 4. On both of them, DBT and C_1 to C_4 derivatives could be detected.

From Aber-Benoît oyster fingerprint, it was clear that no DBT parent could be detected (Figure 4), whereas the presence of C_1 to C_4-DBT was demonstrated. So MF results exhibited that the fingerprints were more and more complicated when alkyl substituents increased, accordingly with the increasing number of possible isomers.

As sulfur compounds are present in minor quantities in a very complex mixture, MF fingerprints may be difficult to explain because some interferences can occur from other compounds especially from the UCM. So, Ogata and Mikaye improved the method by using a technique named "Intensity matching".[8] Mass spectra of sulfur compounds are characterized by the importance of M^+ and $(M+2)^+$ ions. This is due to the presence of isotopic ^{34}S in the molecule. The isotopic ratio can be calculated and is well-defined. On displaying M^+, $(M+1)^+$ and $(M+2)^+$ ions, Ogata,[8] found possible to discard interferences. This technique was used with success on oil samples, one example is given on Figure 5. $(M+2)^+/M^+$ and $(M+1)^+/M^+$ ratio of C_1 and C_2-DBT were determined and were similar to those calculated theoretically. In case of oyster samples, the interpretation was sometimes difficult, particularly when the UCM was important.

4. DBT derivatives as pollution markers

The above fingerprinting techniques were used for monitoring the *Amoco Cadiz* oil spill on oyster areas. The prime idea was that Aber-Benoît samples would be the most polluted whereas those from Carantec area would be only slightly polluted. Binic samples would be considered as controls. So, pollution would decrease with a greater distance from the *Amoco Cadiz* grounding site. In fact, our results were different. They exhibited the persistence of the UCM and DBT derivatives in Aber-Benoît and Carantec samples as estimated, but MF and FPD fingerprints of Binic oysters exhibited the presence of numerous alkylated DBT (Figure 4). They were supposed to be coming from *Tanio* spillage. However, from

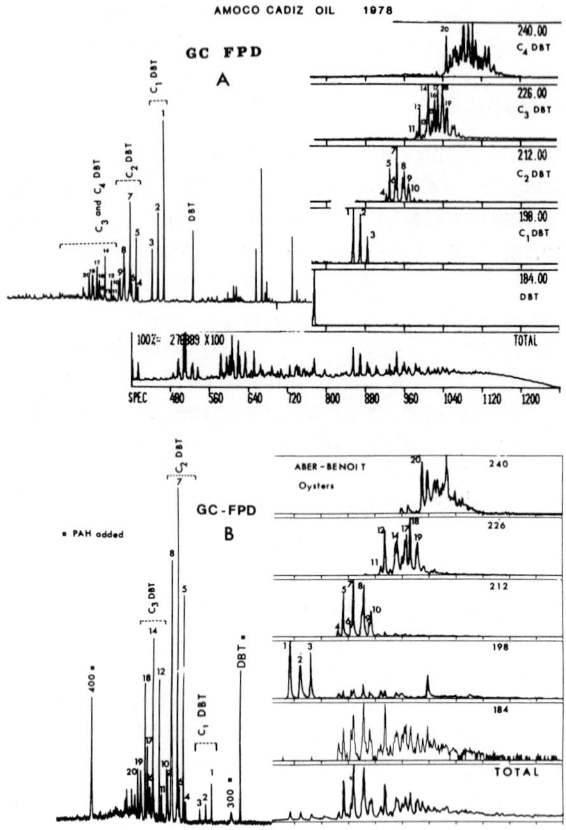

FIGURE 4 GC-FPD and MS fingerprints of aromatic fractions of *Amoco Cadiz* crude oil (A) and oysters collected on Aber-Benoît site in May 1978 and fractionated on LiChrosorb-NH_2 (three ring compounds) (B): the numbers indicated are those of Table I. Total ionic current and molecular ions of DBT and alkylated DBT are presented.

monomethyl DBT relative ratio measurements on FPD fingerprints, it seemed to us that in fact these samples would have been polluted with *Amoco Cadiz* oil (Table II).

DISCUSSION

Literature gave only few results dealing in the use of DBT derivatives as oil pollution markers in biota. Some were concerning artificial pollution of shell-fishes, short-necked clams exposed to crude oil suspension in the

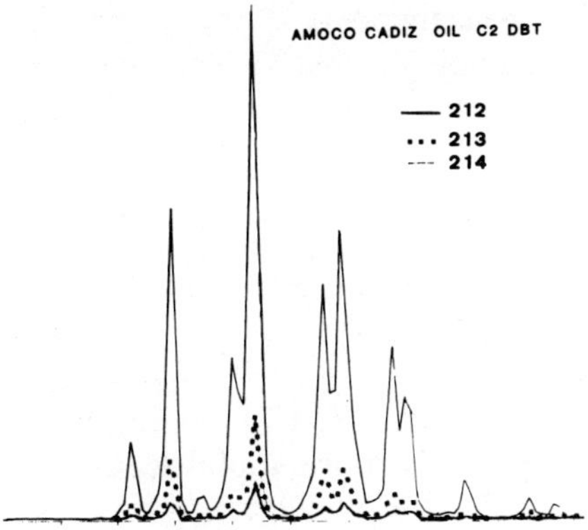

FIGURE 5 Intensity matching method applied to C_2-DBT of *Amoco Cadiz* oil.

TABLE II

Peak height relative ratio of monomethyl-DBT detected by GC-FPD in oil and oyster samples.

Peak height ratio of C_1 DBT	1/3	1/2	2/3
Amoco Cadiz oil	3.2	2.2	1.4
Tanio oil	7.9	6.0	1.3
Aber-Benoît oysters	3.1	2.3	1.3
Carantec oysters	2.9	2.2	1.4
Binic oysters	3	2.3	1.3

laboratory.[7,8] This paper exhibited the ability of using DBT derivatives as pollution markers in oyster samples, three years after an oil spillage by means of GC-FPD and MF techniques. The persistence of such compounds in the marine environment was illustrated.

As sulfur compounds are present as traces in mixtures, the coupling of a high efficiency chromatographic system with sulfur specific detector is essential for the characterization of DBT derivatives and provides a very useful method for rapid screening. However FPD presents one serious drawback consisting in the phenomenon of response quenching due to the presence of non-sulfur compounds within the same peak. So this technique can be improved by coupling GC-FPD with HPLC. Prefractionation of samples on Lichrosorb-NH_2 allows the isolation of three-ring compounds and the identification of DBT derivatives on GC-FPD.

Retention index measurements of these compounds allow interlaboratory comparisons. Problems encountered are due to the numerous possible isomers and to the absence of standard references; only a few of them as benzothiophene, DBT, naphthobenzothiophenes are available.

High efficiency GC coupled to computerized MF allows the obtention of good fingerprints: ions displayed are $m/z=184+14n$ ($n=0$ to 4). Intensity matching method is not very useful for it is too dependent on the type of mass spectrometer used.

For monitoring oil pollution, GC-FPD is more dedicated to routine analysis than GC-MS which is more sophisticated and time-consuming. Moreover, it seems that GC-FPD might allow the determination of the origin of an oil spillage by comparisons of the relative ratio of the three monomethyl DBT and probably C_2 and C_3-DBT in oil and biota samples.

Although the environmental importance of DBT derivatives is not completely clear actually, especially their toxicity, they are likely to receive more attention in the future.[19]

Acknowledgements

The authors acknowledge Mrs. Y. Dreano and Y. Ferec for technical assistance. This work was made possible by the partial financial support of the Centre National pour l'Exploitation des Océans (grants no. 81/6486 and 81/6480).

References

1. *The International Mussel Watch, Report of a Workshop*; National Academy of Sciences (Washington, DC, 1980) p. 49.
2. M. Dastillung and P. Albrecht, *Marine Poll. Bull.* **7,** 13 (1976).
3. J. Albaiges and P. Albrecht, *Int. J. Environ. Anal. chem.* **6,** 171 (1979).
4. J. Albaiges and J. Bordon, *J. Chromatogr.* **204,** 491 (1981).
5. F. Berthou and M. P. Friocourt, *J. Chromatogr.* **219,** 393 (1981).
6. A. Nakamura and T. Kashimoto, *Bull. Environm. Contam. Toxicol.* **20,** 248 (1978).
7. M. Ogata and Y. Miyake, *J. Chromatogr. Science* **18,** 594 (1980).
8. M. Ogata and Y. Miyake, *Water Research* **15,** 257 (1980).
9. P. D. Boehm, D. L. Fiest and A. Elskus, *Proceedings of the International Symposium:* Amoco Cadiz, *Fates and Effects of the Oil Spill, Brest (France).* (CNEXO, Paris, 1981), pp. 159–173.
10. J. Ducreux and M. Marchand, *Proceedings of the International Symposium:* Amoco Cadiz, *Brest (France).* (CNEXO, Paris, 1981), pp. 175–216.
11. J. L. Laseter, G. C. Lawler, E. B. Overton, J. R. Patel, J. P. Holmes, M. I. Shields and M. Maberry, *Proceedings of the International Symposium:* Amoco Cadiz, *Brest (France).* (CNEXO, Paris, 1981), pp. 633–644.
12. M. P. Friocourt, Y. Gourmelun, F. Berthou, R. Cosson and M. Marchand, *Proceedings of the International Symposium:* Amoco Cadiz, *Brest (France).* (CNEXO, Paris, 1981), pp. 617–632.

13. F. Berthou, Y. Gourmelun, Y. Dreano and M. P. Friocourt, *J. Chromatogr.* **203,** 279 (1981).
14. K. Grob, G. Grob and K. Grob, Jr., *J. High Resol. Chromatogr. & Chromatogr. Commun.* **2,** 677 (1979).
15. M. P. Friocourt, F. Berthou, Y. Dreano and M. Marchand, *Analusis* **9,** 468 (1981).
16. M. L. Lee, D. L. Vassilaros, C. M. White and M. Novotny, *Anal. Chem.* **51,** 768 (1979).
17. S. A. Wise, S. N. Chesler, H. S. Hertz, L. R. Hilpert and W. E. May, *Anal. Chem.* **49,** 2306 (1977).
18. C. Willey, M. Iwao, R. W. Castle and M. L. Lee, *Anal. Chem.* **53,** 400 (1981).
19. M. M. Quirk, R. L. Patience, J. R. Maxwell and R. E. Wheatley, *Analytical Techniques in Environmental Chemistry* (ed. J. Albaiges). Pergamon Press, Oxford, 1980, pp. 23–31.
20. M. L. Lee, M. V. Novotny and K. D. Bartle, *Analytical Chemistry of Polycyclic Aromatic Compounds* (Acad. Press, New York), 1981, pp. 448–449.

Determination of Trace Hydrophobic Volatiles in Aqueous Media by a Technique of Multiple Stripping and Trapping in a Closed Circuit†

JOSEF DROZD AND JOSEF NOVÁK

Institute of Analytical Chemistry, Czechoslovak Academy of Sciences, 611 42 Brno, Czechloslovakia.

(*Received November 26, 1981*)

The double-sampling method of quantitative headspace-gas analysis was assayed by analysing model water-gas systems with benzene, toluene, ethylbenzene, *n*-decane, and *n*-dodecane as analytes. With this method, two headspace samples are withdrawn successively from the system and analysed by gas chromatography. The total initial contents of the analytes in the system are calculated by virtue of the decrease in their concentrations in the gaseous phase, brought about by the withdrawal of the first headspace sample. Combined with Grob's closed-loop strip/trap technique, the method provides for the determination of ppb (10^9) concentrations of the above hydrocarbons in water with an average relative error of several percent.

KEY WORDS: Hydrocarbons, hydrophobic volatiles, water, headspace-gas analysis, gas chromatography.

INTRODUCTION

Headspace gas chromatography is mostly employed to measure the so-called characteristic profiles of volatile components of condensed materials. With such applications, the only requirement is that the working conditions be kept invariant during the whole series of

†Presented at the Workshop on the Chemistry and Analysis of Hydrocarbons in the Environment, Barcelona, November 1981.

measurements. Less attention has been paid to the relatively more difficult problem of absolute quantitation in headspace-gas analysis. One of the ways to obtaining absolute quantitative data by headspace gas chromatography is multiple extraction of the material analysed with a gas and gas chromatographic analyses of the extracts. This approach has several versions. The first work on this topic was published by McAuliffe.[1] With his version, a sample of the material to be analysed is placed in an injection syringe and extracted repeatedly by equal volumes of pure gas, each extract being completely pushed out by the syringe piston and analysed by gas chromatography. Plotting the logarithms of the contents of volatiles in the individual extracts against the serial number of extraction steps yields straight lines from the parameters of which the initial contents of the volatiles in the condensed sample can be calculated. Another version, consisting in the successive withdrawal of two headspace-gas samples from a closed gas-condensed phase system, analysis of the samples for the contents of volatiles by gas chromatography, and calculation of the total initial contents of the volatiles in the system from the results of the two headspace-gas analyses, was theoretically outlined by Novák.[2] Recently, Kolb *et al.*[3] suggested a procedure based on the repetitive extraction of a sample of condensed material with a gas, gas chromatographic determination of the contents of volatiles in each extract, and calculation of the initial contents of the volatiles in the condensed sample by summing up the contents found in the individual headspace-gas samples. The linear dependence of the log of headspace-gas contents of volatiles on the number of extractions was used to provide a sufficient amount of data by extrapolation.

The aim of this work is to prove experimentally the concept of the double-sampling method.[2] Model gas-aqueous liquid systems with known trace contents of hydrocarbons were analysed. The Grob's [4–7] closed-loop strip/trap method in both the conservation and equilibration modes of trapping[8] was employed.

PRINCIPLE OF QUANTITATION WITH THE DOUBLE-SAMPLING METHOD

The quantitation with the double-sampling method of headspace-gas analysis is based on the following analyte mass balance:

$$W_i = W_{iG0}[(KV_L/V_G)+1] \qquad (1)$$

$$W_i - w_{iG0} = W_{iG1}[(K'V_L/V_G)+1] \qquad (2)$$

where W_i is the total initial mass of analyte in the system, W_{iG0} and W_{iG1} are the masses of analyte in the gaseous phase of the system before and after the withdrawal of the first headspace sample, w_{iG0} is the mass of analyte in the first headspace sample, K and K' are the liquid/gas partition coefficients of the analyte in the system before and after the withdrawal of the first headspace sample, and V_L and V_G are the volumes of the condensed and gaseous phases of the system, respectively. Provided that $K' = K$, which mostly is the case, Eqs. 1 and 2 yield

$$W_i = \frac{w_{iG0}}{1-(W_{iG1}/W_{iG0})} \tag{3}$$

Further, if equal volumes of the headspace gas are sampled and analysed under constant conditions, the ratio W_{iG1}/W_{iG0} in Eq. 3 is equal to the ratio of the peak areas or peak heights of the analyte in the chromatograms of the second and the first headspace sample; with the arrangement employed in this work, the W_{iG1}/W_{iG0} ratio is given by the ratio of the analyte peak areas or peak heights in the chromatograms of the second and the first concentrate released from the analyte-enrichment trap, respectively. The quantity w_{iG0} (the amount released from the trap after the first run) is determined by direct calibration of the gas chromatograph.

EXPERIMENTAL

Analytical-grade benzene, toluene, ethylbenzene, *n*-decane, and *n*-dodecane (Fluka AG, Buchs, Switzerland) were employed as model analytes. Analytical-grade acetone (Lachema, Brno, Czechoslovakia) was used as a solvent to prepare standard solutions of the above hydrocarbons. The model systems were prepared with distilled water as the liquid phase, which was always boiled up just before use. Blank experiments, carried out at sensitivities ten-times higher than those set in actual analyses, gave zero values.

Instrumentation

A laboratory-made arrangement for stripping and trapping in a closed circuit was employed. A flow diagram of the arrangement is shown in Figure 1. The main component of the set-up is a drop-ball valve/stainless steel bellows pump (1) actuated with a cam driven by an electric motor (2). The liquid analysed is placed in a 100-ml glass vessel (4) provided with

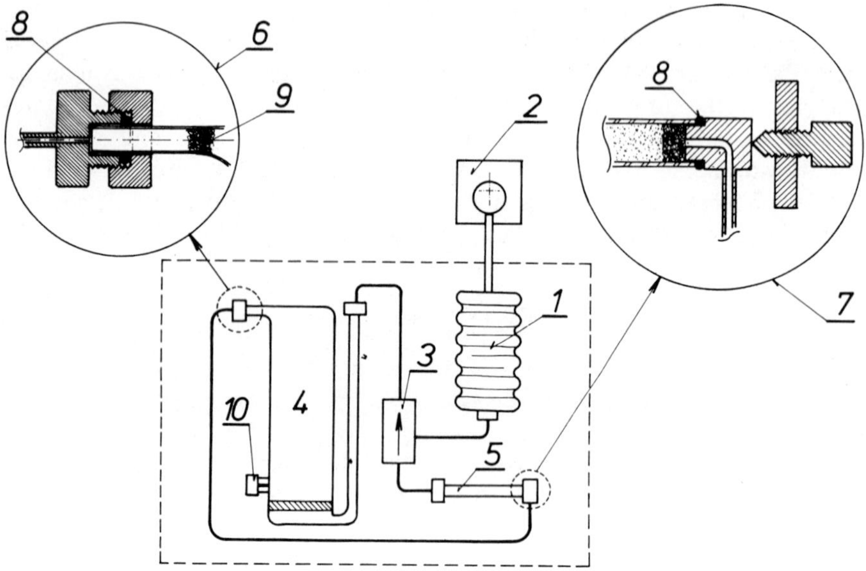

FIGURE 1 Flow diagram of the arrangement for stripping and trapping in a closed circuit.

a sintered-glass frit at the bottom. The gaseous phase of the system is recycled via the valve (3), vessel 4 and trap 5, the latter being a 6 cm/3 mm I.D. glass tube packed with Tenax GC 30/60 mesh (Applied Science Laboratories, State College, PA, U.S.A.). The sorbent was fixed in the tube by two plugs of quartz wool. The individual components of the set-up were interconnected by 2 mm O.D./1 mm I.D. stainless steel capillary. The capillary-to-glass parts connections are shown in detailed representations 6 and 7 in Figure 1. Silicone rubber and/or PTFE rings were used as gaskets (8). A quartz woolplug (9) was placed in the outlet tube of vessel 4 in order to prevent the entrainment of droplets of the liquid phase to the other parts of the circuit. Near the bottom of the vessel there is an inlet port with septum through which samples and/or standards are introduced. The arrangement is fixed to a console and can be immersed into a thermostatting bath.

The analytes captured in the trap are thermally desorbed, and the concentrate obtained is purged by a stream of the carrier gas into the gas chromatograph. The desorption/purge arrangement is illustrated schematically in Figure 2. Oven 2 with controlled temperature is placed around trap 1 while the four-port stopcock 3 is set such as to shortcircuit the trap and direct the carrier gas (4) to the chromatographic column (5).

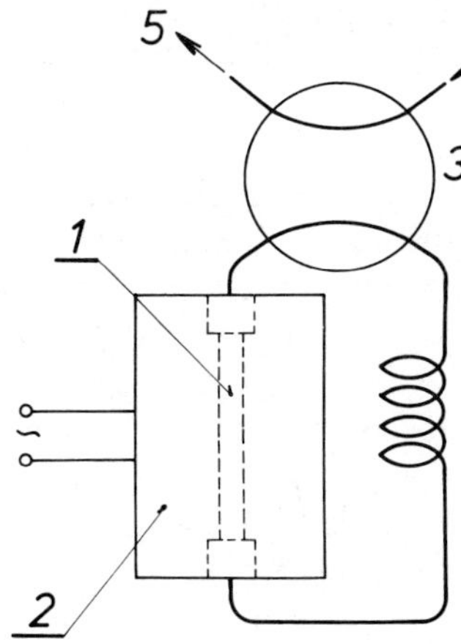

FIGURE 2 Schematic representation of the arrangement for the thermal desorption of captured analytes from the trap and the introduction of the concentrate into the gas chromatograph.

After the trap is brought to the desired temperature, stopcock 3 is turned over to the position allowing the carrier gas to pass via the trap and purge the desorbed analytes into the gas chromatograph. The stopcock and the interconnecting capcillaries are also heated in order to prevent the concentrated analytes from spurious sorption. The outlet of stopcock 3 was connected to column 5 by means of an injection needle pierced through the septum of the inlet port of the gas chromatograph in such a manner that other samples could simultaneously be introduced through the septum by an injection syringe.

The gas chromatographic analyses were carried out on a Shimadzu GC-4A instrument (Shimadzu Seisakusho, Kyoto, Japan) equipped with a flame ionization detector and a 1 m/3 mm I.D. stainless steel column packed with Chromaton N 80/100 mesh (Lachema, Brno, Czechoslovakia) coated with 20% (by wt.) of Carbowax 20 M (Carlo Erba, Milan, Italy). The column and the sample inlet port were kept at 70 and 140°C, respectively. The flow rate of the carrier gas (nitrogen) was 35 ml/min. Before its use, the Tenax trap was kept at 250°C for several hours under a nitrogen stream. An Infotronics CRS-101 integrator (Infotronics, Shannon Airport, Ireland) was used to measure the peak areas.

Procedure

Standard solutions of the model hydrocarbons were prepared by adding known volumes of each of them with a 10-μl Hamilton syringe (Hamilton Micromeasure AG, Bonaduz, Switzerland) to some amounts of acetone in 10-ml measuring flasks and then making up the volumes to the mark with acetone. 0.5-μl volumes of the standard solutions were charged with a 1-μl Hamilton syringe into the gas chromatograph, and the peak areas obtained, A_i^*, were taken as measures of the "given" masses of analytes, W_i^*. Employing the very same 1-μl syringe, 0.5–1 μl of the standard solutions were then introduced into the stripping vessel with 50 ml of distilled water, covering a range of about 1–10 ppb (10^9) of the model hydrocarbons in water. In order to bring the whole circuit into a stationary state before the analysis proper, an empty tube was connected instead of the trap and the pump put on running for 10 min. After that, two traps, each with 16.3 ± 0.1 mg of Tenax GC, were successively connected into the circuit, and the gaseous phase was circulated at a rate of 70 ml/min for 10 min in each run. The whole circuit was thermostatted at 40 ± 0.1°C by means of an ultrathermostat (type 410, VEB Prüfgeräte, Medingen, DDR). The desorptions of the analyte (analytes) captured in the traps were carried out by heating the latter to 160 (and/or 200)°C for 3 min. After each analysis, the vessel was emptied and rinsed with distilled water, and then the whole circuit was purged with a stream of dry nitrogen (about 0.5 1/min), with the pump being running. After 15 minutes, acetone only was detected in blanks.

RESULTS AND DISCUSSION

The quantities W_i^* and A_i^* are related to each other by

$$W_i^* = kA_i^* \tag{4}$$

where k is a calibration factor. With regard to Eq. 3, the "found" mass of analyte, W_i, is given by

$$W_i = kA_i = \frac{W_{it0}}{1-(W_{it1}/W_{it0})} = \frac{kA_{it0}}{1-(A_{it1}/A_{it0})} \tag{5}$$

where A_i is a hypothetical peak area, representing the "found" mass of analyte, W_{it0} and W_{it1} are the masses of analyte in the concentrates released from the traps after the first and the second run, and A_{it0} and

A_{it1} are the respective peak areas. By means of Eqs. 4 and 5, the absolute and relative error can be expressed as

$$W_i - W_i^* = k\left[\frac{A_{it0}}{1-(A_{it1}/A_{it0})} - A_i^*\right], \tag{6}$$

$$\frac{W_i - W_i^*}{W_i} = 1 - \frac{A_i^*[1-(A_{it1}/A_{it0})]}{A_{it0}}, \tag{7}$$

respectively.

The analyte concentrations in the standard solutions were not known precisely, and the results of the analyses were calculated in terms of peak areas by Eqs. 6 and 7. With this way of processing the data, the effects of inaccuracies connected to the preparation of standard solutions were eliminated. For convenience, approximate analyte masses and concentrations were calculated from the measured volumes, densities, and peak areas of the analytes (cf. Table I).

The contents of the analytes in the model systems were too low to be determinable by direct analysis of headspace samples. Therefore, an analyte-enrichment technique had to be employed. We have chosen

TABLE I

Results of the headspace determination of traces of hydrocarbons in model gas-water systems by the closed-loop double-strip/trap method

Analyte	given (ng/50 ml)	given (ppb)	found (ng)	error (ng)	error (%)	stand. deviat. of the mean (ng)	number of measurements
(A)							
Benzene	85.3	1.7	84.0	−1.3	−1.5	2.51	7
(B)							
Benzene	87.9	1.8	90.6	+2.7	+3.1	2.54	5
Toluene	433.5	8.7	418.0	−15.5	−3.6	17.9	5
(C)							
Benzene	439.5	8.8	442.3	+2.8	+0.6	6.07	8
Toluene	86.7	1.7	87.7	+1.0	+1.2	1.56	10
Ethylbenzene	173.4	3.5	169.8	−3.6	−2.1	2.88	11
(D)							
Benzene	87.9	1.8	119.1	+31.2	+35.3	7.65	4
Toluene	433.5	8.7	420.9	−12.6	−2.9	4.21	4
Ethylbenzene	173.4	3.5	181.9	+8.5	+4.9	3.40	4
n-Decane	73.0	1.5	72.7	−0.3	−0.4	1.78	4
n-Dodecane	149.8	3.0	102.8	−47.0	−31.4	1.75	4

Grob's technique of stripping and trapping in a closed circuit, as it appears to constitute a most suitable way of concentrating headspace samples in quantitative headspace trace analysis.

Systems with benzene alone, benzene and toluene, benzene, toluene, and ethylbenzene, and all the three aromatics together with *n*-decane and *n*-dodecane in water were analysed by the above method. The results are summarized in Table I. The "given" and "found" data, obtained by the direct analyses of charges of the standard samples and by the closed-loop double-strip/trap analyses of the model systems, have been calculated by virtue of Eqs. 4 and 5, respectively. The errors, calculated by Eqs. 6 and 7, show the accuracy of the results obtained by the closed-loop double-strip/trap method, the precision of the results being characterized by the standard deviation of the mean.

The results obtained with the systems containing benzene alone, benzene together with toluene, and all the three aromatics (section A, B, and C in Table I) can be considered as excellent, the relative errors and relative standard deviation of the means amounting to some percents only. It is pertinent to note that benzene is equilibrated in the system under the conditions employed. As the retention volume of benzene on 16.3 mg of Tenax GC at 40°C is about 400 ml[9], a two-multiple of the retention volume of benzene approximately is drawn through the trap in each strip/trap run. Under these conditions, the frontal zone of benzene breaks completely through the trap, and the whole circuit becomes practically equilibrated with respect to benzene before the run is finished. About 60 and 23% of the total initial mass of benzene in the system are captured in the trap during the first and the second strip/trap run, respectively, The retention volume of toluene is about 1000 ml under the given conditions, so that its zone breaks through the trap only partially or not at all; about 82 and 14% of the initial mass of toluene are transferred to the trap in the first and the second run. Toluene as well as the other higher boiling hydrocarbons were trapped in a conservation regime. About 99% of ethylbenzene are trapped in the first run, so that practically the entire content of ethylbenzene is recovered in two runs.

At a given flow rate of stripping gas, the fraction of the total initial amount of analyte which is transferred to the trap during a strip/trap run is a function of the water/gas partition coefficient of the analyte and the time of stripping/trapping with conservation trapping and a function of both the water/gas and trapping sorbent/gas partition coefficients of the analyte and the amount of the sorbent in the trap with equilibration trapping.[8] As the water/gas partition coefficients of homologous hydrocarbons decrease with increasing carbon number,[10] the respective rates of stripping increase with increasing carbon number. This is evidenced by the percentage recoveries in the two successive runs, quoted above for the three homologues of aromatics.

Except for benzene and *n*-dodecane, the results of the simultaneous determination of all the three aromatics together with *n*-decane and *n*-dodecane (section D in Table I) are as accurate as those obtained with the aromatics alone. There are different possible explanations for the 35.3-% positive error with benzene and 31.4-% negative one with *n*-dodecane. An unknown peak which was partially overlapped with that of benzene appeared in the chromatograms in this series of analyses, which could cause an error in the determination of benzene peak area. Another possible cause of the errors may be competitive adsorption of the higher hydrocarbons in the trap, interfering variably with equilibration of benzene. The negative error with *n*-dodecane may be due to irreversible adsorption of the latter in the trap. However, an increase of the desorption temperature to 200°C had no effect on the results, and no rests of *n*-dodecane were released from the trap by repeating the desorption procedure. It is also possible that the errors with benzene and *n*-dodecane are interrelated.

CONCLUSIONS

The double-sampling method of quantitative headspace-gas analysis gives current results. Combined with the closed-loop strip/trap technique, the method provides for the determination of ppb (10^9) concentrations of hydrocarbons in water with an average relative error of several percent. With certain systems, mutual interference of the analytes may cause larger errors. This aspect of the method deserves further investigations.

References

1. C. McAuliffe, *Chem. Technol.* **1,** 46 (1971).
2. J. Novák, Quantitative analysis by gas chromatography, *Marcel Dekker, Inc., New York,* 1975, p. 130.
3. B. Kolb, P. Pospisil and M. Auer, *J. Chromatogr*, **204,** 371 (1981).
4. K. Grob, *J. Chromatogr.* **84,** 255 (1973).
5. K. Grob and G. Grob, *J. Chromatogr.* **90,** 303 (1974).
6. K. Grob, K. Grob, Jr. and G. Grob, *J. Chromatogr.* **106,** 299 (1975).
7. K. Grob and F. Zürcher, *J. Chromatogr.* **117,** 285 (1976).
8. J. Novák, J. Goliáš and J. Janák, in Trace organic analysis: A new frontier in analytical chemistry, H. S. Hertz and S. N. Chesler, Editors, U. S. Department of Commerce, National Bureau of Standards, Washington, D.C., 1979, p. 739.
9. J. Novák, J. Vejrosta, M. Roth and J. Janák, *J. Chromatogr.* **199,** 209 (1980).
10. J. A. Jonssön, J. Vejrosta and J. Novák, in preparation.

Some Applications of Computerized GC-MS to the Determination of Biogenic and Anthropogenic Organic Matter in the Environment†

BERND R. T. SIMONEIT

School of Oceanography, Oregon State University, Corvallis, Oregon 97331, USA

(*Received January 9, 1982*)

This is an overview of the application of organic mass spectrometry and ancillary techniques to the analysis of organic matter in environmental research. Such organic matter is usually analyzed in terms of gas, bitumen (lipids), and "kerogen", with asphaltenes and humic substances for some samples. This approach is illustrated with some examples and the origin, the environmental history and the nature of secondary products of this organic matter can be evaluated by using the data derived from both specific molecular and bulk chemical (also physical) analyses. Evaluations of production and fluxes and cross-correlations can thus be made by the application of the same separation and analytical procedures to samples from different environmental compartments (eg., biota, atmosphere, hydrosphere, lithosphere, etc.).

INTRODUCTION

Organic mass spectrometry (MS) has been a tool of analytical chemistry for about 20 years, and it has been applied to environmental research over the past decade, mainly in conjunction with gas chromatography (GC). New mass spectrometric techniques such as chemical ionization MS, high resolution GC-MS, high resolution mass spectrometry, field ionization and desorption MS, tandem mass spectrometry (MS-MS) and pyrolysis GC-

---

†Presented at the Workshop on the Chemistry and Analysis of Hydrocarbons in the Environment, Barcelona, November 1981.

MS, all with the associated online computers and processors are providing new ways to tackle the analytical problems associated with the analysis of complex organic mixtures. The overall field of mass spectrometry has been extensively reviewed.[1-9]

Here I illustrate the utility of some of these various instrumental methods in environnmental research with some examples of typical data and interpretations. The organic matter of environmental samples can be divided into the following categories:

Volatiles	Lipids (bitumen)	Fulvic/humic acids (asphaltenes)	Humin (kerogen)
C_1-C_{15}, CO_2 terpenes, etc.	C_8-C_{40+}	Macromolecular, M.W. $\sim 10^3 - > 10^6$	Macromolecular (>fulvic/humic acids)
Minor amt. of total org. C	Variable amount	Variable amount	Major amount

These categories are based upon the usual operational limits imposed by the organic chemical separation and identification procedures.

Mass spectrometric instrumentation

Low resolution mass spectrometric requirements for most organic compounds of ecochemical significance can be satisfied with either a magnetic or quadrupole mass spectrometer of current design and construction. The quality of the data from either are quite comparable.

Electron impact (EI) is by far the most common method used for sample ionization in mass spectrometry. The usual 70 eV energy causes extensive molecular fragmentation with a high yield of ions and to date the best sensitivity, whereas both EI at lower voltages ($\sim$15 eV or less) and chemical ionization are lower energy processes resulting in less fragmentation and thus stronger molecular ion signals, but overall lower sensitivity.

The most commonly utilized inlet system to the mass spectrometer is the gas chromatograph,[8] however, other methods as for example batch inlets, direct insertion probes, liquid chromatographs and pyro-probes have also been used. The application of the GC/MS technique to environmental research has been extensive and has been amply reviewed.[10-12]

High resolution mass spectrometry (HRMS) has been applied primarily to group type analyses to determine H to C and heteroatomic composition (eg., 13, 14). GC-HRMS is beginning to be applied and has great potential in the use of accurate mass elemental composition chromatograms (vis a vis nominal mass chromatograms at low resolution) of characteristic fragment ions.[15]

Most MS techniques utilize dedicated computer systems and numerous such systems have been assembled for acquisition and processing of mass spectrometric data. They all basically consist of the same programs for data processing and searching of data files.[16,17] The commercial systems are also similar. Most systems have the capability of plotting mass spectra with optional background deletion, scale expansion and scan summing, plotting of total ion current traces and mass chromatograms for GC-MS analyses, listing of prominent ions and optionally, data file searching.

Applications

Analyses of volatile compounds are usually accomplished by GC and confirmed by GC-MS.[14,18-21]

The lipids of environmental samples are extracted with solvents and commonly separated by thin-layer or column chromatography after esterification of free fatty acids (Figure 1). The discrete fractions are then

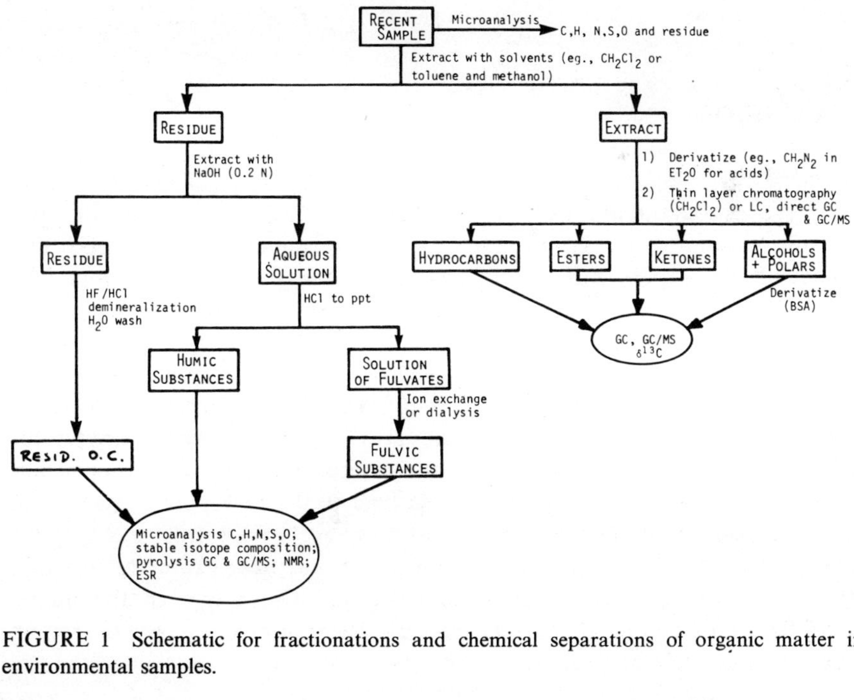

FIGURE 1 Schematic for fractionations and chemical separations of organic matter in environmental samples.

analyzed by GC and GC/MS (eg., 22). It should be noted that lipids of such recent samples compared to geologically ancient samples contain more labile and functionalized compounds (eg., olefins).

Argentation-TLC can be utilized to separate saturated from olefinic compounds and complex fractions can be further separated by molecular sieving or urea (also thiourea) adduction into normal and branched-cyclic components. Other ancillary techniques which can be applied to lipid analyses are chemical alteration (eg., hydrogenation), stable isotope ratios, UV-visible spectrophotometry, infrared spectrophotometry and nuclear magnetic resonance spectroscopy.

Humic and fulvic substances can be extracted from environmental samples with aqueous sodium hydroxide after lipid removal (eg., 23, 24). The residual organic matter, the humin, is concentrated in samples with a high mineral content by treatment with hydrochloric and hydrofluoric acids (eg., 25, 26). Humin (also humates) can be characterized mainly by bulk properties and to a limited extent by chemical degradation (cf., Figure 1). These analyses consist of elemental composition (eg., H/C, O/C, N/C, etc.), stable isotope composition (eg., $\delta^{13}C$, $\delta^{34}S$, $\delta^{15}N$, δD, etc.), electron spin resonance and nuclear magnetic resonance spectrometry[24] and maceral description. Pyrolysis GC and pyrolysis GC-MS have also proven to be of utility for analysis of humin and humates, since they provide a fingerprint pattern and are rapid analytical methods.[27,28]

DISCUSSION

Lipid data

Three types of data are usually monitored in GC-MS analyses of lipid fractions: homologous compound series, molecular markers and the unresolvable complex mixture of branched and cyclic compounds (the hump). Both homologies and molecular markers originate from primary biological precursors with some diagenetic alteration (eg., 29) and some examples of the major classes of such hydrocarbon compounds are summarized in Figure 2. Examples of typical GC analyses of some environmental samples are shown in Figure 3. They are total hydrocarbons in oysters from the Arabian Gulf, illustrating a comparison of the natural biogenic components (resolved peaks in Figure 3a) with the degraded petroleum residues in a contaminated oyster (major hump in Figure 3b).[30] The major biogenic components in these examples are C_{21} and C_{25} polyolefins present in both fractions. The origin of the major hump from petroleum was confirmed by the composition of the minor molecular marker compounds.[30]

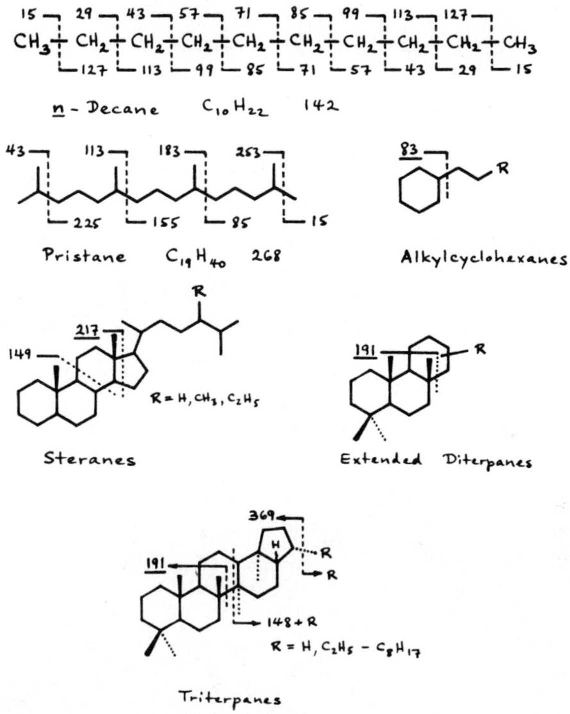

FIGURE 2 Chemical structures of some common hydrocarbon modecular markers with their major mass spectrometric fragmentation patterns.

Some salient features of the data of a typical GC-MS analysis are shown in Figure 4 (total hydrocarbons of a marine sediment of recent age, contaminated by petroleum seepage—Santa Barbara Basin).[31] The total ion current trace is shown in Figure 4a, where the dominant peaks are *n*-alkanes and a bimodal hump is evident above the base line. The molecular markers are distinguished by plotting key ions (eg., m/z 99 for *n*-alkanes, Figure 4b, m/z 191 for extended diterpanes and triterpenoids, Figure 4c, or m/z 217 for steranes, Figure 4d—cf., structures in Figure 2), then analyzing and interpreting the corresponding mass spectra of the various ion maxima. The basic procedures for the interpretation of mass spectra are found in the works by McLafferty,[32] Biemann[33] and Silverstein *et al.*[34]

Some examples of hydrocarbon distributions extracted from aerosol samples are shown in Figure 5.[35] Comparisons are made between rural and urban aerosols and natural plant wax.[35,36] The *n*-alkanes range from

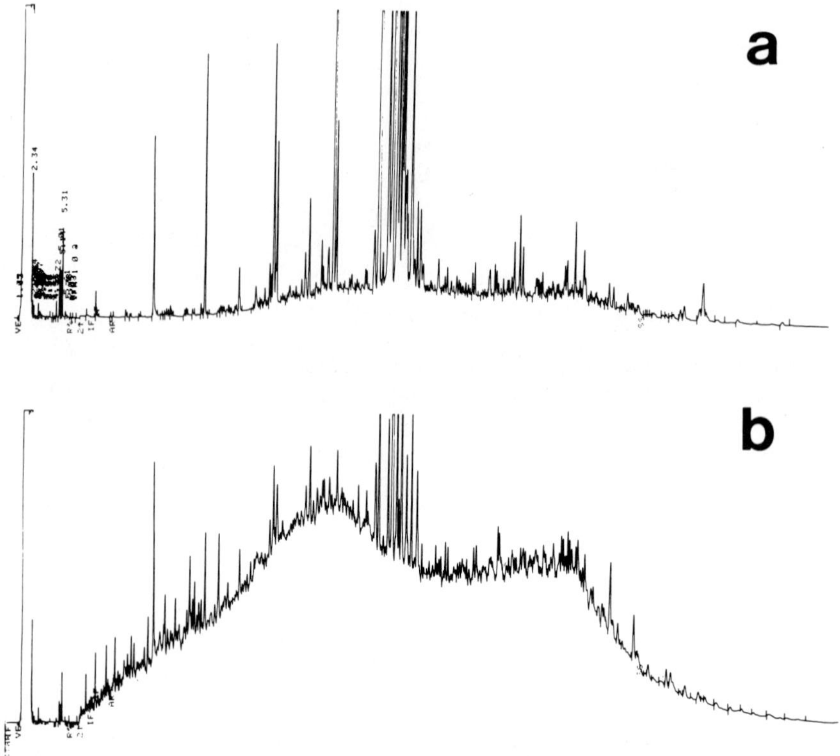

FIGURE 3 Gas chromatograms (FID) of fractions I (primarily hydrocarbons) of extracts of oysters: (a) from Ras az Zaur, (b) from Mina Abdullah[30].

about C_{15} to C_{34} with an odd carbon number predominance (CPI > 1.2[36]) for the rural samples from Lake Tahoe (during summer) and Mt. Lassen (Figure 5a–c). The homologs > n-C_{23} are believed to be derived chiefly from waxes of vascular plants.[36] These distributions compare quite favorably with the n-alkane distribution of the mixed grass wax (Figure 5d). The dominance of n-C_{29} and n-$C_{27} \simeq n$-C_{29} may indicate a mixed origin from forest and grassland.

The ratios of branched and cyclic hydrocarbons to the n-alkanes can be used as an approximate measure of the level of petroleum contamination (e.g. 31). Values of 10–200 have been observed for sediment lipids contaminated with petroleum seepage, whereas the natural hydrocarbons derived from vascular plants exhibit no hump and this ratio is < 1.0 for aerosols over oceanic areas.[36] For these samples the hump: n-alkane ratio ranges from 1.4 to 3.4, indicating some contamination by petroleum

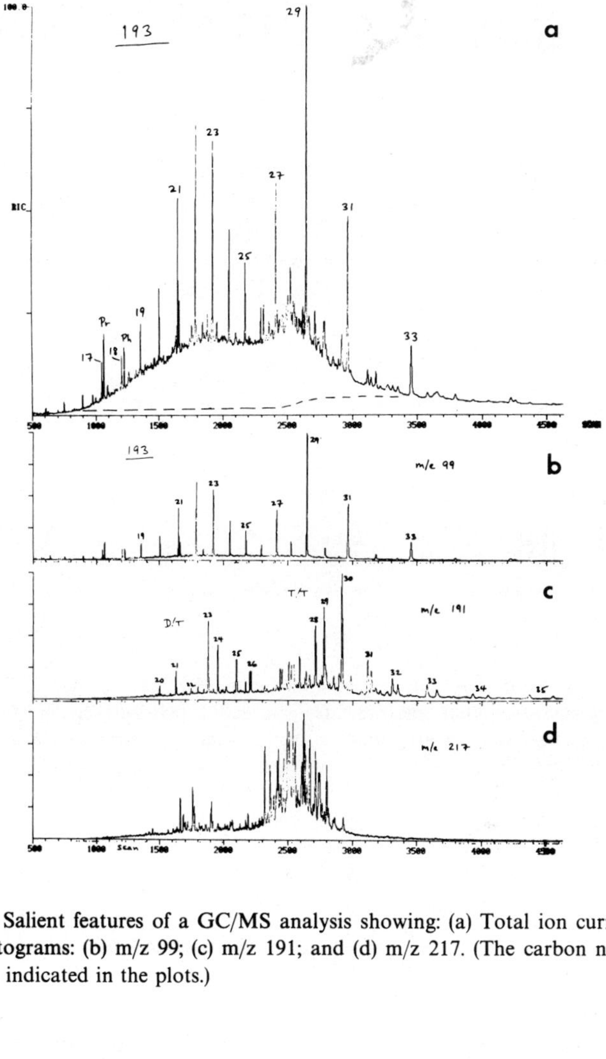

FIGURE 4 Salient features of a GC/MS analysis showing: (a) Total ion current trace and mass chromatograms: (b) m/z 99; (c) m/z 191; and (d) m/z 217. (The carbon numbers of the homologs are indicated in the plots.)

residues. The hydrocarbons of the sample from Mt. Lassen consist essentially of *n*-alkanes from only natural sources.

Most petroleums contain *n*-alkanes ranging from C_1 to about C_{35} with no carbon number predominance (CPI $\simeq 1.0$) and isoprenoid hydrocarbons (e.g. phytane and pristane). The C_1-C_{20} homologs are usually lost due to evaporation (according to their boiling points) in the environment and/or in the experimental procedure, or were separated by distillation in the refining processes. Thus, absolute concentrations of the homologs $< C_{20}$ (Figure 5) are not accurate, but qualitative comparisons can be made.

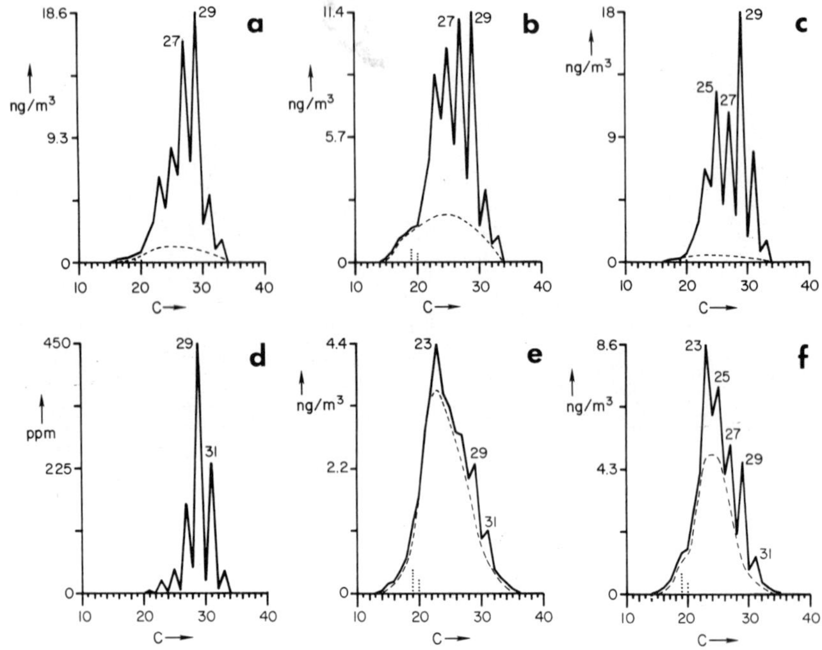

FIGURE 5 Distribution diagrams for *n*-alkanes extracted from various aerosol and plant wax samples[35] (height of dotted lines indicates isoprenoid hydrocarbons and the dashed line envelope indicates the approximate contribution of *n*-alkanes from petroleum residues): (a) Sugarpine Point State Park, Lake Tahoe, night, summer; (b) Sugarpine Point State Park, day, summer; (c) Battle Creek Meadow Ranch, Mt. Lassen, summer; (d) grass epicuticular wax; (e) Sugarpine Point State Park, winter; (f) Sierra Ski Ranch, Lake Tahoe, summer.

Pristane and phytane are diagenetic products of phytol and are not primary consistuents of most terrestrial organisms. Their presence in these hydrocarbon fractions, coupled with the presence of a broad hump, indicate petroleum contamination. An estimate of the *n*-alkane component derived from petroleum can be made from these data (smooth envelopes in Figure 5 indicated by the dashed lines).[35] At Sugarpine Point State Park and the Mt. Lassen area, a diurnal variation is evident: the contribution from petroleum residues is less during the night (Figure 5a) than during the day (Figure 5b). The sample from the Mt. Lassen area (Figure 5c) exhibits the lowest level of such contamination. The influence of heavy vehicular traffic is illustrated with the sample from Sierra Ski Ranch (Figure 5f) collected during summer, and can be compared with the distribution in Figure 5a. The level of contamination at Sierra Ski Ranch is significantly higher. Seasonal variation is illustrated with the samples from Sugarpine Point State Park collected during winter (Figure 5e) and summer (Figure 5a). The contamination by petroleum residues is greater during winter. It should also be noted that the amount of natural plant wax is lower in the winter. The presence of petroleum contamination is also confirmed by the molecular markers (triterpanes, steranes and extended tricyclic diterpanes) in the hydrocarbon fractions.[35,36]

Some typical examples of homologous compounds in lipid fractions from sediments, geologically recent and ancient, are shown in Figure 6 to illustrate patterns of their distributions. Sample 36-330-10-1, 95–102 cm, which is of Jurrasic age from the South Atlantic, exhibits *n*-alkanes maximizing at *n*-C_{17} and *n*-C_{19} (Figure 6a) and even carbon numbered *n*-fatty acids with a maximum at *n*-C_{16} (Figure 6e). Both of these distributions are indicative of primary bacterial lipid residues and there are only traces of allochthonous homologs $>C_{21}$. These lipids are of a marine origin.[22,29,37]

Sample ML71-2-23, 8.0 m is from Mangrove Lake, Bermuda, of recent origin in a highly productive sapropelic environment.[29] The *n*-alkanes (Figure 6b) show maxima at *n*-C_{17} from primary algal synthesis, at *n*-C_{22} from microbial degradation of algal detritus and a minor one at *n*-C_{29}, with a strong odd carbon number predominance $>C_{26}$, derived from

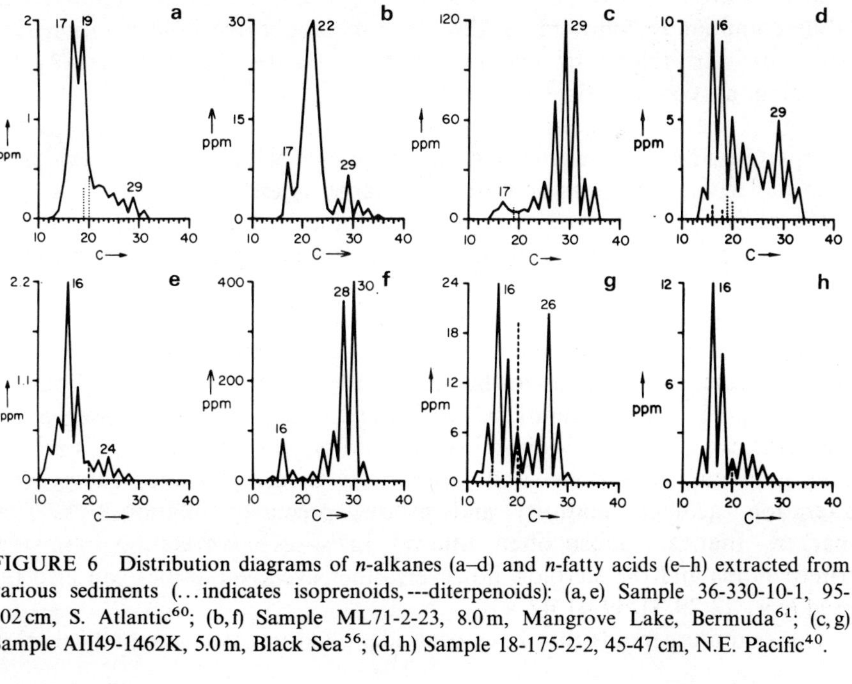

FIGURE 6 Distribution diagrams of *n*-alkanes (a–d) and *n*-fatty acids (e–h) extracted from various sediments (... indicates isoprenoids, --- diterpenoids): (a, e) Sample 36-330-10-1, 95-102 cm, S. Atlantic[60]; (b, f) Sample ML71-2-23, 8.0 m, Mangrove Lake, Bermuda[61]; (c, g) Sample AII49-1462K, 5.0 m, Black Sea[56]; (d, h) Sample 18-175-2-2, 45-47 cm, N.E. Pacific[40].

higher plant wax.[22,31,38] The *n*-fatty acids (Figure 6f) exhibit a bimodal distribution where the predominant maximum at C_{28} and C_{30} is of a plant wax origin and the minor homologs $> C_{20}$ are of a marine (or lacustrine) derivation.

The Black Sea is a sink for terrigenous lipids from plant wax as indicated by the example (Figure 6c, g). The *n*-alkanes exhibit a maximum at n-C_{29} with a strong odd-to-even carbon number predominance, typical of higher plant wax.[22,29] This is corroborated by the *n*-fatty acids with their maxima at n-C_{16} and n-C_{26} and strong even carbon number predominance, where the homologs $> C_{20}$ are of an allochthonous terrigenous origin.[22,29]

Sample 18-175-2-2, 45–47 cm is an example of a reducing micro-environment in the northeastern Pacific Ocean, where the *n*-alkanes (Figure 6d) exhibit a strong even-to-odd carbon number predominance for the homologs $< C_{24}$ and the distribution follows that of the *n*-fatty acids $< C_{24}$ (Figure 6h). These even *n*-alkanes ($< C_{24}$) may be derived from reduction of fatty acids or of olefins from fatty alcohols.[39,40] The *n*-alkanes $> C_{24}$ with the odd carbon number predominance and maximum at n-C_{29} are derived from vascular plant wax.

It should be emphasized that these distributions are found only under ideal environmental conditions, precluding the occurrence of extensive biodegradation. Biodegradative and oxidative alterations first remove the labile compounds, followed by the other metabolizable lipid components, which in turn leads to changes in the relative distributions of the homologous compounds such as for example *n*-alkanes or fatty acids. Therefore, homolog distributions in environmental samples should be interpreted with caution and such data coupled with supportive evidence from molecular indicators and other available parameters.

Petroleum products and coal extracts exhibit hydrocarbon distributions where, for example, the *n*-alkanes exhibit essentially no carbon number predominance and range from $< C_{10}$ to C_{35+} (eg., 39, 41). The contents of fatty acids, ketones and alcohols are usually low in petroleums.

Molecular markers or indicators are organic compounds with specific chemical structures, which can be correlated with an origin from biogenic precursors and which cannot be synthesized by abiogenic processes. Such markers can thus be utilized as indicators, for example of sources, environmental conditions and alteration, diagenetic and catagenetic alteration, geologic maturity and product-precursor relationships. The markers that are most often utilized in source correlations are the triterpenoids and the steroids, however, other cyclic molecules can also be used (e.g., 22, 29, 31, 39, 41, 42, 43).

The sesquiterpenoids that have been identified in environmental samples (mainly sediments) are primarily cadalene, with various tetrahydro analogs, and they are of both a marine (algal) and terrigenous origin.[31]

The diterpenoids that have been characterized fit into two classes, those derived from terrigenous sources[40,44] and extended tricyclic diterpanes of a probable marine origin. The terrigenous diterpenoids consist of a large number of compounds and the most abundant analogs are dehydroabietic acid; dehydroabietin; dehydroabietane; simonellite; retene; tetrahydroretene; fichtelite; pimanthrene; 17-nordehydroabietane; iosene; 13-methylpodocarpa-8,11,13-triene; and norsimonellite.[40] The extended tricyclic diterpanes are found in shales, petroleum, and recent sediments. Most of their occurrences were coupled with the presence of the 17α(H)-hopane series of triterpanes, which are geologically mature, i.e., old.[31,45,46] These extended tricyclic diterpanes range from $C_{19}H_{34}$ to $C_{26}H_{48}$ and sometimes to $C_{29}H_{54}$ with very similar distributions. Some distributions are shown as examples in Figure 7. The samples from the Southern California Bight (Figure 7c, d) are contaminated by petroleum seepage, indicating that the extended tricyclic diterpanes are syngenetic with the 17α(H)-hopane series. However, the same extended tricyclic diterpanes in a recent sample from Walvis Bay (Figure 7b) must be derived from a marine source since no petroleum contamination is evident.

The triterpenoids found in many environmental samples are of a microbial origin and are usually comprised of the hopane and to a lesser extent the moretane series. Some sedimentary sinks can also receive triterpenoid residues from terrigenous sources. The common homologs consist of trisnorhopane (I†, R=H), norhopane (I, R=C_2H_5), hopane (I, R=C_3H_7) and extended hopanes ranging from C_{31} to C_{35} (II) with minor amounts of the corresponding moretanes (III). The 17β(H) stereochemistry predominates in viable and recent, immature samples[49,50] and various triterpenes such as diploptene (IV), 17β(H)-hop-21(22)-ene (V) hop-17(21)-ene (VI) are also present.[51] The examples of recent sediments from the Gulf of California and Walvis Bay (Figure 7a and b, also cf. Figure 4c with 7a) illustrate this point in that only 17β(H)-hopanes and hopenes are present in the latter. The extended 17β(H)-hopanes occur as only the 22S diastereomer in recent samples as the direct markers of biosynthesis. In ancient sediments or crude oil, the hopanes occur as the 17α(H)-stereomers and the extended 17α(H)-hopanes are found as 22R and 22S diastereomeric pairs, where full maturity is indicated by an S/R ratio of about one.[31,49,51] The samples from the Southern California Bight (Figure 7c and d) show this type of distribution, where in sample 193 only petroleum residues and matured 17α(H)-hopanes are found and in sample

†These chemical structures are given in Appendix I.

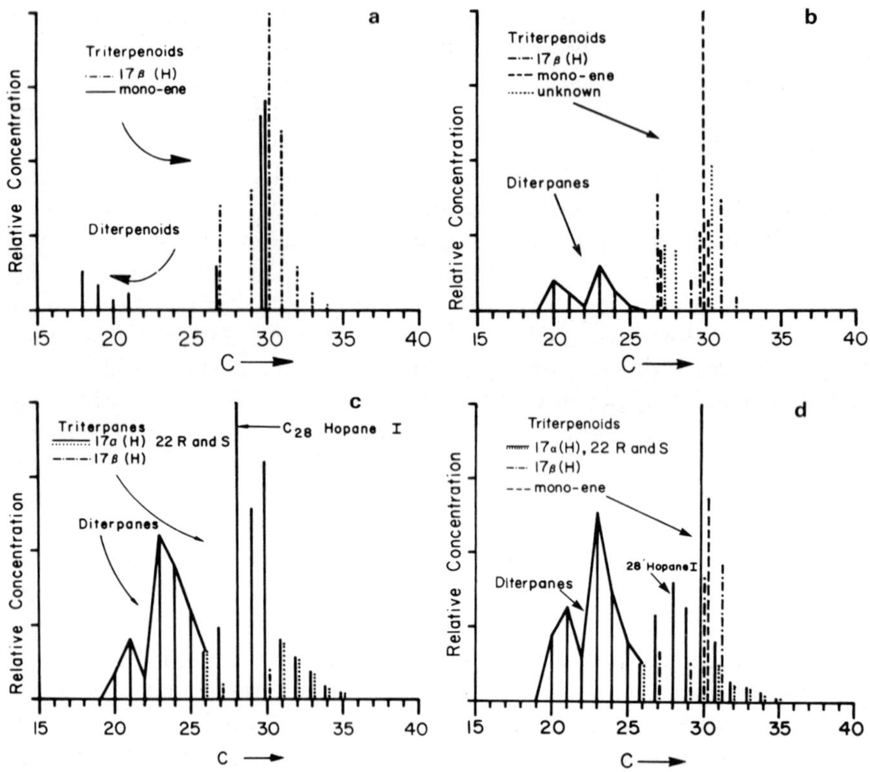

FIGURE 7 Relative distribution histograms for diterpenoids and triterpenoids (based on the m/z 191 mass chromatograms or gas chromatographic response) of some examples. The R and S diastereomers are also indicated and C_{28} Hopane I is 17α(H), 18α(H), 21β(H)-28, 30-bisnorhopane. (a) Sample 30G-I (102-105 cm), Guaymas Basin, Gulf of California[37]; (b) Sample AII 93/3-18 (197-201 cm), Walvis Bay, SW Africa[31]; (c) Sample 193 (25–31 cm), Santa Barbara Coastal Area, Southern California Bight[31]. (d) Sample 575 (25–31 cm), Tanner Basin, Southern California Bight[31].

575, a mixture of Recent and ancient triterpenoids and allied lipids are present. Also, a $C_{28}H_{48}$ triterpane is a major analog in the Southern California Bight sediments and it has been identified as 17α(H), 18α(H), 21β(H)-28, 30-bisnorhopane (VIII).[31,52] This compound appears to be a specific marker for Southern California peteroliferous residues, as it is not commonly found in most other areas.[31,35]

Steroidal compounds are common in environmental and geological samples and they undergo complex diagenetic reactions yielding various series of hydrocarbons, alcohols and ketones (VIII).[43,47,48,53,54] The saturated hydrocarbons are comprised of the steranes [VIII, R,R′,R″=H, eg., cholestane, 5α(H) *vs.* coprostane, 5β(H)], diasteranes (IX) and traces of other isomers, and the 5α stereomers usually predominate over the 5β. Various series of 4-methylsteroids have also been identified in some samples.[55,56] Recently, more detailed analyses of steroid products from environmental samples have revealed a more complex diagentic pattern and a miriad of compounds (eg. 43, 47, 48). The general concensus is that the steroid nucleus is limited in defining the biological source organisms since the steroid skeletal structures are ubiquitous in the biosphere. It is the constitution of the steroid side chain that may prove to be more characteristic for genetic correlations.

Other minor components that can also be utilized as molecular markers are for example tetraterpenoids, *iso*- and *anteiso*-alkanes or fatty acids, hydroxy fatty acids, tetrapyrrole pigments and aromatic hydrocarbons.

Humin or protokerogen data

Humin (protokerogen) and to some extent humic and fulvic substances are complex mixtures of high molecular weight moieties of various, essentially undefined structures and compositions.[22,29,67] The end member products of humin are coals for terrigenous and alginites for marine origins. Most protokerogens of the aquatic realm are admixtures of all input sources (eg., 37, 58).

Since protokerogen (also humic substances) is an invaluable endogenous environmental marker for the origin and nature of the bulk of the organic matter, it is important to know its constitution. Protokerogen and to some extent humic and fulvic substances are commonly characterized by their bulk chemical and physical properties. Some of these are the elemental composition, stable isotopic compositions, electron spin resonance and nuclear magnetic resonance spectroscopy, pyrolysis GC and pyrolysis GC/MS fingerprinting (Figure 1).[25,36,39,58,59] Some protokerogens have also been chemically degraded for analysis of the resultant components.[58] Bulk chemical parameters can then be correlated to indicate trends.

Protokerogen is also sensitive to thermal stress (eg. the normal geothermal gradient due to depth of burial[37,58] or from transients such as intrusives[25,29] or combustion).[36] This catagenesis results in the generation of bitumen (also gases) from the protokerogen, which in turn becomes more aromatized. This phenomenon can be followed by monitoring various bulk properties as decribed above. Catagenesis can also be simulated by subjecting organic matter to pyrolysis and analyzing the bitumen that forms by GC and GC/MS (eg., 28). Such bitumen can be characterized in terms of homologous and marker compounds as described above. Carbonaceous detritus that has been under severe thermal stress does not yield liquid products on pyrolysis.[28]

CONCLUSIONS

The applications of mass spectrometry with its ancillary techniques and the other methods of organic geochemistry are of great utility for the investigation of organic matter in environmental research. The characterization of gas, lipids and protokerogen (also humic and asphaltic substances) of environmental samples and of secondary products can provide information about the origin, environmental history, toxicity potential and results of degradation of the organic matter. Thus it is recommended to carry out detailed analyses of all the fractions of organic matter in the same manner in order to facilitate the interpretation and intercorrelation of data from various types of environmental samples.

Acknowledgements

I thank the National Science Foundation (Grant ATM81-18101) for partial financial assistance.

Appendix I

Chemical Structures Cited

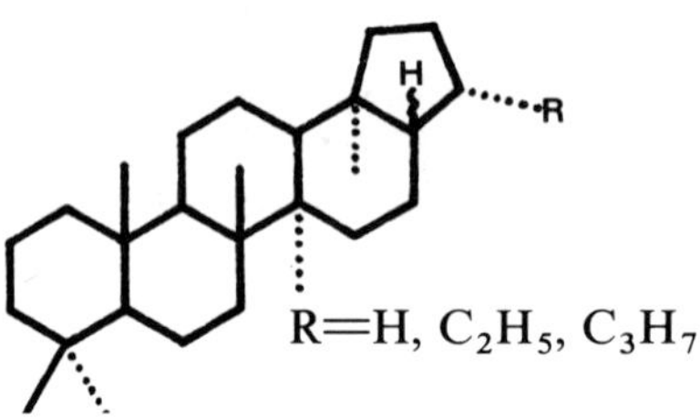

I. Hopanes

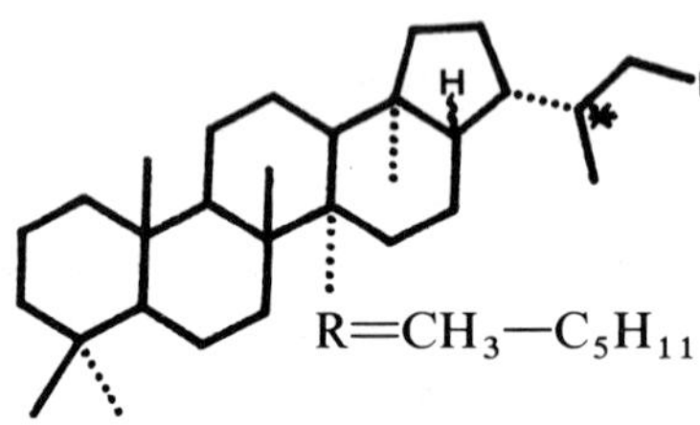

II. Extended Hopanes

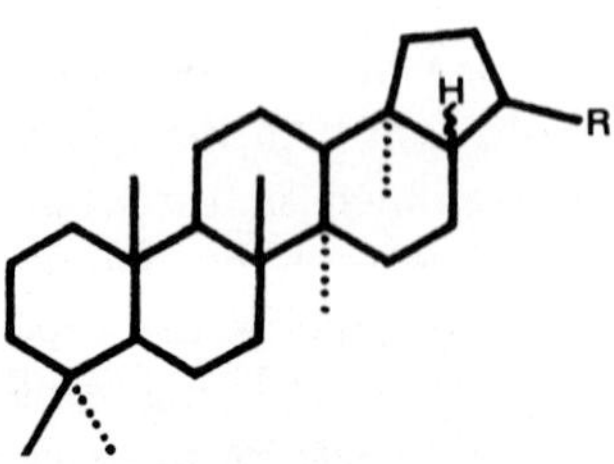

III. Moretanes

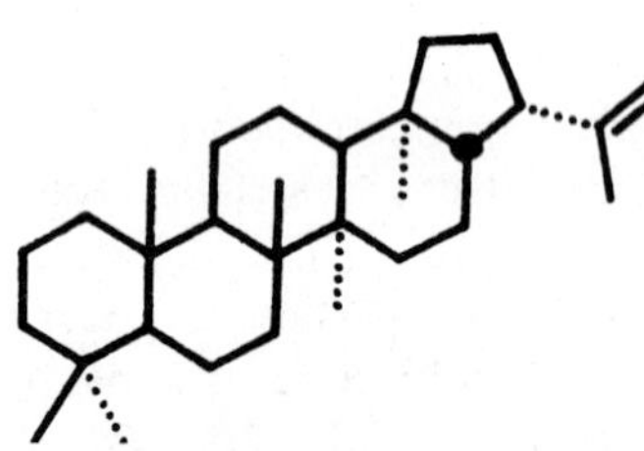

IV. Diploptene

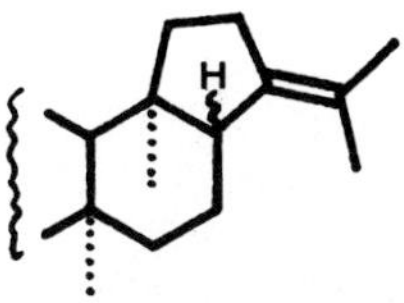

V. Hop-21(22)-ene

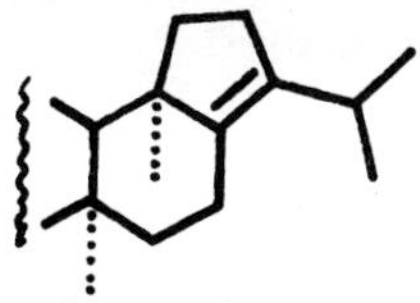

VI. Hop-17(21)-ene

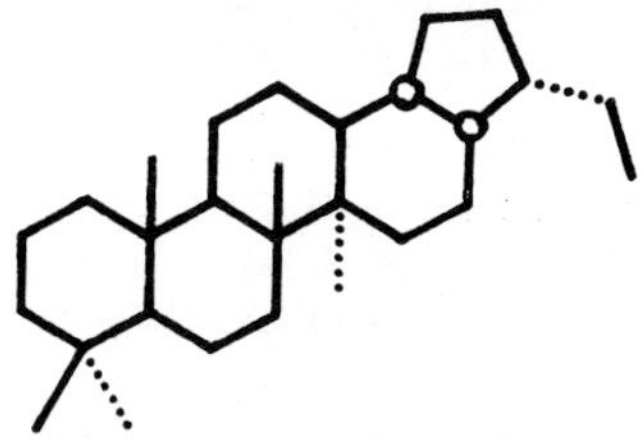

VII. 17α, 18α, 21β(H) -28,30-Bisnorhopane

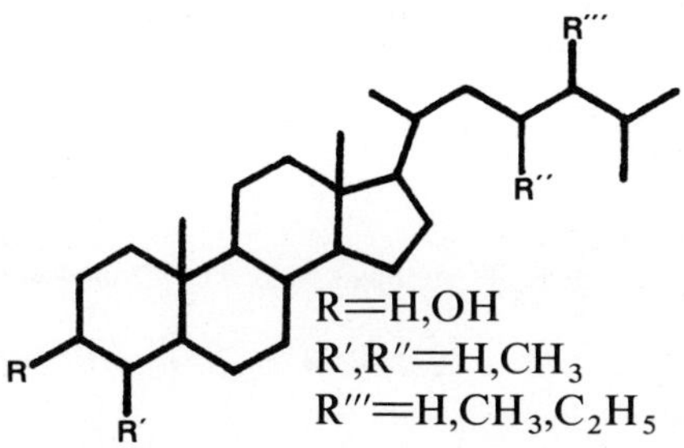

VIII. Steroidal derivatives

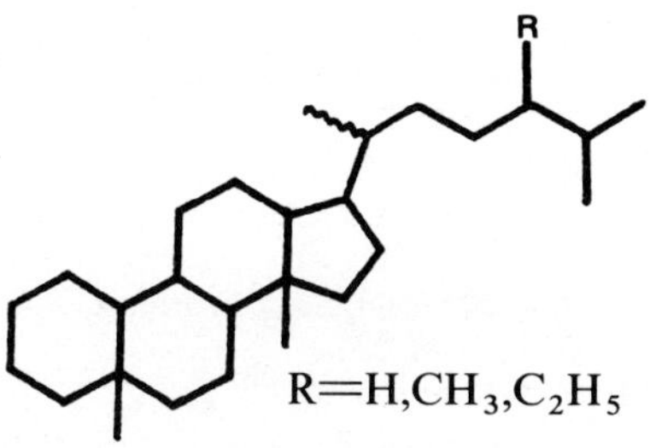

IX. Diasteranes

References

1. A. L. Burlingame, R. E. Cox and P. J. Derrick, *Anal. Chem.* **46,** 248R (1974).
2. A. L. Burlingame, B. J. Kimble and P. J. Derrick, *Anal. Chem.* **48,** 368R (1976).
3. A. L. Burlingame, C. H. L. Shackleton, I. Howe and O. S. Chizhov, *Anal. Chem.* **50,** 346R (1978).
4. A. L. Burlingame, T. A. Baillie, P. J. Derrick and O. S. Chizhov, *Anal. Chem.* **52,** 214R (1980).
5. R. A. W. Johnstone, *Mass Spectrometry, Vol. 3,* The Chemical Society, London (1975).
6. R. A. W. Johnstone, *Mass Spectrometry, Vol. 4,* The Chemical Society, London (1977).
7. R. A. W. Johnstone, *Mass Spectrometry, Vol. 5,* The Chemical Society, London (1979).
8. W. H. McFadden, *Techniques of Combined Gas Chromatography/Mass Spectrometry: Applications in Organic Analysis* J. Wiley and Sons, New York (1973).
9. J. Freudenthal, *Adv. Mass Spectrom. Vol. 7B,* Applied Science Publishers, Barking (1978).
10. A. Alford, *Biomed. Mass Spectrom.* **2,** 229 (1975).
11. A. Alford, *Biomed. Mass Spectrom.* **4,** 1 (1977).
12. A. Alford, *Biomed. Mass Spectrom.* **5,** 259 (1978).
13. F. W. Bowes, M. J. Mulvihill, B. R. T. Simoneit, A. L. Burlingame and R. W. Risebrough, in *Dioxin, Toxicological and Chemical Aspects* F. Cattabeni *et al.*, eds. (Spectrum Publications, New York, 1978), pp. 79–98.
14. B. R. T. Simoneit, D. H. Smith and G. Eglinton, *Arch. Env. Contamin. Toxicol.* **3,** 385 (1976).

15. A. L. Burlingame, B. J. Kimble, E. S. Scott, D. M. Wilson, M. J. Stasch, J. W. de Leeuw and L. H. Keith, in *Identification and Analysis of Organic Pollutants in Water* L. H. Keith, ed., Ann Arbor Science Publishers. 587 (1976).
16. R. A. Hites and K. Biemann, *Anal. Chem.* **42,** 855 (1970).
17. H. S. Hertz, R. A. Hites and K. Biemann, *Anal. Chem.* **42,** 681 (1971).
18. R. A. Rasmussen, *J. Air Poll. Contr. Assoc.* **22,** 537 (1972).
19. H. B. Singh, L. J. Salas, H. Shigeishi and E. Scribner, *Science* **203,** 899 (1979).
20. W. A. Lonneman, S. L. Kopczynski, P. E. Darley and F. Sutterfield, *Environ. Sci. Techn.* **8,** 229 (1974).
21. A. P. Altshuller, W. A. Lonneman, F. D. Sutterfield and S. L. Kopczynski, *Environ. Techn.* **5,** 1009 (1971).
22. B. R. T. Simoneit, in *Chemical Oceanography* J. P. Riley and R. Chester, eds., Academic Press, New York, 223 (1978).
23. D. H. Stuermer and G. R. Harvey, *Nature* **250,** 480 (1974).
24. D. H. Stuermer, K. E. Peters and I. R. Kaplan, *Geochim. Cosmochim. Acta* **42,** 989 (1978).
25. B. R. T. Simoneit, S. Brenner, K. E. Peters and I. R. Kaplan, *Geochim. Cosmochim. Acta* **45,** 1581 (1981).
26. R. P. Philp, M. Calvin, S. Brown and E. Yang, *Chem. Geol.* **22,** 207 (1978).
27. S. Larter and A. G. Douglas, in *Environmental Biogeochemistry and Geomicrobiology* W. E. Krumbein, Ed., Ann Arbor Science Publishers, 373 (1978).
28. D. van de Meent, S. C. Brown, R. P. Philp and B. R. T. Simoneit, *Geochim. Cosmochim. Acta* **44,** 999 (1980).
29. B. R. T. Simoneit, in *The Impact of the Treibs' Porphyrin Concept on the Modern Organic Geochemistry*, A. Prashnowsky, ed., BJM Universität, Würzburg, 133 (1981).
30. V. C. Anderlini, L. Al-Harmi, R. W. Risebrough, B. de Lappe, W. Walker II, B. R. T. Simoneit and A. S. Newton, *Mar. Poll. Bull.* **12,** 57 (1981).
31. B. R. T. Simoneit and I. R. Kaplan, *Mar. Environ. Res.* **3,** 113 (1980).
32. F. W. McLafferty, *Interpretation of Mass Spectra* Second Edition, W. A. Benjamin, Inc., Reading, Mass. (1973).
33. K. Biemann, *Mass Spectrometry, Organic Chemical Applications* McGraw-Hill, New York (1962).
34. R. M. Silverstein, G. C. Bassler and T. C. Morrill, *Spectrometric Identification of Organic Compounds* Third Edition, John Wiley and Sons, New York (1974).
35. B. R. T. Simoneit, M. A. Mazurek and T. A. Cahill, *J. Air. Pollut. Control. Assoc.* **30,** 387 (1980).
36. B. R. T. Simoneit, in *Advances in Organic Geochemistry 1979* A. G. Douglas and J. R. Maxwell, eds., Pergamon Press, Oxford, 343 (1980).
37. B. R. T. Simoneit, M. A. Mazurek, S. Brenner, P. T. Crisp and I. R. Kaplan, *Deep-Sea Res.* **26A,** 879 (1979).
38. R. W. Johnson and J. A. Calder, *Geochim. Cosmochim. Acta* **37,** 1943 (1973),.
39. B. P. Tissot and D. H. Welte, *Petroleum Formation and Occurrence: A New Approach to Oil and Gas Exploration* Springer-Verlag, Berlin, 538 pp. (1978).
40. B. R. T. Simoneit, *Geochim. Cosmochim. Acta* **41,** 463 (1977).
41. J. M. Hunt, *Petroleum Geochemistry and Geology* Freeman, San Francisco, 617 pp. (1979).
42. W. K. Seifert and J. M. Moldowan, *Geochim. Cosmochim. Acta* **42,** 77 (1978).
43. A. S. Mackenzie, R. L. Patience, J. R. Maxwell, M. Vandenbroucke and B. Durand, *Geochim. Cosmochim. Acta* **44,** 1709 (1980).
44. B. R. T. Simoneit and M. A. Mazurek, *Atmosph. Environ.* in press (1982).
45. C. M. Ekweozor and O. P. Strausz, in *Advances in Organic Geochemistry 1981* M. Bjorøy, ed., in press.

46. F. R. Aquino Neto, J. Connan, J. M. Trendel, A. Restle and P. Albrecht, in *Advances in Organic Geochemistry 1981* M. Bjorøy, ed., in press.
47. C. Lee, R. B. Gagosian and J. W. Farrington, *Geochim. Cosmochim. Acta* **41,** 985 (1978).
48. C. Lee, J. W. Farrington and R. B. Gagosian, *Geochim. Cosmochim. Acta* **43,** 35 (1979).
49. M. Dastillung and P. Albrecht, *Mar. Poll. Bull.* **7,** 13 (1976).
50. G. Ourisson, P. Albrecht and M. Rohmer, *Pure Appl. Chem.* **51,** 709 (1979).
51. P. W. Brooks, G. Eglinton, S. J. Gaskell, D. J. McHugh, J. R. Maxwell and R. P. Philp, *Chem. Geol.* **20,** 189 (1977).
52. W. K. Seifert, J. M. Moldowan, G. W. Smith and E. V. Whitehead, *Nature* **271,** 436 (1978).
53. S. J. Gaskell, *The environmental geochemistry of sterols* Ph.D. Thesis, University of Bristol, England (1974).
54. W. Y. Huang and W. G. Meinschein, *Geochim. Cosmochim. Acta* **43,** 739 (1979).
55. G. Mattern, P. Albrecht and G. Ourisson, *Chem. Comm.* 1570 (1970).
56. B. R. T. Simoneit, *Deep-Sea Res.* **24,** 813 (1977).
57. M. Schnitzer and S. U. Khan, *Humic Substances in the Environment* Marcel Dekker, New York, 327 pp. (1972).
58. B. Durand, ed., *Kerogen, Insoluble Organic Matter From Sedimentary Rocks* Editions Technip, Paris (1980).
59. B. R. T. Simoneit, S. Brenner, K. E. Peters and I. R. Kaplan, *Nature* **273,** 503 (1979).
60. B. R. T. Simoneit, in *Initial Reports of the Deep Sea Drilling Project, Vol. 50* Y. Lancelot, E. Winterer, *et al.*, U.S. Government Printing Office, Washington, D.C., 637 (1980).
61. P. G. Hatcher, B. R. T. Simoneit, F. T. Mackenzie, A. C. Neumann, D. C. Thorstenson and S. M. Gerchakov, *Org. Geochem.* in press (1982).

Part II
CHEMICAL ASPECTS
(Distribution, Contamination)

The Environmental Chemistry of Volatile Aromatic Hydrocarbons†

ERNEST MERIAN

International Association of Environmental Analytical Chemistry, and Swiss Association for Environmental Research, Im Kirsgarten 22, CH-4106 Therwil, Switzerland

(*Received December 2, 1981*)

The approximately annually produced 3,000 million tons of crude oil contain about 40 million t C_6-C_8-aromatics, and about 30 million t naphthalenes. In refineries some higher quantities are additionally produced by disproportionating, cracking or by other processes. Minor shares are isolated, and used in the chemical industry. Most of these valuable amounts are however burnt, some parts are transformed during combustion, or are lost.

It is estimated that about 13 million t C_6-C_8-aromatics, and about 3 million t naphthalene and higher alkylated volatile aromatic hydrocarbons enter annually the environment, mostly from the oil industry and from automobile exhaust. About 15% is introduced into the sea. The bigger share is first introduced into the atmosphere, toluene and xylenes also in considerable amounts from solvent losses. Some naphthalene and chlorinated benzenes enter the environment as insecticides.

Global cycles are discussed. It seems that C_6-C_8-aromatics evaporate or stay rather in the atmosphere, and are transformed or degraded there, whereas naphthalenes accumulate rather in the sea, particularly in fish, other aquatic animals or sediments. Of course the local concentrations may vary widely. Also, one should not forget, that it takes some weeks to mix thoroughly the atmosphere of one hemisphere, 1 to 2 years between the hemispheres, 1,000 years to mix thoroughly oceans, and even longer soils, according to their local composition.[1]

Air contains for instance about 0,001–0,003 mg benzene/m^3 in remote rural areas. But in urban streets or in specific industrial sites the concentration may be several hundred times higher. Transport behaviour of volatile aromatic hydrocarbons, their degradation and overall environmental fate are also discussed, to make clear, where emphasis for further studies is relevant.

KEY WORDS: Global cycles, environmental sources, environmental fate, volatile aromatic hydrocarbons, benzene, toluene, xylenes, naphthalenes.

---

†Presented at the Workshop on the Chemistry and Analysis of Hydrocarbons in the Environment, Barcelona, November 1981.

INTRODUCTION

To get an idea, where relevant problems occur and where emphasis for future studies is necessary, it is important to have a better understanding of global and local cycles of volatile aromatic hydrocarbons, of their transport behaviour, their degradation and their overall environmental fate. From many informations in the literature (which are not always comparable) it was tried to get the orders of magnitude of the global cycles. This should be a first approach for further discussion. It would be in the interest of everyone involved, if the experts in the field—especially also from the oil industry—could verify, improve and precise these data, which were estimated by the author by extrapolation. Also, global cycles can of course not reflect local situations.

It can be assumed reasonable to differentiate between three differently behaving groups of compounds:

—Benzene is the most volatile, most water soluble and most toxic of the aromatic one- or two-ring hydrocarbons.

—Toluene, xylenes, ethylbenzene, styrene have a medium volatility, a medium water solubility, are relatively rapidly decomposed, and are not very toxic. Both groups are relatively easily evaporated from water into air.

—Napthalene, alkylated naphthalenes have a low volatility, a low water solubility, and are therefore not decomposed very rapidly. They tend to be precipitated from the air into the water, where they are agglomerating, sedimenting or accumulating in living organisms. Naphthalenes are about 2 to 100 times more toxic to fish and crustacea than C_6-C_8-aromatics.[2]

GLOBAL IMPACTS OF PETROLEUM PRODUCTS IN GENERAL

Unfortunately most of the valuable compounds mentioned are burned. Besides losses into the environment one finds decomposition products of volatile aromatics, including pyrolytic compounds (also additional volatile aromatics from pyrolysis of other compounds). Korte and Boedefeld have already worked out[3] a material balance of global natural gas and liquid petroleum operations. Of the about 1,600 million t of natural gas produced in 1974 around 82% were burned and about 9% were flared, and only the rest was fractionated or used as such. The liquids concentrate of natural gas may contain up to 4% benzene.[4]

Of the about 2,890 million t of crude oil produced in 1974 about 2,400 million t of fuels were produced by refineries and about 100 million t were already burned or lost before this stage. This is a total of about 88%, whereas about 12% or 300 million t were non-fuels, such as naphthas, bitumen, petroleum coke, lubricants and white spirits. Losses into the environment amounted to about 2.4% or 70 million t of the total annual production, which figure does not include losses and decomposition products from fuel distribution and during combustion. Of the 2,400 million t of fuels, about 840 million t were residual fuel oil, about 640 million t gas/diesel oil and about 600 million t motor gasoline.[3] The latter contains about 2% benzene, about 14% toluene, about 24% C_8-aromatics and about 12% C_9-C_{10}-benzols.

Crude oil contains already between 16.5 and 21.9% aromatics,[3] which corresponds to about 500 million t. Of this amount 5 million t may be natural benzene, 12 million t toluene, 22 million t xylenes and 30 million t naphthalenes.[5] Additionally roughly 500 million t of volatile aromatics are produced in the petrochemical industry by refining and cracking of oil fractions. The losses into the environment are evaporated, precipitated, biodegradated or abiotically decomposed. A rather comprehensive review of the fate of hydrocarbons in the marine environment has been given by the National Academy of Sciences[6] and by Exxon.[7] Annually about 5–10 million t of hydrocarbons get into the sea.[7,8] The aromatic content of tanker spills is no more than 20%, but it is relatively persistent.[5] Marine organisms contain 1–30 ppm of hydrocarbons, which are not differentiated.[6] In mussels of non-polluted water 0.2–0.6 ppm were found, in mussels at polluted Danish coasts, 10–150 ppm.[9] Relatively good information about metabolism mechanisms of organic hydrocarbons in marine organisms are available.[5,10]

BENZENE IN THE ENVIRONMENT

One can estimate that worldwide about 5 million t of benzene are of natural origin, about 26 million t are additionally produced by refining and cracking of oil fractions, and 1 million t are produced from coal tar.[5] Only part of it is isolated as pure benzene for further use. One estimates that 10 million t[5] to more than 20 million t[11] are contained in motor-spirits. From Figure 1 it can be concluded that annually about 15 million t of benzene are used as basis product in the chemical industry, about 13 million t are destroyed through burning and about 4 million t are lost into the environment, mainly into the atmosphere. If the biological and abiotic degradation should take on the average a few days, one can calculate that there must be globally always about 100,000 t of benzene in the environment. This amount is distributed very irregularly. More relevant is of course the individual local concentration.

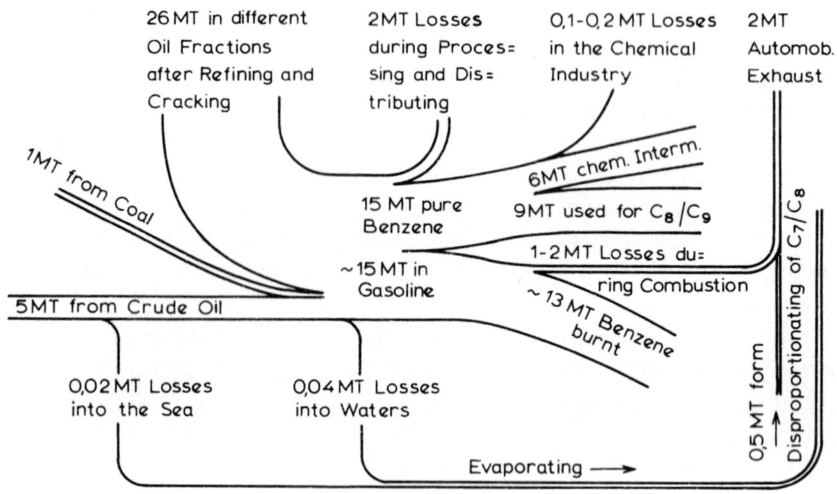

FIGURE 1 Global annual cycle of benzene.

It has been estimated that a human individual ingests about 250 μg benzene per day from food (for instance eggs, rum), about 310 μg from 10 cigarettes and about the same amount over a 24 hour period, if the inhaled air contains 12–43 μg/m^3 benzene,[4,12] which may correspond to concentrations found in cities and industrial areas. A town inhabitant resorbs therefore daily the order of 1/2 mg benzene. Benzene occurs in natural amounts of 0.001 to 0.4% in crude oil, it is a constituent of combustion products, formed from forest fires and smouldering wood, it may be released from plants into the air, and it is therefore also a natural constituent of foods of plant and animal origin.[4] It seems that the natural benzene content of the air is dependent on latitude, for instance in Alaska one finds 0.1 ppb, in the Amazonas region up to 14 ppb, perhaps also due to plant emissions[12] but perhaps also because of the difference of temperature and therefore of the different distribution between air, soil and water. The global industrial production of benzene is at least five times higher, compared to the natural input.

In earlier years occupational intoxications from inhalation of benzene occurred relatively frequently—including cases of leukemias—when benzene was still used as a solvent.[5] For the present time it has been calculated that the number of leukemia cases per annum in the general population of the U.S.A. due to atmospheric benzene could be about 90.[13] This is about 1% of the total leukemia deaths in the United States, based on 1973 vital statistics.[13] Since the recognition of the dangers[5,12] threshold values were fixed, for instance a "Technical Guiding Concentration TRK" of 8 ppm in the Federal Republic of Germany.[14,15] This corresponds to $26\,mg/m^3$.

In the environment such high concentrations have only been found near some specific industrial sites.[16] In streets the immission loads are in the order of 0.04 to $0.1\,mg/m^3$.[5,12] Surface and/or ground waters contain about $1\,\mu g/l$ benzene,[17] rarely up to $100\,\mu g/l$.[12] It evaporates easily,[4,5] and the half life of benzene in water is only about 37 minutes at room temperature.[4,12] In the ice free port of Valdez at the end of the Trans-Alaska Pipeline (around the freezing point) benzene, toluene and xylenes do however not evaporate any more rapidly. Occasionally exceedances of the monthly average and the 24-hr maximum have been observed.[18] Although benzene is relatively stable—as compared to other aromatic hydrocarbons—Korte and Klein[19] believe that the time of 50% mineralization in the atmosphere is likely to be about two days. Benzene is however abiotically decomposed[3,20] only in presence of ultraviolet light and nitric oxides and/or at catalytic surfaces of solid particles. Benzene is degraded by a number of organisms.[5] Profile analytical tests show that benzene is not accumulated to a critical extent in algae and in fish.[19] On the other side, 50 ppm or lower concentrations of benzene or toluene in sea water reduce heart rates and respiration rates (oxygen consumption) of mussels.[21] Benzene and other volatile aromatics are even more persistent in soil.[5]

TOLUENE, XYLENES, ETHYLBENZENE AND STYRENE IN THE ENVIRONMENT

From Figure 2 it can be taken that the cycles of C_7- and C_8-aromatic hydrocarbons are about three times as important in weight as the one of benzene. About 34 million t are of natural origin in crude oil, about 62 million t are added anthropogenically in petrochemistry,[5] and about 8 million t are synthetically produced from benzene. About 8 million t are used as solvents, about 7 million t as basis products for other chemical compounds, and about 10 million t for polymers (with a higher life time). However about 70 million t are destroyed through burning, and about 6 million t of toluene and about 3 million t of xylenes are lost into the environment, mainly into the atmosphere.

Since C_7- and C_8-aromatic hydrocarbons are less toxic than benzene, less efforts for evaluations have been made.[5] They are not carcenogenic, and their threshold and MAK-values (maximum concentrations at working places) are fixed at 100 to 200 ppm (400 to $800\,mg/m^3$).[14] Toluene

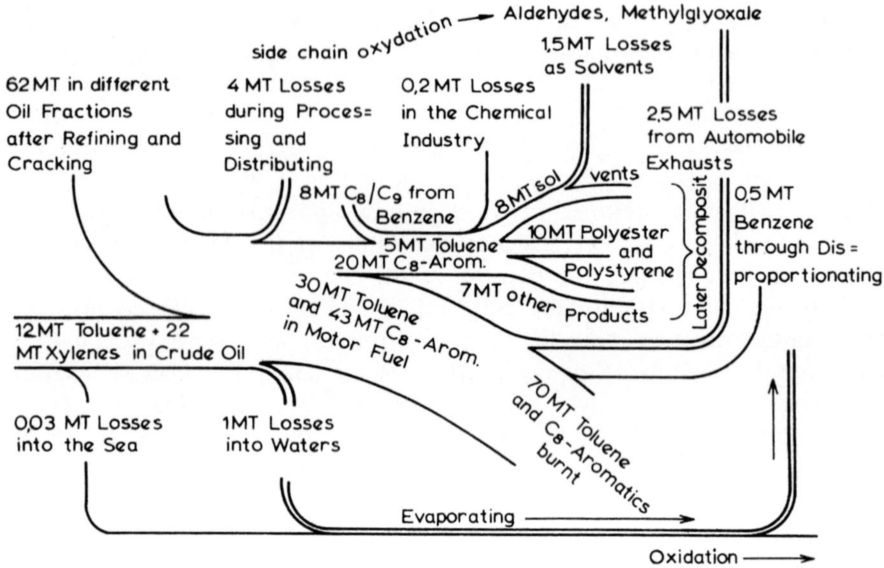

FIGURE 2 Global annual cycle of toluene and C_8-benzenes.

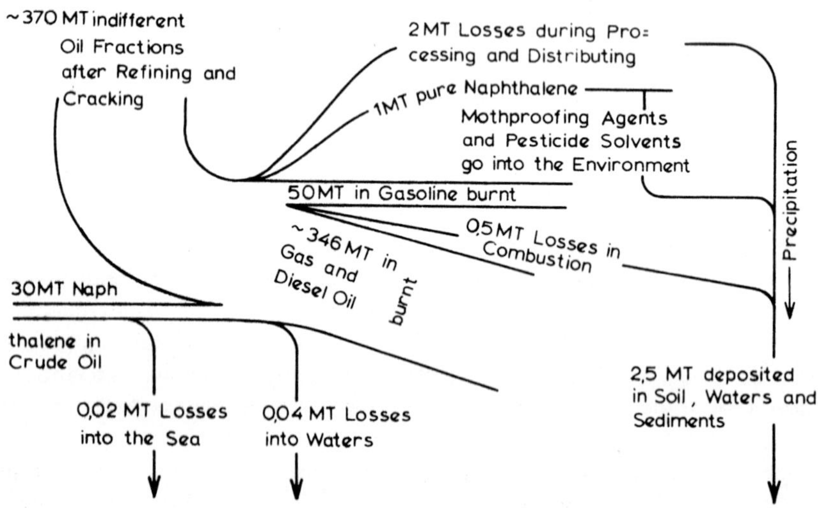

FIGURE 3 Global annual cycle of higher alkylated benzenes and naphthalenes.

and xylenes smell more than benzene, and have therefore somewhat lower odour threshold in the order of 1 ppm, and lower.[5] They may also be sources for eye-irritant decomposition products (smog!) after photochemical reactions, especially in presence of nitrogen oxides.

Since there are also side-chain reactions possible, the life times are probably lower than the one of benzene. When automobile exhaust gases are exposed to ultraviolet light in presence of nitric oxides, 68% of the toluene and 84% of the xylenes decompose within six hours.[5,20] Nevertheless the immission concentrations of toluene and xylenes are higher in streets, as compared to benzene, probably because gasoline contains also much higher amounts. One may find about 0.05 to 0.2 mg/m^3 toluene and about 0.02 to 0.15 mg/m^3 xylenes (lower vapour pressure!) in the atmosphere of streets, and about 2 $\mu g/l$ Toluene, about 3 $\mu g/l$ xylenes, and about 1–2 $\mu g/l$ ethylbenzene in surface water.[5,17] Cigarette smoke contains also about twice as much toluene than benzene.[12]

HIGHER ALKYLATED BENZENES, NAPHTHALENES AND ALKYLATED NAPHTHALENES IN THE ENVIRONMENT

From Figure 3 one can see the amounts involved, which are even higher as compared to benzene, toluene and xylenes. About 30 million t of natural naphthalenes are annually introduced into the system with crude oil. To this anthropogenicly some 370 million t higher alkylated benzenes and naphthalenes are additionally added from refining and cracking process.[5] The approximately 400 million t are mainly parts of diesel fuel and gas oil.[3,5] About 1 million t of naphthalene is used in the chemical industry or as insecticide, about 396 million t are destroyed through burning, and about 3 million t are lost into the environment, mainly at the end of the cycles into the sea, its sediments and its organisms.

These compounds have about 100 times lower vapour pressure, about five to ten times lower water solubility, and a somewhat higher density and partition coefficient between octanol and water than toluene and the xylenes.[5] They are relatively often precipitated—first from the air into the water, and then from the water into the sediments—and tend to agglomerate into particles or into organisms. Especially crustaceans accumulate naphthalenes very rapidly in the cephalothorax.[8] In cold zones sea organisms (for instance flat fishes) are more exposed to risk, because they can take up more naphthalenes, and because metabolism is slower.[10] As already mentioned naphthalenes are about 2 to 100 times more toxic than benzene and its alkylated derivatives to fish and crustacea.[2] Also for man the maximum allowable occupational concentrations of naphthalenes are about 4 to 10 ppm (20 to 50 mg/m^3), that is to say, these volatile aromatics are about 20 times more toxic than toluene and the xylenes.[15] Diphenyl has even a MAK-value of 0.02 ppm.[15]

SUMMARY

Volatile aromatic compounds may have locally adverse effects. Global pollution and exposition to plants, animals and men are probably not too important, apart from higher molecular compounds (C_{10} and more), which may accumulate. As with metals some smaller amounts occur naturally, but locally anthropogenic loads may be much larger. Apart from possible local risks, it is unfortunate that of about 556 million t of volatile aromatic hydrocarbons about 475 million t are burned and about 17 million t are lost into the environment.[5] Table I gives a summary of the estimations of the orders of magnitude, where the losses occur:

TABLE I

Global losses of volatile aromatics into the environment

	Benzene	Toluene	Xylenes and other C_8 and C_9's	Higher alkylated benzols and naphthalenes
Losses into the sea	0.07 MT	0.5 MT	0.6 MT	0.1 MT
Losses of refineries into the air	2 MT	2.5 MT	1.5 MT	2 MT
Evaporation of gasoline	0.1 MT	0.05 MT	0.02 MT	
Automobile exhaust	2 MT	2 MT	0.5 MT	0.5 MT
Diesel exhaust				0.2 MT
Solvent losses		1 MT	0.5 MT	
Losses in the chemical industry	0.1 MT	0.1 MT	0.05 MT	0.01 MT
Total	4.3 MT	6.2 MT	3.2 MT	2.8 MT

References

1. F. Korte, Proceedings, The scientific basis for the assessment of hazards from chemicals associated with the environment, SECOTOX, London, Sept. 1981.
2. J. M. Neff, *Polycyclic Aromatic Hydrocarbons in the Aquatic Environment: Sources, Fates and Biological Effects*, Elsevier's Applied Science Publishers Ltd., Barking, Essex, U.K., especially pages 199–204 (1979).

3. F. Korte and E. Boedefeld, Ecotoxicological review of global impact of petroleum industry and its products, *Ecotoxicology and Environmental Safety* **21,** 55–103 (1978).
4. A. Kluge, Benzene in the environment, Workshop on Benzene, Interpretation of Data and Evaluation of Current Knowledge, Vienna June 1980, Minutes of the Forum für Wissenschaft, Wirtschaft und Politik e. V., Bonn, page 2 (1981).
5. E. Merian and M. Zander, Volatile aromatics, Chapter in the Handbook of Environmental Chemistry (Editor: O. Hutzinger). Volume 3, Part B, Springer-Verlag, Berlin, 117–161 (1982).
6. National Academy of Sciences, *Vapor-phase Organic Pollutants*, Washington, D.C., U.S.A. (1976).
7. Exxon Background Ser. 12, *Fate and Effects of Oil in the Sea*, New York (1978), and *Petroleum in the Marine Environment*, Nat. Academy of Sciences.
8. I. B. Tarshis, Uptake and depuration of petroleum hydrocarbons by crayfish, *Arch. Environm. Contam, Toxicol.* (Springer) **10,** Nr. 1, 79–86 (1981).
9. K. Jensen, Levels of hydrocarbons in mussels and surface sediments from Danish Coastal areas, *Bull. Environm. Contam. Toxicol.* **26,** Nr. 2, 202–206 (1981).
10. U. Varanasi, D. J. Gmur and W. L. Reichert, Effect of environmental temperature on naphthalene metabolism by juvenile starry flounder, *Arch. Environm. Contam. Toxicol.* (Springer) **10,** Nr. 2, 203–214 (1981).
11. P. Brackley, The technology and uses of benzene in the petroleum industry, Workshop on Benzene, Interpretation of Data and Evaluation of Current Knowledge, Vienna June 1980, Minutes of the Forum für Wissenschaft, Wirtschaft und Politik e. V., Bonn, page 3 (1981).
12. DGMK-Project, Evaluation of benzene, toxicity in man and animals, Deutsche Gesellschaft für Mineralölwissenschaft und Kohlechemie e. V., Hamburg, 174–176 (1980); see also J. H. Sterner *et al.*, Assessment of health effects of benzene germane to low-level exposure, E.P.A. 600/1 (1978).
13. R. E. Albert *et al.*, Carcinogen Assessment Group's final report on population risk to ambient benzene exposure, E.P.A. (1978); see also J. H. Sterner *et al.*, Assessment of health effects of benzene germane to low-level exposure, E.P.A. 600/1 (1978).
14. MAK-Werte, Maximale Arbeitsplatzkonzentrationen, dfg-Mitteilung XVI (1980).
15. E. Merian, Legal limits and controls for volatile aromatic hydrocarbons, *Chemosphere* **9,** Nr. 5/6, N3–N6 (1980).
16. E. D. Pellizari and E. Sawicki, The measurements of carcinogenic vapors in ambient atmospheres, E.P.A. 600 (1977).
17. E.P.A., Information System EPA-R2-73-277, Corvallis, and information system, Monitoring to detect previously unrecognized pollutants in surface water, 560/6-77-015, Washington, D.C., U.S.A. (1977).
18. R. J. Fritz, Environmental problems associated with the Alaskan pipeline, Proceedings, Environmental Impact Analysis and Risk Acceptance, International Association of Environmental Coordinators, Brussels, Oct. 1981.
19. F. Korte and W. Klein, Degradation of benzene in the environment, Workshop on Benzene, Interpretation of Data and Evaluation of Current Knowledge, Vienna June 1980, Minutes of the Forum für Wissenschaft, Wirtschaft und Politik e. V., Bonn, pp. 4/5 (1981).
20. E. Shinoyama, *J. Japan Soc. Air. Pollution* **10,** 264 (1975).
21. T. D. Sabourin and R. E. Tullis, *Bull. Environm. Contam. Toxicol.* **26,** 729–736 (1981).

Microbial Degradation of Hydrocarbons†

DAVID T. GIBSON

Center for Applied Microbiology and Department of Microbiology, The University of Texas at Austin, Austin, Texas 78712, U.S.A.

(*Received November 27, 1981*)

The filamentous fungus, *Cunninghamella elegans*, oxidizes naphthalene, biphenyl, benzo[a]-pyrene, benzo[a]anthracene and 3-methylcholanthrene to the same metabolic products that are formed by mammals. The initial oxidation products are apparently arene oxides that can isomerize to phenols or undergo enzymatic hydration to form *trans*-dihydrodiols. In the case of benzo[a]pyrene evidence was obtained for the formation of (+)-9α,10α-epoxy-7β,8α-dihydroxy-7,8,9,10-tetrahydrobenzo[a]pyrene which has been implicated as the ultimate carcinogenic benzo[a]pyrene metabolite formed by higher organisms. In contrast, bacteria oxidize naphthalene, biphenyl, benzo[a]pyrene, benzo[a]anthracene and 3-methylcholanthrene to dihydrodiols in which the hydroxyl groups have a *cis* relative stereochemistry. Arene oxides have not been detected as intermediate products. Many monocyclic aromatic hydrocarbons are also oxidized through *cis*-dihydrodiols by bacteria and the enzyme catalyzing the dioxygenation of toluene to *cis*-toluene dihydrodiol has been shown to be a multienzyme system composed of a flavoprotein and two iron sulfur proteins. The differences in the metabolism of aromatic hydrocarbons by bacteria, fungi and mammals may represent a fundamental difference between procaryotic and eucaryotic cells and is almost certainly related to the physiological functions of the different enzyme systems.

KEY WORDS: Aromatic hydrocarbons, metabolism, bacteria, fungi, mammals, epoxide formation.

INTRODUCTION

Studies on the environmental distribution of aromatic hydrocarbons were initiated almost 25 years ago when Kern[20] reported the presence of chrysene in soil samples. Since that time improved analytical techniques

†Presented at the Workshop on the Chemistry and Analysis of Hydrocarbon in the Environment, Barcelona, November 1981.

have led to the realization that a single sediment sample can contain thousands of aromatic compounds.[2,16] The ubiquitous nature of this class of compounds means that they come into contact with all forms of life and thus studies on the microbial degradation of aromatic hydrocarbons have importance in our understanding of reactions that may be occurring in the environment. However, interest in the metabolism of aromatic hydrocarbons goes back over 200 years to Dr. John Hill's observations on the carcinogenic properties of snuff,[23] and Sir Percivall Pott's reports on the incidence of scrotal skin cancer in English chimney sweeps.[22] It is now known that most aromatic hydrocarbons are toxic, and in some cases, carcinogenic for experimental animals. The virtual explosion in studies on the metabolism of polycyclic aromatic hydrocarbons by mammals that has occurred in the past decade has been occasioned, in no small part, by the growing realization that 60–90% of human cancers are caused by exposure to environmental chemicals.[17] In contrast, studies on the microbial degradation of aromatic hydrocarbons are much less extensive.[12] It is the purpose of this brief chapter to summarize some of the results obtained in our own laboratory on the degradation of aromatic hydrocarbons by bacteria and fungi. Where possible the results are related to comparable studies with higher organisms.

INITIAL REACTIONS IN THE OXIDATION OF AROMATIC HYDROCARBONS

The oxidation of aromatic hydrocarbons by all forms of life involves the enzymatic incorporation of atmospheric oxygen into the substrate. However, the purpose and mechanism of this reaction differs in mammals and bacteria. Mammals have evolved a sophisticated battery of enzymes to convert lipophilic compounds to water-soluble metabolites that can be excreted from the body *via* the kidneys. As will be described below, fungi utilize similar reactions although the physiological function of these enzymes in eucaryotic microorganisms has not been firmly established. In contrast, many bacteria utilize the enzymatic incorporation of molecular oxygen into aromatic hydrocarbons to prepare the molecule for fission of the aromatic nucleus. Subsequent degradation of the ring-fission products leads to small molecules that can be used for the synthesis of cellular constituents.

MAMMALIAN AND FUNGAL OXIDATION OF BENZO[a]PYRENE

The initial reactions in the oxidation of benzo[a]pyrene by mammals are shown in Figure 1. Quinones are apparently formed from 6-

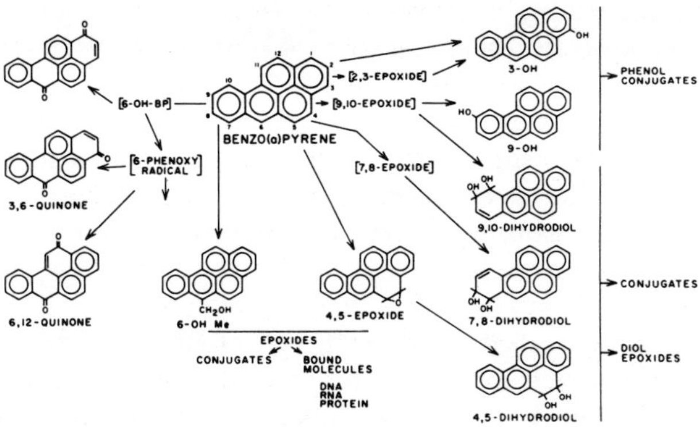

FIGURE 1 Products formed from benzo[a]pyrene by mammals.

hydroxybenzo[a]pyrene while epoxides are formed at the 4,5-, 7,8-, 9,10- and probably 2,3-positions. Interest in benzo[a]pyrene epoxides stems from the fact that they are electrophilic metabolites that can react with cellular nucleophiles. They are formed by the enzymatic incorporation of one atom of molecular oxygen into benzo[a]pyrene by microsomal cytochrome P-450. The oxides are substrates for a second enzyme, epoxide hydrolase, which catalyzes the incorporation of water into the molecules to form dihydrodiols that have a trans relative stereochemistry. It is now known that the enantiomeric specificity of the enzymes involved in benzo[a]pyrene metabolism plays a role in the carcinogenic properties of benzo[a]pyrene. For example, the (−)-enantiomer of *trans*-7,8-dihydroxy-7,8-dihydrobenzo[a]pyrene (BP-7,8-dihydrodiol) is much more potent as a carcinogen and mutagen than the (+)-enantiomer. This is because BP-7,8-dihydrodiol is oxidized to (+)-9α,10α-epoxy-7β,8α-dihydroxy-7,8,9,10-tetrahydrobenzo[a]pyrene (Diol epoxide-2) which has been identified as the ultimate carcinogenic metabolite formed from benzo[a]pyrene. Diol epoxide-2 is an extremely reactive electrophile that can react with DNA to form covalent adducts. Details of the mammalian metabolism of benzo[a]pyrene may be found in a recent review by Gelboin.[11] The reactions leading to the formation of diol epoxide-2 are shown in Figure 2.

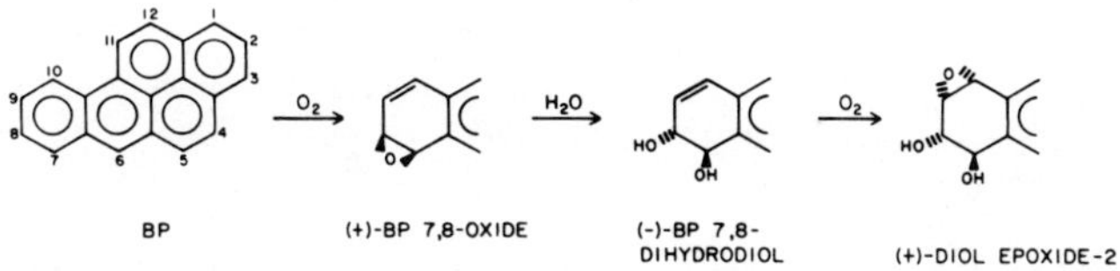

FIGURE 2 Reactions leading to the formation of (+)-Diol epoxide-2 in mammals.

Five years ago we initiated studies to see if eucaryotic microorganisms would oxidize aromatic hydrocarbons. A screen of 86 species of fungi belonging to 64 genera revealed that 47 species from 34 genera oxidized naphthalene to the same spectrum of metabolites that are formed by mammals.[8] All organisms tested from the order *Mucorales* oxidized naphthalene, and *Cunninghamella elegans* was chosen for further investigation. This organism oxidized naphthalene (Figure 3) and biphenyl (Figure 4) to the same products that are formed from these substrates by mammals.

When *C. elegans* was incubated with benzo[a]pyrene several metabolic products were detected by thin-layer chromatography. Fractions were isolated from the thin layer chromatograms and analyzed further by high pressure liquid chromatography. The results showed that benzo[a]pyrene was oxidized to 3-hydroxybenzo[a]pyrene, 9-hydroxybenzo[a]pyrene, benzo[a]pyrene 1,6-quinone, benzo[a]pyrene 3,6-quinone, *trans*-9,10-dihydroxy-9,10-dihydrobenzo[a]pyrene and *trans*-7,8-dihydroxy-7,8-dihydrobenzo[a]pyrene. In addition, sulfate and glucuronide conjugates of hydroxylated metabolites were also detected. These reactions, which are remarkably similar to those observed in higher organisms, are summarized in Figure 5.

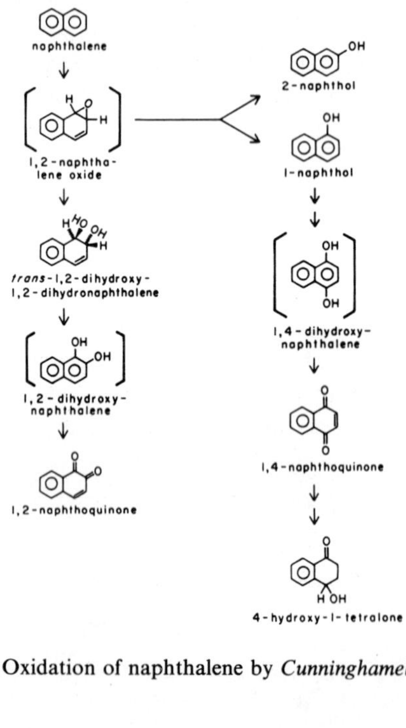

FIGURE 3 Oxidation of naphthalene by *Cunninghamella elegans.*[7]

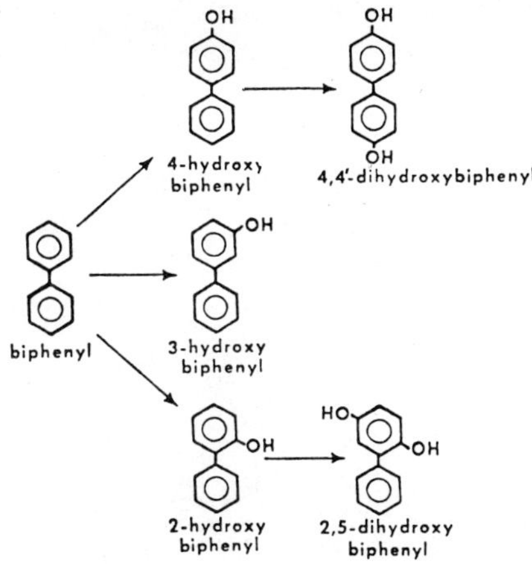

FIGURE 4 Oxidation of biphenyl by *Cunninghamella elegans*.[9]

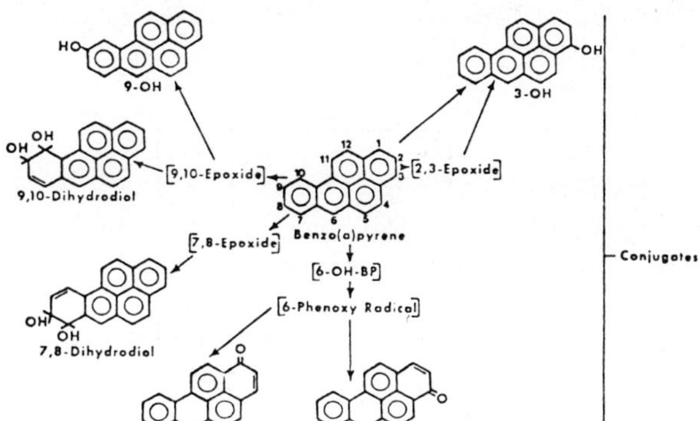

FIGURE 5 Oxidation of benzo[a]pyrene by *Cunninghamella elegans*.[6]

Since *C. elegans* oxidizes benzo[a]pyrene to *trans*-7,8-dihydroxy-7,8-dihydrobenzo[a]pyrene we decided to investigate the possibility that the fungus produces benzo[a]pyrene 7,8-diol-9,10-epoxides which have been implicated as the probable ultimate carcinogenic benzo[a]pyrene metabolites. Two diastereometric diol epoxides, each of which can exist in two enantiomeric forms, are known. Since the diol epoxides are unstable they are isolated and identified as their hydrolysis products (Figure 6). When polar products formed from benzo[a]pyrene by *C. elegans* were analyzed by high pressure liquid chromatography a complex spectrum of

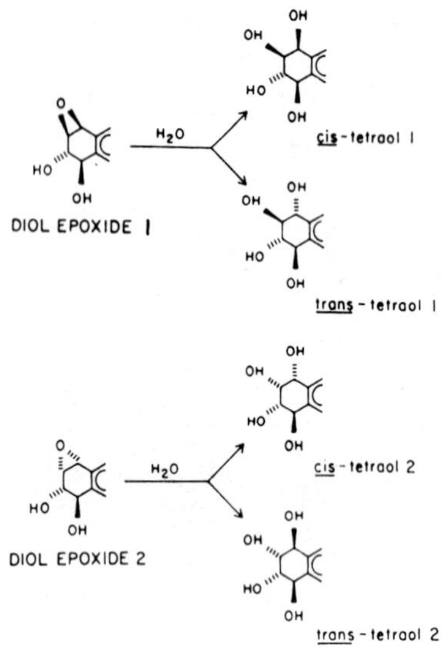

FIGURE 6 Hydrolysis of diastereomeric benzo[a]pyrene diol epoxides to tetraols. Absolute stereochemistry is not implied.

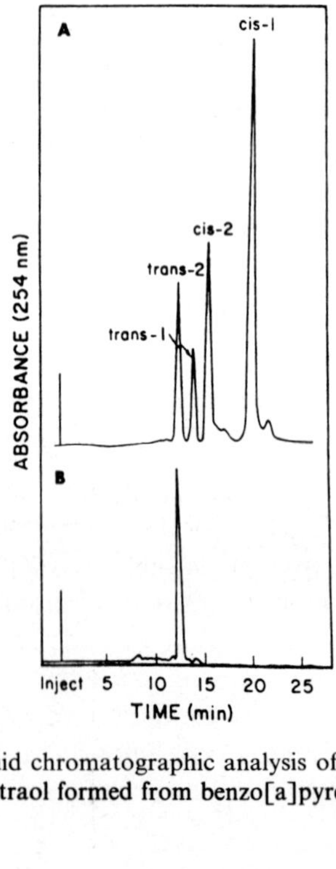

FIGURE 7 High-pressure liquid chromatographic analysis of the tetraols formed from Diol epoxide-1 and -2 (A) and the tetraol formed from benzo[a]pyrene by *C. elegans* (B).[5]

metabolites was detected. Most of these products were sulfate conjugates. However, one metabolite gave an identical absorption spectrum to that given by the tetraols formed from the diol epoxides. Since the absorption spectrum did not differentiate between the four possible tetraols shown in Figure 6 the fungal product was analysed by high pressure liquid chromatography. Figure 7 shows that the metabolite had the same retention time as *trans*-tetraol 2 which is formed from diol epoxide-2. The same tetraol was formed from (±)-*trans*-7,8-dihydroxy-7,8-dihydrobenzo[a]pyrene by *C. elegans.* These observations suggested that the organisms forms either (±)-7α,8α-dihydroxy-9α,10α-epoxy-7,8,9,10-tetrahydrobenzo[a]pyrene or one of the two possible enantiomers of this compound. The question of enantiomeric specificity is important since (+) diol epoxide-2 has been implicated as the ultimate chemical carcinogen formed from benzo[a]pyrene. Since *trans*-tetraol 2 could be formed from either (+) or (−) BP-7,8-dihydrodiol (Figure 8) the enantiomeric purity of the *trans*-7,8-dihydroxy-7,8-dihydro diol formed from benzo[a]pyrene was determined. Radioactive 7,8-dihydro-diol formed by *C. elegans* was mixed with racemic 7,8-dihydrodiol and reacted with (−)-menthoxyacetyl chloride. The resulting diastereomers were separated by high pressure liquid chromatography and the radioactivity in each compound was determined. More than 95 percent of the radioactivity cochromatographed with the di-(−)menthoxyacetate formed from (−)-*trans*-7,8-dihydroxy-7,8-dihydrobenzene (Figure 9). These results are almost identical to those reported for mammals which form the more tumorigenic (−)-enantiomer of *trans*-7,8-dihydroxy-7,8-dihydrobenzo[a]pyrene. The reactions used by

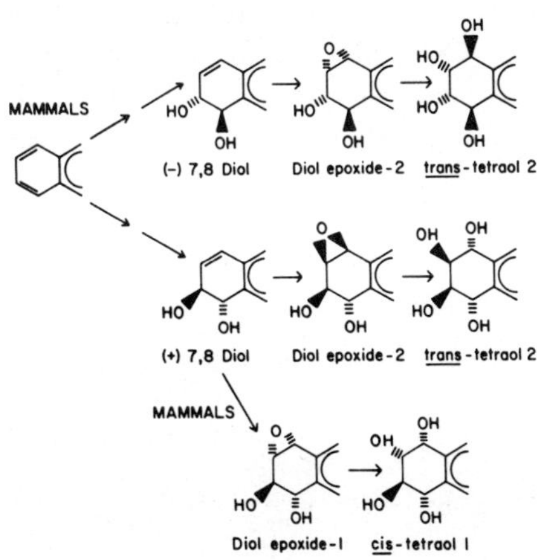

Figure 8 Diol epoxide formation from enantiomers of *trans*-7,8-dihydroxy-7,8-dihydrobenzo[a]pyrene.

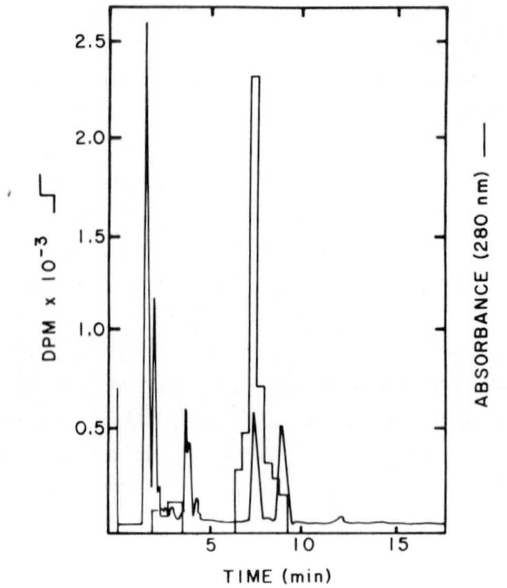

FIGURE 9 Separation of di-(−)menthoxyacetates formed from (±)-BP 7,8-dihydrodiol and the radioactive BP 7,8-dihydrodiol formed from benzo[a]pyrene by *Cunninghamella elegans*. The compound eluting at 7.5 min is the diester formed from (−) BP 7,8-dihydrodiol.[4]

C. elegans to oxidize benzo[a]pyrene at the 7,8-position are shown in Figure 10.

Recently Dodge (1980)[10] and Cerniglia and Gibson *et al.* (1982)[3] have shown that *C. elegans* oxidizes benzo[a]anthracene and 3-methylcholanthrene to the same metabolites that are produced from these substrates by mammals. The only exception, that was also noted with benzo[a]pyrene, is the absence of oxidative attack at the K-region. The physiological function of the enzymes involved in aromatic hydrocarbon oxidation by fungi have not been established. However, their resemblance to the drug-metabolizing systems of higher organisms is certainly remarkable and studies with purified enzyme systems should provide further comparative information.

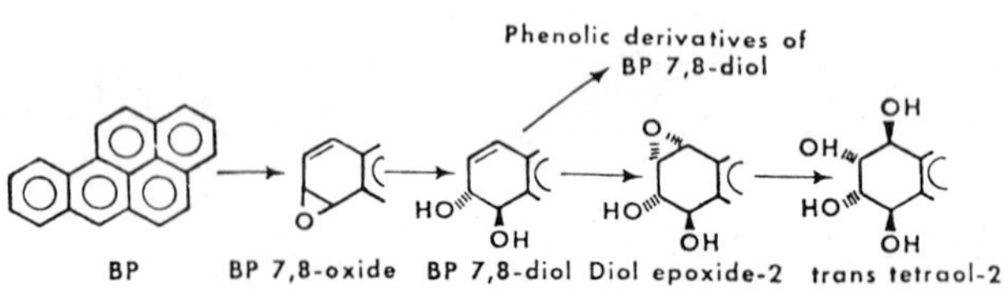

FIGURE 10 Reactions utilized by *C. elegans* to oxidize benzo[a]pyrene to diol epoxide-2. The hydrolysis of Diol epoxide-2 to *trans*-tetraol-2 is non enzymatic.

BACTERIAL OXIDATION OF BENZO[a]PYRENE

Bacteria capable of growth with benzo[a]pyrene have not been isolated. However, cells of a *Beijerinckia* sp. after growth with biphenyl oxidized benzo[a]pyrene to acid products that were not identified.[14] A mutant strain of the same organism that oxidizes biphenyl to *cis*-2,3-dihydroxy-2,3-dihydrobiphenyl,[13] oxidized benzo[a]pyrene to two dihydrodiols which were separated by high pressure liquid chromatography (Figure 11).

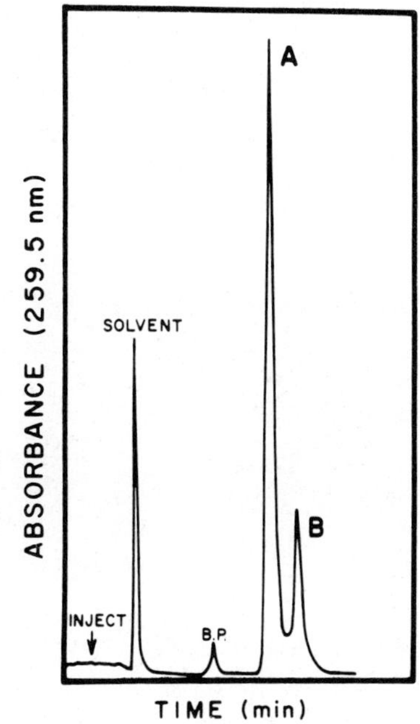

FIGURE 11 Separation of dihydrodiols formed from benzo[a]pyrene by a mutant strain of *Beijerinckia*.

Analysis of the proton magnetic resonance spectrum of each dihydrodiol revealed that *cis*-hydroxylation had occurred at the 7,8- and 9,10-positions. The major product was *cis*-9,10-dihydroxy-9,10-dihydrobenzo[a]pyrene[13] and the initial reactions in the bacterial oxidation of benzo[a]pyrene by *Beijerinckia* are shown in Figure 12. The same mutant strain of *Beijerinckia* oxidized benzo[a]anthracene to three *cis* dihydrodiols at the 1,2-, 8,9- and 10,11-positions. The structures of these metabolites are shown in Figure 13. Anthracene and phenanthrene were oxidized to *cis*-1,2-dihydroxy-1,2-dihydrophenanthrene and *cis*-3,4-dihydroxy-3,4-dihydrophenanthrene respectively.[1,19,21]

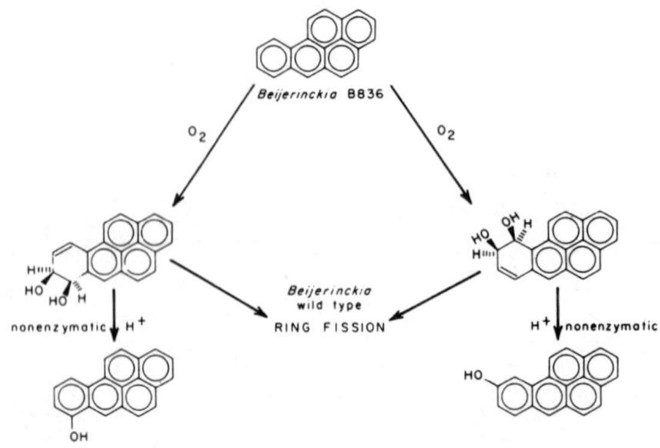

FIGURE 12 Initial reactions in the oxidation of benzo[a]pyrene by a *Beijerinckia* species.

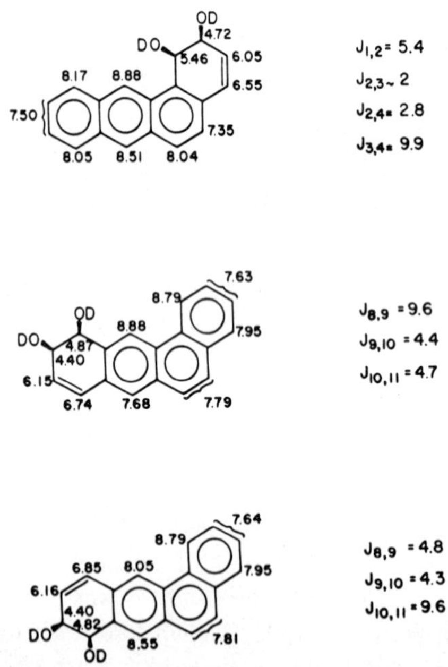

FIGURE 13 *Cis*-Dihydrodiols formed from benzo[a]anthracene by *Beijerinckia* B-836. Numbers refer to chemical shifts observed in the p.m.r. spectrum after exchange of hydroxyl protons with deuterium. The coupling constants $J_{1,2}$, $J_{10,11}$ and $J_{8,9}$ are indicative of a cis relative stereochemistry.

It is now apparent that *cis*-hydroxylation is a common reaction used by different bacterial strains to initiate the oxidation of many aromatic hydrocarbons and related compounds. Some examples are given in Figure 14 and a more complete list has been published.[12] In all cases that we have examined the *cis*-dihydrodiols are oxidized by pyridine-nucleotide dependent dehydrogenases to catechols which are substrates for dioxygenases that cleave the aromatic nucleus.

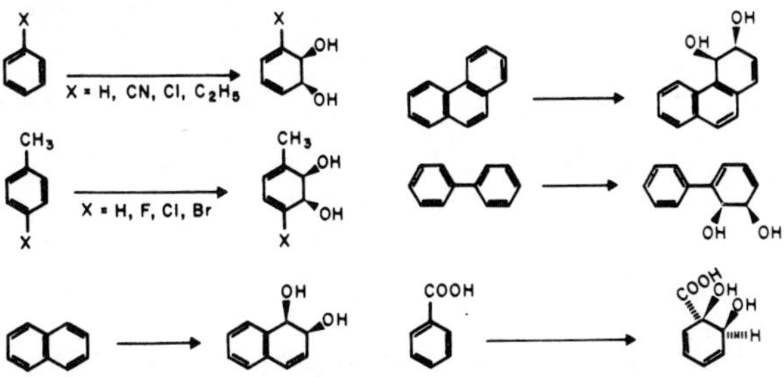

FIGURE 14 Some *cis*-dihydrodiols that are formed by bacteria.

The enzymes that catalyze the formation of *cis*-dihydrodiols incorporate both atoms of molecular oxygen into the aromatic nucleus.[15,18] Recently we have resolved the enzyme that catalyzes the oxidation of toluene to *cis*-1(S),2(R)-dihydroxy-3-methylcyclohexa-3,5-diene (*cis*-toluene dihydrodiol) into three protein components.[26] In the presence of reduced nicotinamide-adenine dinucleotide (NADH) and ferrous iron the three proteins catalyze the reaction shown below. Preliminary studies indicate that electrons are

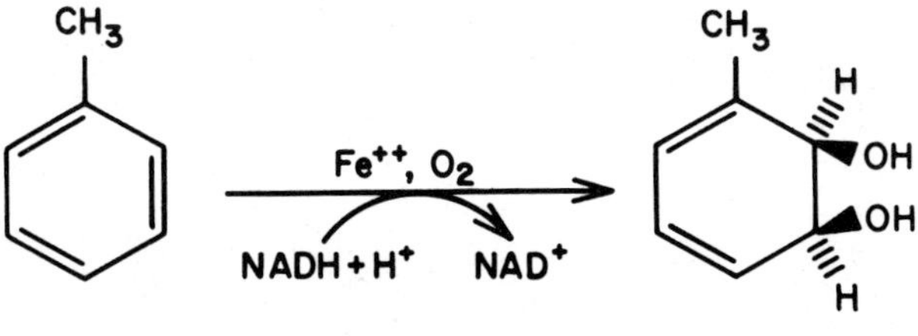

transferred from NADH to a flavoprotein that contains flavin adenine dinycleotide as its prosthetic group.[24] The reduced flavoprotein (reductase$_{TOL}$) transfers electrons to a small iron-sulphur protein (ferredoxin$_{TOL}$) which in turn reduces a large iron sulphur- protein (ISP$_{TOL}$). The oxygenase activity of the toluene dioxygenase system is associated with ISP_{TOL},[25] and the flow of electrons is shown in Figure 15. Future studies will be directed towards the elucidation of the mechanisms of oxygen activation and insertion into the aromatic nucleus.

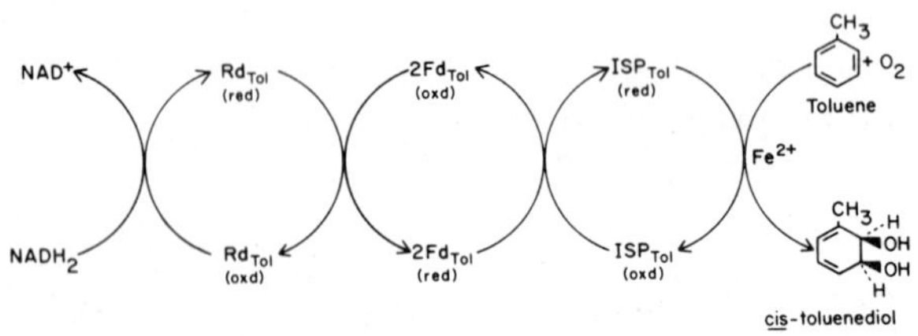

FIGURE 15 Organization of the toluene dioxygenase enzyme system.

CONCLUSIONS

Higher organisms utilize a monooxygenase enzyme system to oxidize aromatic hydrocarbons to reactive arene oxides. The oxides can isomerize to phenols, undergo enzymatic conjugation with glutathione and are substrates for epoxide hydrolase which converts them to trans dihydrodiols. In addition, certain arene oxides can bind to nucleophilic cellular constituents and initiate tumor formation. Many fungal species appear to utilize almost identical reactions for the oxidation of aromatic hydrocarbons. Thus, *Cunninghamella elegans* oxidizes naphthalene, biphenyl, benzo[a]anthracene, benzo[a]pyrene and 3-methylcholanthrene to the same spectrum of metabolites that are formed from these substrates

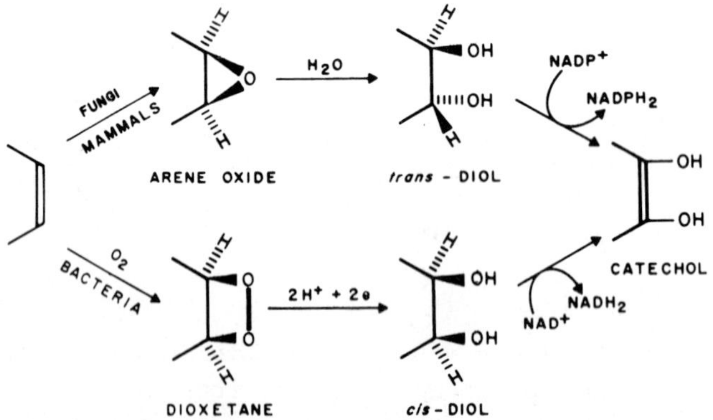

FIGURE 16 Reactions utilized by fungi, mammals and bacteria for the oxidation of aromatic hydrocarbons.

by mammals. The only discernable difference is the absence of K-region oxidation in the polycyclic molecules. In contrast, bacteria utilize a dioxygenase enzyme system to incorporate both atoms of molecular oxygen into aromatic hydrocarbons. The first detectable products are *cis*-dihydrodiols which are oxidized by pyridine nucleotide-dependent dehydrogenases to catechols. The latter compounds are substrates for a different class of dioxygenases that catalyze fission of the aromatic ring. The differences between mammals, fungi and bacteria in terms of the initial reactions involved in the oxidation of aromatic hydrocarbons are shown in Figure 16.

Acknowledgements

The studies described in this paper were supported in part by grants ES-00537 awarded by the National Institute of Environmental Health Sciences, R01 CA-19078 awarded by the National Cancer Institute, F-440 from the Robert A. Welch Foundation and the Exxon Education Foundation.

References

1. M. N. Akhtar, D. R. Boyd, N. J. Thompson, M. Koreeda, D. T. Gibson, V. Mahadevan and D. M. Jerina, *J. Chem. Soc.* Perkin Transactions I, pp. 2506 (1975).
2. M. Blumer and M. Rudrum, *J. Inst. Petr.* (*London*) **56** (548), 99 (1970).
3. C. E. Cerniglia, R. H. Dodge and D. T. Gibson, *Chem. Biol. Interact* **38,** 161 (1982).
4. C. E. Cerniglia, W. Mahaffey and D. T. Gibson, *Biochem. Biophys. Res. Comm.* **94,** 226 (1980).
5. C. E. Cerniglia and D. T. Gibson, *J. Biol. Chem.* **255,** 5159 (1980).
6. C. E. Cerniglia and D. T. Gibson, *J. Biol. Chem.* **254,** 12174 (1979).
7. C. E. Cerniglia and D. T. Gibson, *App. Environ. Microbiol.* **34,** 363 (1977).
8. C. E. Cerniglia, R. L. Hebert, P. J. Szaniszlo and D. T. Gibson, *Arch. Microbiol.* **117,** 135 (1978).
9. R. H. Dodge, C. E. Cerniglia and D. T. Gibson, *Biochem. J.* **178,** 223 (1979).
10. R. H. Dodge, *Ph.D. Dissertation.* The University of Texas at Austin, Austin, Tx. 78712 U.S.A. (1981).
11. H. V. Gelboin, *Phys. Reviews* **60,** 1107 (1980).
12. D. T. Gibson, Biodegradation of aromatic petroleum hydrocarbons, in *Fate and Effect of Petroleum Hydrocarbons.* D. A. Wolfe (ed.), Pergamon Press, New York, pp. 36–46 (1977).
13. D. T. Gibson, V. Mahadevan, D. M. Jerina, H. Yagi and H. J. C. Yeh, *Science* **189,** 295 (1975).
14. D. T. Gibson, Microbial degradation of polycyclic aromatic hydrocarbons, in *Proceedings of the Third International Biodegradation Symposium*, J. M. Sharpley and A. M. Kaplan (eds.), Applied Science Publishers Ltd., London, pp. 57–66 (1975).
15. D. T. Gibson, G. E. Cardini, F. C. Masales and R. E. Kallio, *Biochemistry* **9,** 1631 (1970).
16. W. Giger and M. Blumer, *Anal. Chem.* **46,** 1663 (1974).
17. J. Higginson and C. S. Muir, in *Cancer Medicine*, J. F. Holland and E. Frei III (eds.), Lea and Febiger, Philadelphia.

18. A. M. Jeffrey, H. J. C. Yeh, D. M. Jerina, T. R. Patel, J. F. Davey and D. T. Gibson, *Biochemistry* **14,** 575 (1975).
19. D. M. Jerina, H. Selander, H. Yagi, M. C. Wells, J. F. Davey, V. Mahadevan and D. T. Gibson, *J. Amer. Chem. Soc.* **98,** 5988 (1976).
20. W. Kern, *Helvetica Chim. Acta* **30,** 1595 (1947).
21. M. Koreeda, M. N. Akhtar, D. R. Boyd, J. D. Neill, D. T. Gibson and D. M. Jerina, *J. Org. Chem.* **43,** 1023 (1978).
22. P. Pott, *Nat. Cancer Inst. Monog.* **10,** 7 (1963).
23. D. E. Redmond Jr., *New England J. Med.* **282,** 18 (1970).
24. V. Subramanian, T. N. Liu, W. K. Yeh, M. Narro and D. T. Gibson, *J. Biol. Chem.* **256,** 2723 (1981).
25. V. Subramanian, T. N. Liu, W. K. Yeh and D. T. Gibson, *Biochem. Biophys. Res. Comm.* **91,** 1131 (1979).
26. W. K. Yeh, D. T. Gibson and T. N. Liu, *Biochem. Biophys. Res. Comm.* **78,** 401 (1977).

Biogeochemistry of Aromatic Hydrocarbons in the Benthos of Microcosms†

J. W. FARRINGTON,‡ B. W. TRIPP,‡ J. M. TEAL,‡ G. MILLE,§ K. TJESSEM,¶ A. C. DAVIS,‡ J. B. LIVRAMENTO,‡ N. A. HAYWARD‡ and N. M. FREW‡

‡*Woods Hole Oceanographic Institution, Woods Hole, MA 02543 U.S.A.*
§*Faculté des Sciences et Techniques, Marseilles, France*
¶*Norsk Hydro, Bergen, Norway*

(*Received February 22, 1982*)

Investigations into the fate of petroleum compounds in the marine environment were carried out using experimental microcosms of two sizes and designs. Aromatic hydrocarbons or No. 2 fuel oil were spiked to the water of a 13 m^3 continuous flow system and to a 228 l recycled flow system. The transport and alteration of this oil was traced in the sediment and benthic organisms (*Glycera americana*, *Crepidula sp.*, and *Nephtys incisa*) of these microcosms. Measurable contamination was found in both sample types. The aromatic hydrocarbon distribution, including relative isomeric distribution (e.g., C_2-phenanthrenes) was found to be different in sediment and in organisms from that which was originally introduced to the experimental microcosm. Differences in isomer distribution between *Glycera* and *Crepidula* were also detected. Based on the experimental data: molecular weight and specific isomeric form, biochemical processes, solubility, and particle adsorption/desorption influence the fate of petroleum compounds in benthic ecosystems.

KEY WORDS: Aromatic hydrocarbons, benthic ecosystem, microcosms, sediment, isomeric distribution.

INTRODUCTION

The fate of petroleum compounds in the aquatic environment has been the subject of oil spill studies and laboratory experiments.[1,2,3,4,5] Major concerns with both chronic and acute inputs of petroleum and other fossil fuel compounds to aquatic ecosystems include: the transport of these

†Presented at the Workshop on the Chemistry and Analysis of Hydrocarbons in the Environment, Barcelona, November 1981.

compounds to sediments, accumulation in sediments and benthic organisms, and long-term effects on benthic ecosystems[6,7,8,9,10] (and references cited previously).

One means of investigating these concerns is the use of large scale and medium scale microcosms. The need for and experimental use of such microcosms has been described in papers in Giesy[11] and Grice and Reeves.[12] We have participated in a series of experiments involving chronic oil additions to the MERL (Marine Ecosystems Research Laboratory, located at the Graduate School of Oceanography, University of Rhode Island) microcosms. These experiments were designed to study the fate and effects of chronic low-level additions of No. 2 fuel oil to a shallow water estuarine ecosystem. Grassle *et al.*[13] have shown that chronic oil additions had an adverse effect on the benthic microcosm community. The simulated oil pollution resulted in a highly significant decline in the number of macrofaunal and meiofaunal individuals in the oil-dosed microcosms. This effect could be seen in both the numbers of individual macrofaunal species and in all metazoan meiofaunal groups. Quantification of the extent of incorporation of fuel oil compounds into the benthos was of interest for this reason.

We have also been conducting studies of the biogeochemistry of hydrocarbons using a smaller scale experimental benthic chamber of our own design.[14,15] We report here some of the results of one of our benthic chamber experiments as well as the results of our analyses of the benthic organisms from the MERL experiments. The goal of these studies was to understand the biogeochemical processes that transport individual petroleum compounds to and within benthic ecosystems.

EXPERIMENTAL

The MERL microcosms experiment used six cylindrical 13 m^3 simulated ecosystems tanks (1.8 m diameter × 5.5 m) containing 30 cm of natural silt-clay sediment collected from mid-Narragansett Bay. The water column in the ecosystem tanks was pumped from Narragansett Bay and was maintained at a flow rate of 20 l per hour with a turnover time of 30 days. The dominant macrofauna in the benthos were *Nucula annulata* and *Mediomastus ambiseta*. The operation of the microcosms and the details of the No. 2 fuel oil experiment are given in Gearing *et al.*[16] and Elmgren *et al.*[17]

A water-accommodated No. 2 fuel oil fraction was added to the water column of the MERL tanks. Details of the chemical composition of the No. 2 fuel oil, the addition procedures for the experiment and the biogeochemistry of hydrocarbons in the water column and sediments are given in Gearing *et al.*,[16,18] Wade and Quinn[19] and Gearing and Gearing.[20,21] A summary of information relevant to this paper is presented in Table 1.

TABLE I

Summary of additions of No. 2 fuel oil to the MERL experimental microcosm and resulting concentrations

First oil addition	February 14, 1977 (7–045)[a]
Last oil addition	August 1, 1977 (7–213)
Number of additions (approx. twice weekly)	42
Total No. 2 fuel oil added per tank	108 g.
Range of water column concentrations for total No. 2 fuel oil	11–636 μg/liter
Average water column concentration ± standard deviation	182 ± 105 μg/liter

[a]Code is: year—Julian date.

At the end of five and one-half months of semi-weekly oil additions and two months recovery period, all sediments in the three control and three experimental tanks were sieved through 0.64 cm sieves and the larger animals retained for enumeration, weighing, and ecological studies.[13] *Glycera americana*, a carnivorous polychaete, was the only infaunal species collected in sufficient quantities for both ecological studies and chemical analyses. *Crepidula sp.*, the common slipper shell, was collected from the wall of the tanks at the sediment water interface, and were also analyzed for hydrocarbon concentrations. Two chronic oil addition MERL experiments were conducted although data are presented only on the benthic animals from the first experiment. Animals from the second experiment were not analyzed.

The benthic chamber experiments were shorter-term (approximately 30 days versus 300) than the MERL experiments discussed above and focused on assessing shorter term interactions between infauna and their habitat in determining the distribution of aromatic hydrocarbons in benthic ecosystems.

These experiments, presented schematically in Figure 1, used an enclosed recirculating chamber that entrains 228 l of water and recycles this over 0.37 m^2 of sediment at controlled velocity. Silt-clay sediment was collected by grab sample from Buzzards Bay (Station BC; 41°27.2′N × 70°52.0′W) and the oxic top centimeter was separated from the

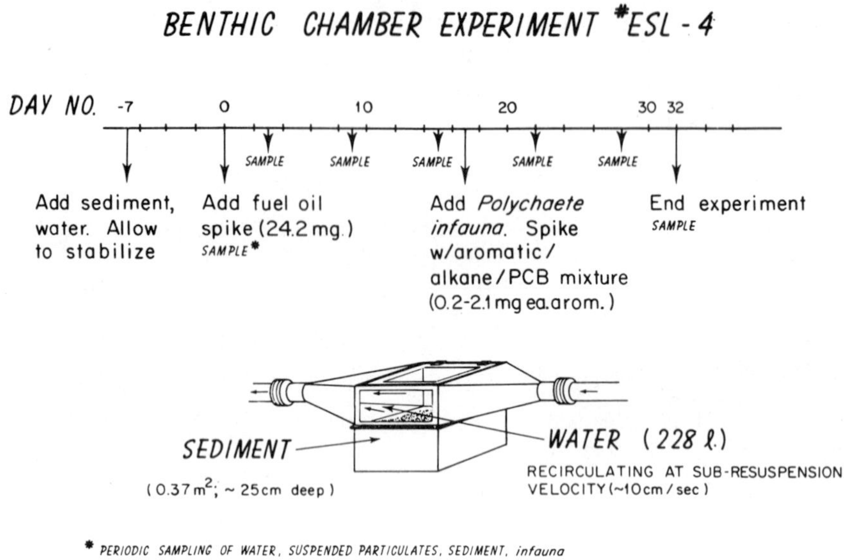

FIGURE 1 Schematic representation of a fuel-oil aromatic hydrocarbon experiment in the benthic chamber. Polycheate infauna and a hydrocarbon spike were added approximately halfway through the 32-day experiment.

deeper anoxic sediment. Both parts of the sediment were frozen and thawed to kill macrofauna and then the sediment was loaded into the chamber with the oxic surface sediment layered on the top.

Sea water from Vineyard Sound, Massachusetts was added to fill the chamber and the system was operated for a one-week stabilization period before the start of the experiment. The seawater was drained and fresh seawater was added from Vineyard Sound and the chamber was operated continuously at 6–9 cm/sec bulk water flow which was below the resuspension velocity for surface sediments in the system. Temperature was monitored throughout the experiment and was initially 18°C dropping to 15°C by day 15 and remaining at 15°C for the remainder of the experiment. The dissolved O_2 level was usually maintained at 6–9 ppm by periodic additions of oxygen although there was one twenty-four hour period where a low of 3–4 ppm O_2 was experienced. The general condition of the experiment and in particular the sediment-water interface was monitored each day by visual observation. The water in the chamber was initially spiked with No. 2 fuel oil (106 μg/l) by injecting a hexane/ethanol solution (24.2 mg No. 2 fuel oil/7 ml of solution) downstream of the sediment in the chamber with water flowing. No visible slick was observed but a slight fuel oil odor was detected after the spike. The hydrocarbon-PCB spike was added on day 17 in the same manner to obtain between 1 and 10 μg/L of each individual hydrocarbon compound. The biogeochemistry of PCBs is the subject of a related investigation and will be discussed elsewhere.

Nephtys incisa were collected from the same Buzzards Bay station as sediment by grab sampling and sieving through a 1 mm mesh screen. The sediment in the chamber was divided by a teflon sheet parallel to the water flow and nearly flush with the sediment interface and a total of 75 *Nephtys* were placed at the sediment–water interface in one-half of the chamber on the same day as collection (day 17).

Samples of water, suspended particulates, and sediment cores were taken throughout the experiment, but only data on the sediment and *Nephtys* will be presented here. Cores of sediment in the chamber were obtained by pushing a 5.8 cm diameter polypropylene core liner into the sediment, inserting a stopper and removing the core. An inverted glass jar was placed in the core hole to prevent slumping or trapping of suspended particulates. Cores were sectioned at once using a stainless steel spatula and then frozen until analysis. All samples discussed in this paper were sampled from the half of the chamber containing polychaetes.

Nephtys were sampled on day 22 by removing exposed or partially exposed individuals with forceps. On day 32, samples of sediment were taken for sediment and pore water analyses. The remaining sediment in the box was sieved through a 1 mm screen to collect *Nephtys*. The samples of *Nephtys* analyzed were: day 22, 5 each—2.2 g wet weight total; day, 32, 50 each—15.9 g wet weight total; day 0, control from station BC-Buzzards Bay, 4–8 each—2.5 g wet weight.

CHEMICAL ANALYSES

Details of the analytical methods are given in Farrington *et al.*[22] and Teal *et al.*[6] Briefly, animals were extracted by aqueous alkaline digestion of wet tissue, extractable lipids were partitioned into diethyl ether, and then separated into alkane/cycloalkane and aromatic hydrocarbon fractions by silica gel column chromatography. Sediments were Soxhlet extracted with toluene–methanol, partitioned into hexane, cleaned of sulfur and polar organics on a small copper–alumina column and then fractionated on silica gel. The hydrocarbons were analyzed by glass capillary gas chromatography and by computerized glass capillary gas chromatography–mass spectrometry. Quantification of aromatic hydrocarbons was by GC-flame ionization detector signal or by GC–MS when aromatic hydrocarbons could not be reliably quantified by GC–FID due to interference from alkenes and naphtheno-aromatics. Internal standards of hexamethylbenzene, hexaethylbenzene and anthracene were used to correct for losses during extraction and analysis.

RESULTS

MERL

The results of hydrocarbon analyses and related data for the benthic organisms *Glycera americana* and *Crepidula sp.* are presented in Tables II, III and Figure 2. Our gas chromatographic data of the alkane–cycloalkane and the aromatic fractions show that the hydrocarbon distribution in the *Glycera* and *Crepidula* is indicative of a partially weathered and degraded No. 2 fuel oil. Gearing *et al.*[16,18] and Wade and Quinn[19] have presented

TABLE II

Summary of benthic animal samples taken from the MERL tanks for hydrocarbon analysis and gravimetric results from column chromatographic separation of lipid extract

	No. of individuals	Total wet wt. (gm)	Dry wt. (gm)	UCM-GC[a] (μg/gm dry wt.)	
Glycera americana					
7-271-08[c]	1	1.05	0.205	f1	<3[b]
(Control)				f2	<3
7-271-02-114	12	12.3	2.00	f1	32.1
(Oil added)				f2	30.3
7-271-09	5	6.55	1.28	f1	62.9
(Oil added)				f2	68.3
Crepidula sp.					
7-277-08	9	10.3	1.28	f1	5.03
(Control)				f2	<3
7-277-02-113	11	8.3	1.03	f1	33.7
(Oil added)				f2	7.49
7-277-07-113					
7-237-07-116	8	5.5	0.682	f1	<3
(Oil added)				f2	<3
7-271-09-115	18	18.5	1.75	f1	57.3
(Oil added)				f2	16.9

[a]Estimate of total petroleum concentration. The f_1 fraction contains alkane and cycloalkane; the f_2 fraction contains aromatics.

[b]Small sample size made measurement of <3 μg/gdw uncertain.

[c]Sample code is explained as follows; e.g., 7-271-08—Year (1977)—Julian date (271)—tank No. (08).

TABLE III

Results of the analyses of benthic animal samples for selected aromatic hydrocarbons (10^{-9} gm/gm dry wt.)

	Glycera americana			*Crepidula sp.*			
	Control	Oil tanks		Control	Oil tanks		
	7-271-08[b]	7-277-02	7-271-09	7-271-08	7-277-02	7-277-07	7-277-09
Naphthalene	—[a]	—	—	—	—	—	—
C_1-Naphthalenes	—	—	—	—	—	11	8
C_2-Naphthalenes	—	14[c]	560	—	—	20	19
C_3-Naphthalenes	—	140	2000	—	—	—	24
C_4-Naphthalenes	—	990	2960	—	—	—	—
Dibenzothiophene	—	140	124	—	—	—	—
C_5-Naphthalenes	—	650	1550	—	—	—	—
C_6-Dibenzothiophene	—	300	628	—	—	—	—
Phenanthrene	—	28	113	41	78	56	68
C_1-Phenanthrene	—	377	1280	35	260	170	238
C_2-Phenanthrene	—	1540	2460	26	430	215	777
C_3-Phenanthrene	—	2200	4600	8	300	148	863
C_4-Phenanthrene	—	780	1780	—	160	64	378
C_5-Phenanthrene	—	740	260	—	48	—	133
Fluoranthene	76.2	73.2	127	53	38	24	26
Pyrene	122	227	514	70	83	49	136

[a]None detected. Detection limit greater than 0.5×10^{-9} g.
[b]Sample code is explained as follows; e.g., 2-271-08—Year code (1977)—Julian date (271)—tank No. (08).
[c]Data are good to only two significant figures for several of the compounds due to GC/MS quantitation limitations.

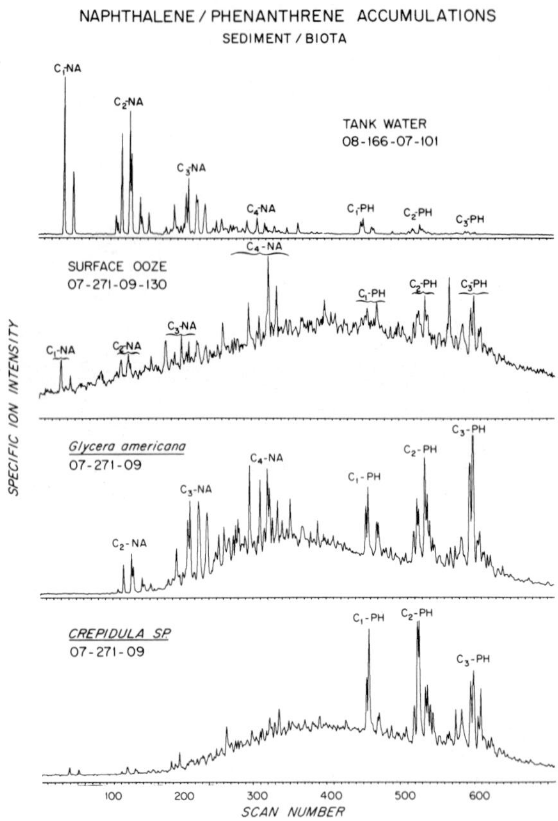

FIGURE 2 Selected ion plot (Σ m/z 192, 156, 170, 192, 206, 220) for samples taken from the MERL experimental microcosm showing the relative differences in the naphthalene and phenanthrene series between samples. Key: NA—naphthalenes, PH—Phenanthrenes.

gas chromatographic data of hydrocarbons in the sediment and water column showing the presence of a similarly degraded No. 2 fuel oil.

Concentrations of No. 2 fuel oil compounds in organisms from the experimental tanks were at least an order of magnitude greater than the control tank organisms. The single *Glycera* analyzed from the control tank had petroleum hydrocarbon concentrations at low or non-detectable levels, similar to those we have reported for other polychaete samples from similar depositional environments.[14,22] These latter data give us more confidence in the analyses of the single available *Glycera* control sample.

The concentrations of aromatic hydrocarbons are also at least an order of magnitude greater for experimental tank samples than for control tank samples, with the exception of fluoranthene and pyrene. Those two aromatic hydrocarbons were present at trace levels in the No. 2 fuel oil but are among the predominant hydrocarbons in the assemblage of combustion product hydrocarbons found in sediments of the northeast U.S. coast including Narragansett Bay and Buzzards Bay sediments used in our experiments.[9,23,24,25] Fluoranthene and pyrene were also found in similar concentrations in samples of mixed polychaete and *Nephtys sp.* taken from sediments in the New York Bight.[23] Some elevation of pyrene concentrations in a few samples from the experimental tanks is noted (Table III) and we suggest that this may be due to pyrene contributed by the repeated No. 2 fuel oil additions. Phenanthrene is also a major component of combustion product hydrocarbons in sediments but was not detected in control tank *Glycera.* Presumably this is due to uptake, release and/or metabolic processes which result in greater or longer retention of fluoranthene and pyrene.

The C_1- through C_5-phenanthrenes were present in much lower concentrations than corresponding alkyl naphthalenes in the MERL No. 2 fuel oil and in the oil–water dosing mixture (Figure 2). However, in the *Glycera* and *Crepidula,* we found equal or greater concentrations of phenanthrenes relative to naphthalenes (Figure 2). This is probably the result of a combination of preferential loss of the lower molecular weight naphthalenes to the atmosphere and to dissolution, more rapid microbial degradation of naphthalenes[16,18] and selective uptake or release by the *Glycera* and *Crepidula.* The combination of these factors made the C_1 through C_5 phenanthrenes more readily available to the organisms over the long-term and, once taken up, more strongly retained. We have recently reported similar results for No. 2 fuel oil aromatics taken up by *Mytilus edulis* as a result of a small oil spill.[22] These results are also consistent with short-term laboratory aquaria studies using individual aromatic hydrocarbons.[1]

There are definite differences in the relative abundance of individual aromatic hydrocarbons when comparing sediment, *Glycera* and *Crepidula* from the same tank on the same date (Figure 2). We think this is due to the different feeding strategies and habitats of the two kinds of organisms. *Glycera,* burrowing through the sediments, is in intimate contact with No. 2 fuel oil compounds adsorbed to the sediments or present in interstitial water. These sediments retained the lower molecular weight C_1 to C_4 naphthalenes much longer than did the overlying water column after addition of the oil dispersion ceased.[18] The filter feeder *Crepidula* was primarily in contact with the water, even though these animals lived within a few centimeters of the sediment-water interface. We theorize that after the oil additions ceased, naphthalenes were less available to or more readily released by the *Crepidula* because of their exposure to cleaner water, thus explaining the observed difference between the aromatic hydrocarbon distribution of the two species.

Alternate explanations include the differential induction of aryl hydrocarbon hydroxylase or mixed function oxidases for the two species or differences in membrane selectivity controlling the passage of aromatic hydrocarbons between the animals and their habitat. There is evidence to suggest that biochemical processes do play an important role in governing the differences between the aromatic hydrocarbon distributions in the two species. Figure 3 presents a magnification of mass chromatograms from Figure 2 and it is readily apparent that within the C_2-phenanthrenes the

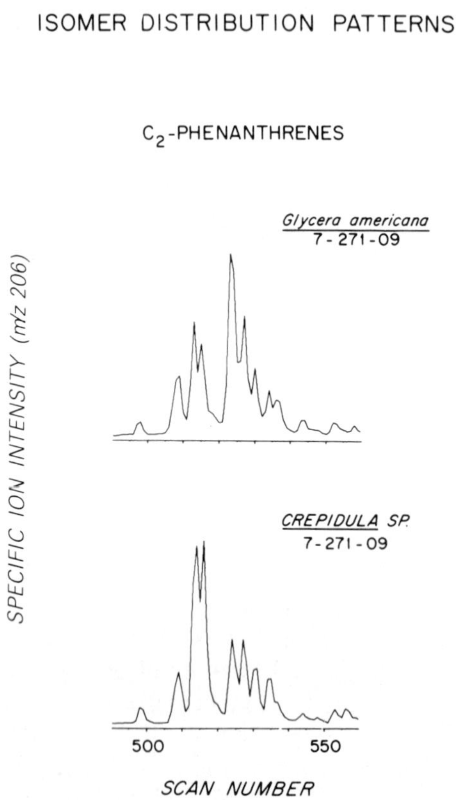

FIGURE 3 Expanded ion plots from Figure 2, highlighting differences in the isomer distribution pattern between species.

relative distribution of isomers differs in these species. The C_2-phenanthrene isomeric distributions in the water-accommodated No. 2 fuel oil are also different than those shown in Figure 3. We noted a similar phenomena in a study of the release of No. 2 fuel oil aromatics from *Mytilus*.[22]

Differences are also observed between the aromatic composition of surface sediment and *Glycera*. These differences could be related to biochemical processes associated with *Glycera* itself or with its prey.

These MERL results demonstrate: (i) Chronic additions of No. 2 fuel oil resulted in measurable contamination of benthic animals; (ii) The No. 2 fuel oil compounds found in the animals resembled those of a weathered and microbially degraded oil; (iii) Phenanthrenes predominated over naphthalenes in the samples despite their lower relative abundance in the original No. 2 fuel oil–water dispersion; (iv) Differences in isomeric distributions within the alkylphenanthrenes were found when comparing *Glycera*, *Crepidula* and surface sediments, thus biochemical processes probably control aromatic hydrocarbon distributions within the organisms.

BENTHIC CHAMBER EXPERIMENTS

The MERL oil experiments were several months in duration and involved No. 2 fuel oil which has a restricted range of molecular weights of aromatic hydrocarbons. The benthic chamber experiments involved a shorter time scale and some heavier molecular weight compounds as well as No. 2 fuel oil range aromatic hydrocarbons. We do not report here all aspects of the benthic chamber experiments, but only a subset of data which is both complementary and supplementary to the MERL results discussed above.

Addition of a single spike of aromatic hydrocarbons resulted in measurable contamination of surface sediments and *Nephtys* as little as five days later (Day 22) and the aromatic compounds were still present 15 days (Day 32) after the spike in both animals and sediments (Figure 4). Differences in aromatic hydrocarbon composition were found both within sample types on the two sampling dates (i.e., day 22 vs. day 32) and between sample types on the same day. Data on the abundance of PAH relative to pyrene in the mixture spiked to the chamber, in *Nephtys*, and in surface sediment are presented in Table IV. A comparison of the PAH compositions in *Nephtys* and sediments indicates a selectivity of uptake/release or selective hydrocarbon metabolism by the deposit feeding *Nephtys*. For example, on day 32, 2,3,6-trimethylnaphthalene,

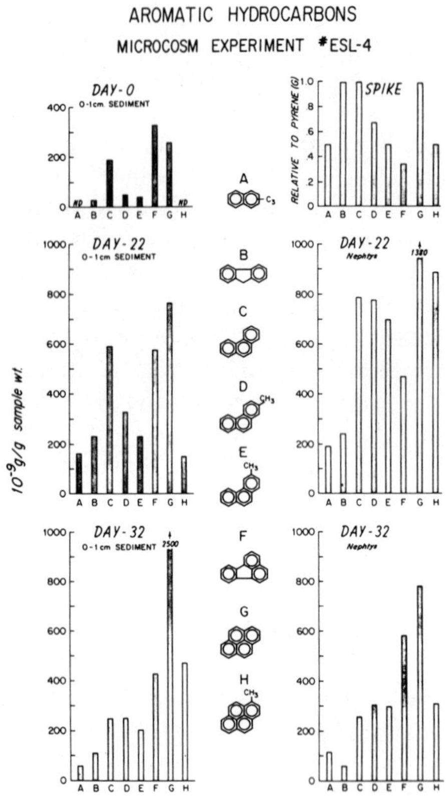

FIGURE 4 Bar graph of results from the benthic chamber experiment demonstrating the differences in relative concentration of selected aromatics found with time and sample type. Sediment—ng/g dry wt. in Nephtys—ng/g wet wt.

phenanthrene, and methylphenanthrenes were present in higher relative abundance in *Nephtys* than in surface sediment. However, there were also lower relative abundances of 2,3,6-trimethylnaphthalene, fluorene, and phenanthrene in sediments and *Nephtys* on day 32 compared to their abundances in the spike. Day 22 values were either the same as day 32 or in between the values for the spike and day 32.

The same general trends as described in the previous paragraph were noted for two earlier benthic chamber experiments, one conducted in a similar manner as described in this paper and a second experiment *in situ* in Buzzards Bay, Massachusetts, U.S.A.[14]

TABLE IV

Abundance of selected aromatic hydrocarbons relative to pyrene in the ESL 4[a] benthic chamber experiment. Ratios are calculated from electronic integration of the gc-fid signal after subtracting background values

	Day 17	Day 22		Day 32	
	Spike	Sediment	Nephtys	Sediment	Nephtys
2,3,6-Trimethyl-Naphthalene	0.53	0.21	0.14	0.05	0.15
Fluorene	1.02	0.30	0.17	0.08	0.07
Phenanthrene	1.05	0.77	0.57	0.20	0.33
1-methylphenanthrene	0.47	0.35	0.51	0.16	0.38
2-methylphenanthrene	0.70	0.43	0.56	0.20	0.39
Fluoranthene	0.33	0.75	0.34	0.35	0.34
Pyrene	1.00	1.00	1.00	1.00	1.00
1-methylpyrene	0.51	0.20	0.30	0.38	0.39

[a]Environmental Systems Laboratory, W.H.O.I.

DISCUSSION

The results of the long-term MERL experiments and the shorter-term benthic chamber experiments demonstrate experimentally that low concentrations of water-accommodated petroleum compounds can be incorporated into sediments and fauna of benthic ecosystems. In both types of experiments, biological and physical processes act in concert to produce differences in the resulting PAH composition of selected fauna and surface sediment. It is important to note that these differences in composition were found in both the longer-term chronic oil addition experiments in MERL and the shorter-term single dose experiments in the benthic chambers.

The actual route or routes of input of petroleum compounds to the fauna cannot be determined from the data presented here. It is possible for these animals to take up hydrocarbons from their food, (ie., particulate matter or prey animals), from the water column directly, or from interstitial water of the benthic ecosystem. Further experimentation is required to determine the relative importance of these various exposure pathways.

The experimental chamber and MERL results agree with field data derived from our observations of the long-term fate of No. 2 fuel oil in marsh sediments[6] and the fate of No. 2 fuel oil taken up by *Mytilus edulis* exposed to a small oil spill.[22] The heavier molecular weight PAH such as the alkyl phenanthrenes that are present in lower abundance relative to the naphthalenes and alkylnaphthalenes in the contaminating oil become the more dominant aromatic compounds in sediment and animal samples as time progresses.

Thus, the concentration of an aromatic compound in the initial composition of released oil is not, by itself, a good guide to the study of chronic long-term biological effects. We recognize that we have not analyzed for the potentially more toxic heteroatom aromatic compounds in these experiments, and attention should be given in the future to these compounds. Furthermore, it is important to recognize that the processes removing PAH from the ecosystem components (e.g., metabolism by animals and bacteria and photochemistry) can produce a number of oxygenated compounds which may be more biologically active than their aromatic hydrocarbon precursors.[25] In fact, an experiment in the MERL microcosms with C-14 labelled benzanthracene clearly demonstrated that benzanthracene metabolites and reaction products were incorporated into the sediments of the benthic ecosystem and could be detected at the end of the 220 day experiment.[26]

We have conducted a similar experiment with C-14 labelled benzanthracene in the benthic chamber for 42 days. Analyses of samples from this experiment are now in progress. We think that further experiments with labelled compounds will provide insight into the routes and rates of reaction and transfer of pollutant organic compounds in benthic ecosystems. Controlled benthic chamber experiments and microcosm experiments of the type at MERL have proven to be well-suited for these types of investigations.

Very little data exists on the relationship between long-term effects on organisms and body burdens of fossil fuel compounds and metabolites. Most toxicity data relates to effects on organisms as a function of concentrations in the water or sediment of their habitat.[1,27] Thus, we cannot extrapolate from concentrations of hydrocarbons observations in the organisms we analyzed to specific short or long-term effects.

It is interesting to note that as a result of chronic exposure concentrations of 1–5 ppm, groups of aromatic hydrocarbons (e.g., C_3 to C_5 naphthalenes and C_1 to C_3 phenanthrenes) accumulated in *Glycera* (Table III). This accumulation occurs despite the fact that polychaetes have been shown to be capable of metabolizing aromatic hydrocarbons and that this metabolic capacity can be induced.[28]

The experiments we have discussed dealt with only one form of pollutant compound addition to an ecosystem; i.e., via the water soluble or accommodation route. Shaw and Wiggs[29] measured hydrocarbons in *Mytilus edulis* (a filter feeder) and *Macoma balthica* (a deposit feeder) from an area in Alaska subjected to both petroleum and coal inputs. They found that hydrocarbons in *Mytilus edulis* reflected petroleum input while hydrocarbons in *Macoma* reflected coal input. We have recently compared PAH data from *Mytilus edulis* and surface sediments from the northeast U.S. coast. Surface sediments near urban harbor areas had a predominance of PAH of a pyrogenic source while *Mytilus* from nearby stations showed significant inputs of petroleum PAH.[23] Our data and that of Shaw and Wiggs[29] raise the interesting question of the relative biological availability of fossil fuel hydrocarbons from various sources. Experimentation on the interactive influences of the physical–chemical form of the pollutant hydrocarbon, feeding strategies of animals, and uptake/release and metabolism are needed if we are to assess the relative importance of various sources of inputs of fossil fuel compounds to the marine environment.

Acknowledgements

Drs. Patrick and Juanita Gearing, James Quinn, Stuart Wakeham and Terry Wade provided valuable discussions and assistance as did the entire MERL staff. C. Winget, H. Clifford, M. Connor, R. Howarth, G. Hampson and the E.S.L.–W.H.O.I. provided assistance with benthic chamber experiments. We gratefully acknowledge financial support from the U.S. Department of Energy (Contract EE-77-S-02-4256), the Bureau of Land Management U.S. Department of the Interior (via Interagency Agreement IA7-20) and U.S. E.P.A. grant R-803902020 via University of Rhode Island M.E.R.L. This is Contribution No. 5094 of Woods Hole Oceanographic Institution.

References

1. D. W. Connel and G. J. Miller. In *CRC Critical Reviews in Environmental Control*, Volume 11, issues 1 and 2, pp. 39–104 and 105–162 (1981).
2. J. M. Neff, *Polycyclic aromatic hydrocarbons in the aquatic environment* (Applied Sciences Publishers, London, 1980).
3. Royal Society, *The Effects of Oil Pollution: Some Research Needs* (The Royal Society, London, 1980).

4. National Academy of Sciences, *Petroleum in the Marine Environment*, Proceedings of a Workshop, May 1973, Washington, D.C. (1975).
5. D. C. Malins, M. M. Krahn, D. W. Brown, W. D. MacLeod, Jr. and T. K. Collier, *Helgoland. Meeresunter.* **33,** 257 (1980).
6. J. M. Teal, K. Burns and J. W. Farrington, *J. Fish. Res. Bd. Can.* **35(5),** 510 (1978).
7. H. L. Sanders, J. F. Grassle, G. R. Hampson, L. S. Morse, S. Garner-Price and C. C. Jones, *J. Mar. Res.* **38(2),** 265 (1980).
8. J. W. Farrington. In *Advances in Chemistry Series No. 185 "Petroleum in the Marine Environment."* L. Petrakis and F. T. Weiss (eds.) (American Chemical Society, Washington, D.C., 1980), pp. 1–22.
9. S. G. Wakeham and J. W. Farrington. In *Contaminants and Sediments: analysis chemistry, biology*, Vol. 2, R. A. Baker (ed.) (1980).
10. J. M. Augenfeld, *Marine Environmental Res.* **3,** 307 (1980).
11. J. P. Giesy, Jr. (ed.) *Microcosms in Ecological Research*, U.S. Department of Energy, Washington, D.C. Symposium Series 52 (CONF-481101) (1980).
12. G. D. Grice and M. R. Reeve, (eds.) *Marine Mesocosms: Biological and Chemical Research in Experimental Ecosystems* (Springer-Verlag, New York, 1981).
13. J. F. Grassle, R. Elmgren and J. P. Grassle, *Mar. Environ. Res.* **9,** 279 (1980–81).
14. J. W. Farrington, J. M. Teal and B. W. Tripp. In: *Marine Pollution Transfer Processes*, M. Waldichuck (ed.). IAPSO Symposium PS-9, Canberra, Australia, December 1979, in press (1981).
15. C. L. Winget, *WHOI Technical Report 78–94* (1978).
16. J. N. Gearing, P. J. Gearing, T. Wade, J. G. Quinn, M. B. McCarthy, J. W. Farrington and R. F. Lee. In: *1979 Oil Spill Conference*, (American Petroleum Institute, Washington, D.C. 1979), pp. 555–564 (1979).
17. R. Elmgren, G. A. Vargo, J. F. Grassle, J. P. Grassle, D. R. Heinke, G. Langlois and S. L. Vargo. In: *Microcosms in Ecological Research*, J. P. Giesy, Jr. (ed.). U.S.D.E. Symposium Series No. 52 (Conf. 781101) (1980).
18. P. J. Gearing, J. N. Gearing, R. J. Pruell, T. L. Wade and J. G. Quinn, *Environ. Sci. Technol.* **14(9),** 1129 (1980).
19. T. L. Wade and J. G. Quinn, *Environ. Res.* **3,** 15 (1980).
20. P. J. Gearing and J. N. Gearing, *Mar. Environ. Res.*, in press (1982).
21. P. J. Gearing and J. N. Gearing, *Mar. Environ. Res.*, in press (1982).
22. J. W. Farrington, A. C. Davis, N. M. Frew and K. S. Rabin, *Marine Biology*, in press (1981b).
23. J. W. Farrington, J. M. Teal, B. W. Tripp, A. C. Davis and J. B. Livramento, *Can. J. of Aquatic Sciences and Fisheries* (submitted) (1981).
24. J. L. Lake, C. Norwood, C. Dimock and R. Bowen, *Geochim. et Cosmochim. Acta.* **43,** 1847 (1979).
25. D. C. Malins and H. O. Hodgins, *Environ. Sci. and Tech.* **15,** 1271 (1981).
26. K. R. Hinga, M. E. Q. Pilson, R. F. Lee, J. W. Farrington, K. Tjessem and A. C. Davis, *Environ. Sci. Technol.* **14,** 1136 (1980).
27. J. W. Anderson. In: *Marine Pollution: Functional Responses*, W. B. Vernberg, A. Calabrese, F. P. Thurberg, F. John Vernberg (eds.). (Academic Press, New York, 1979).
28. R. F. Less, S. C. Singer, K. R. Tenore, W. S. Gardner and R. M. Philpot. In: *Marine Pollution: Functional Responses*, W. B. Vernberg *et al.* (eds.) (Academic Press, New York, 1979).
29. D. G. Shaw and J. N. Wiggs, *Mar. Pollut. Bull.* **11,** 297 (1980).

Appearance and Structure Analysis of Aromatic Oil Residues in Baltic Mussel and Fish†

SEIJA SINKKONEN

University of Jyväskylä, Department of Chemistry, Kyllikinkatu 1-3, SF-40100 Jyväskylä 10, Finland

(*Received November 27, 1981*)

The non-polar aromatic compound fractions from Baltic mussel and fish have been analysed to find oil residue compounds and to study their possible enrichment in food chain.

GC/mass spectrometry was used to study the structure of the compounds and to determinate quantitatively some aromatic compounds or compound groups. In mussel and flat fish samples there were found several polycyclic hydrocarbons, most abundant were the alkylsubstituted forms as in petroleum.

Reversed phase high pressure liquid chromatography with variable wavelength UV-detector was also used to separate and determinate some components in the aromatic fraction.

Statistical treatment of the analysis results was used to differentiate between the compounds of biogenic origin and the compounds originating from oil pollution.

KEY WORDS: Polycyclic hydrocarbons, oil residues, Baltic Sea, GC-MS, HP-LC, mussel, fish.

ANALYSIS PROCEDURE

Detailed extraction, clean up and determination procedures for the analysis of the oil residues has been described elsewhere.[1] The mussel and flat fish samples were digested in alkaline methanol:water and extracted with n-hexane. Fractionation was done by column chromatography using

†Presented at the Workshop on the Chemistry and Analysis of Hydrocarbons in the Environment, Barcelona, November 1981.

SiO_2-Al_2O_3-columns. The non-polar aromatic compound fraction was eluted with dichloromethane after the aliphatic hydrocarbons had been eluted with hexane. Hexaethylbenzene was used as an internal standard in the aromatic fraction.

GC/MS ANALYSES

Contents of the non-polar aromatic fractions from mussel and fish samples were analyzed mainly by gas chromatography/mass spectrometry. Some information was obtained from FID-gas chromatography and reversed phase high pressure liquid chromatography.

Gas chromatograph Varian 3700 connected with a new mass spectrometer system Finnigan MAT 212 was used. The column was 0.3 mm diameter 30 m long glass capillary coated with SE-30. Helium was used as carrier gas 1 ml/min. Temperature program applied started at 80° then +4°/min up to 250°. During the gas chromatography, repeated scans of low resolution mass spectra were done and stored in the disc unit.

Total ion chromatograms obtained are very complex. The oil residue hydrocarbons are partly overlapped by compounds of natural origin.

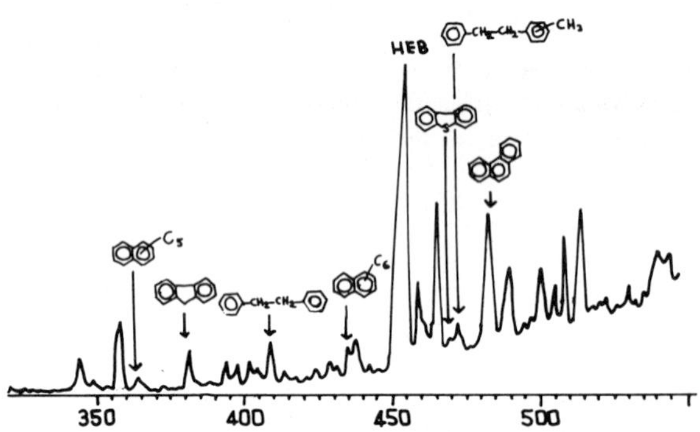

FIGURE 1 Total ion chromatogram of non-polar aromatic compound fraction from a fish sample. The aromatic oil residue hydrocarbons are partly overlapped by compounds from the nature. Some compounds are identified according to their mass spectra and retention times.

The possible oil residue compounds were sought from the total ion chromatogram by selective ion mass chromatography (Figure 2).

Alkyltrimethylbenzenes which are the major components in the non-polar aromatic fractions extracted from sediments and crude oils,[2] were sought by their characteristic m/z-values 77, 91, 105, 133, 147, etc. Their relative amount in mussel and fish oil residues appeared to be much lower

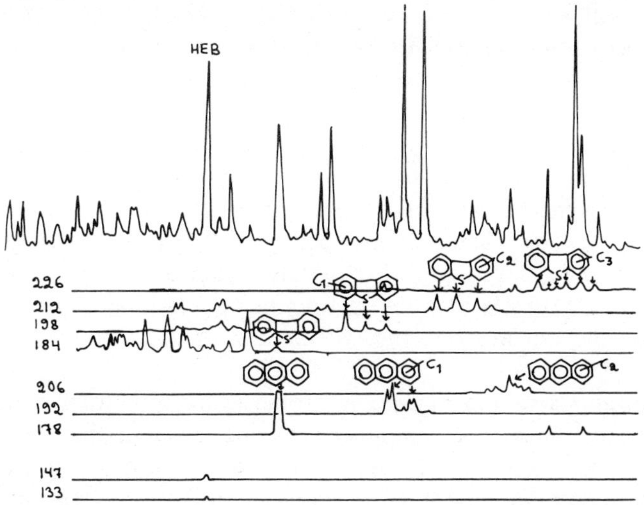

FIGURE 2 Total ion chromatogram and selective ion mass chromatograms to find alkyltrimethylbenzenes (m/z 133, 147), anthracene/phenanthrene (178), methylated phenanthrenes (192, 206), dibenzothiophene (184) and methylated dibenzothiophenes (198, 212, 226) from non-polar aromatic fraction of a fish sample. GC/MS Finnigan Mat 212/SS 188.

than in crude oil. Probably alkyltrimethylbenzenes are metabolized in fish in a relatively short time in contrast to the other oil originating aromatic hydrocarbons. Naphthalene and alkylnaphthalenes were sought by m/z-values of their molecular ions 128, 142, 156, 170, 184 and 198. Naphthalene and monomethylnaphthalenes are so volatile that they are easily lost during the analyses procedure and in addition fishes are able to metabolize naphthalene to 1-naphtyl glucuronic acid, 1,2-dihydro-1,2-dihydroxynaphthalene and 1-naphthol and methylnaphthalenes to corresponding methylated metabolites.[3] On the contrary more alkyl-substituted naphthalenes and trimethylbenzenes could be analyzed without loss and they were abundant in mussel samples. Anthracene/phenanthrene and the corresponding methylated compounds could be detected by m/z-values of their molecular ions 178, 192, 206, 220 and 234. These compounds were found in considerable amounts in many fish and mussel samples but at least some of the alkylated anthracenes/phenanthrenes may be natural compounds and not originating from the oil pollution.

Biphenyl and alkylbiphenyls found with m/z values 154, 168, 182, etc. and fluorenes and alkylfluorenes with m/z values 166, 180, 194, 208, etc. are possible oil pollutants although these compounds obviously in part are of natural origin.

Dibenzothiophene and especially methyl-, dimethyl- and trimethyl-dibenzothiophenes are known to be extremely stable oil residues which cannot be metabolized by various organisms[4] and also could bio-accumulate and be enriched in the food chain. Dibenzothiophene and alkyldibenzothiophenes by m/z values 184, 198, 212, and 226 were detected in several mussel and fish samples. From the polycyclic aromatic hydrocarbons pyrene (m/z of the molecular ion 202), benzo-fluorene (216), chrysene (228), benzo(c)phenanthrene (228), perylene (252), benzo(a)pyrene (252) and coronene (300) sought only pyrene and chrysene were found in a few samples. It is possible that the compounds containing more than four aromatic rings did not move in the GC column used. They were found by HPLC.

Some compounds in the non-polar aromatic fraction were determined quantitatively by mass chromatography using hexaethylbenzene as an internal standard.[1] The results are more relative than absolute while model compounds to determine the response factors were not available.

STRUCTURE DETERMINATIONS

Total mass spectra of the peaks in the total ion chromatogram were run. Usually there were so many compounds in the samples that all of them were not completely resolved, so it was impossible to get mass spectra of pure compounds but there are interfering peaks from the background. Low resolution mass spectra of some scan numbers from the total ion chromatogram in Figure 1 are presented in Figures 3, 4 and 5.

The spectra corresponding scan numbers 380 and 402 are quite pure mass spectra of fluorene and diphenylethane. On the contrary the low resolution mass spectrum of scan number 435 obviously is a mixture of several isomeric C_6-alkylnaphthalenes. The definite structures and locations of aliphatic side chains cannot be determined.

HIGH PRESSURE LIQUID CHROMATOGRAPHY

Some nonpolar aromatic fractions of mussel and fish samples have been analyzed also by high pressure liquid chromatography coupled with a variable wavelength UV-detector. A Pye Unicam LC-3 liquid chromatograph system with Altex pump was used. Helium gas was used to remove air from the eluent system.

Different reversed phase columns were tested for the separation of the aromatic hydrocarbons in the samples with standard solutions. The following standard compounds were used: biphenyl, naphthalene, fluorene,

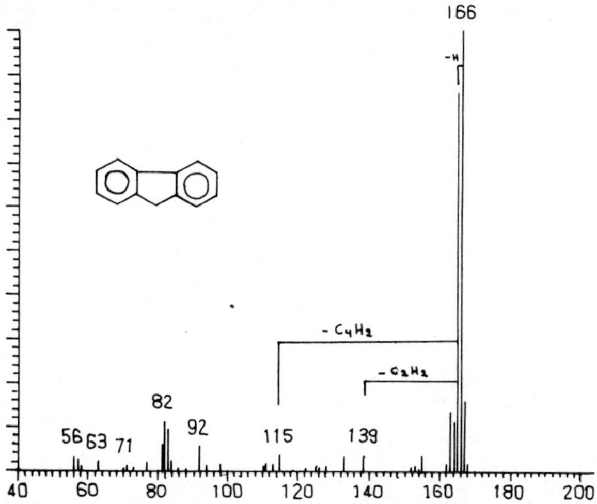

FIGURE 3 Low resolution mass spectrum of scan 380 from the total ion chromatogram in Figure 1. This is fluorene with $M^+ = 166$.

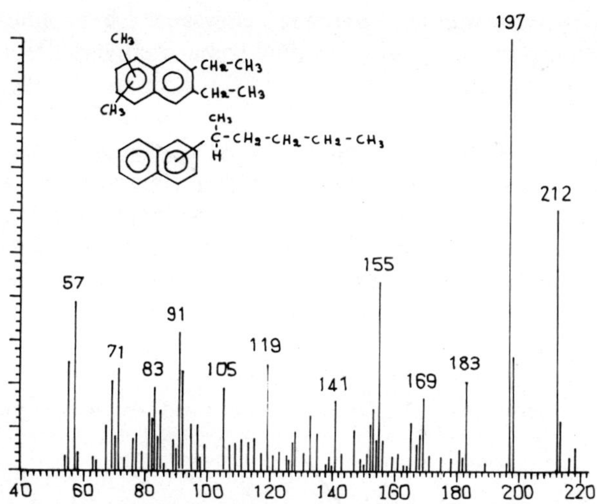

FIGURE 4 Low resolution mass spectrum of scan 435 from the total ion chromatogram illustrated in Figure 1. There seems to be a mixture of several isomeric C_6-alkylnaphthalenes. Some possible structures have been drawn in the figure.

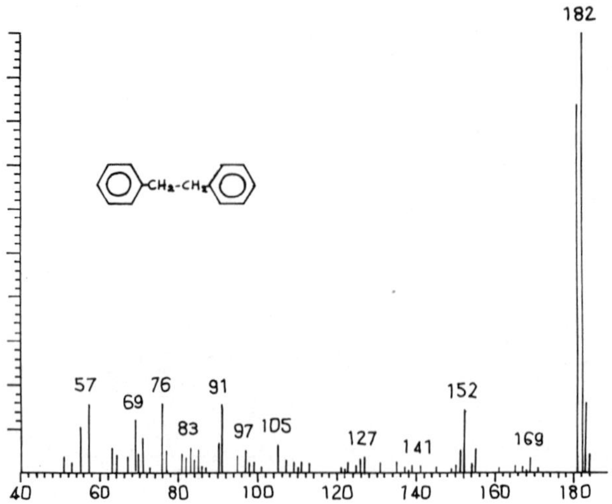

FIGURE 5 Low resolution mass spectrum of scan 402 from the total ion chromatogram illustrated in Figure 1. The compound is diphenylethane with $M^+ = 182$.

anthracene, phenanthrene, pyrene, chrysene, benzo(c)phenanthrene, benzo(a)pyrene, perylene, coronene and some methylated forms of these. The Zorbax ODS-column (25 cm long, ID 4.6 mm, particle size 5.5 μm) proved to be best to the separation of these compounds, thus far.

Methanol:water and asetonitrile:water solutions were used as eluents. The latter eluent was found to have higher separating power for these aromatic hydrocarbons. The water used was distilled and filtered before mixing with the organic solvent. Sensitivity of the UV-detector for the polycyclic aromatic compounds is very good when compared to the FI-detector in gas chromatography. The wavelength 254 nm is best when there is a complex mixture if different hydrocarbons to be measured. However, higher polyaromatic hydrocarbons do have absorption maxima at higher wavelengths and these can be used for their more selective analysis[5].

The complexity of the mixtures of the non-polar aromatic fractions extracted from mussel and fish samples is such that all the compounds cannot be resolved. Some compounds however were identified by comparing retention times and the UV-absorbances at different wavelengths with the corresponding data of standard compounds and by recording mass spectra of some fractions (Figure 6).

The quantitative determination was done by running a standard mixture after each sample run and comparing the heights of the peaks in the sample and in the standard run.

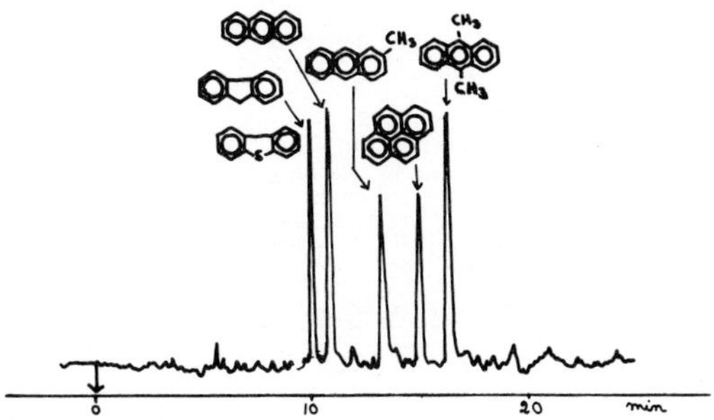

FIGURE 6 A high pressure liquid chromatogram from the non-polar aromatic hydrocarbon fraction of a flat fish from Askö. Philips Pye Unicam LC-3 system with UV-detector. Isocratic run with acetonitrile: water 80:20, ml/min, Zorbax ODS-column (25 cm, ID 4, 6 mm), UV-detector at 254 nm. The compounds are identified by comparing the retention times of the peaks with the retention times of model compounds. The same compounds were identified from the sample also by mass spectrometry.

RESULTS

Quantitative results of ten different compound groups from mussel samples 1979 were treated statistically to differentiate oil components from the naturally in mussel occurring compounds. The Pearson correlations were run. While RES (resolved aliphatic hydrocarbons) and UCM (unresolved complex mixture) had been observed in GC analysis to be dominantly from oil pollution, those of the other results which correlated strongly and significantly with UCM and RES could be considered as oil residues. In Macoma samples dibenzothiophene (DBT), methyldibenzothiophenes (C1DBT), dimethyldibenzothiophenes (C2DBT), anthracene-phenanthrene (A, F), methylphenanthrene (C1F) and trimethylbenzenes (TMB) showed this behavior assumed for dominantly oil residues (Table I).

Linear regression between the analysis results UCM + RES and dibenzothiophenes verify the nature of the latter as dominant oil residues (Figure 7).

The results are useful in detection of the non-visible oil pollution of sea which is apparent from the differences observed between places and times of sampling.

TABLE I

Pearson correlation coefficients (lower triangle) and number of cases (upper triangle) of the analysis results from *Macoma baltica* samples.

	UCM	RES	A, Ph	C1Ph	C2N	TMB	DBT	C1DBT	C2DBT	PRIS	PHY
UCM	19	19	17	16	16	17	13	17	17	19	19
RES	0.796	19	17	16	16	17	13	17	17	19	19
A, Ph	0.966	0.860	17	16	16	17	13	17	17	17	17
C1Ph	0.848	0.904	0.820	16	15	16	12	16	16	16	16
C2N	0.221	0.178	0.385	0.481	16	16	12	16	16	16	16
TMB	0.714	0.797	0.805	0.804	0.538	17	13	17	17	17	17
DBT	0.723	0.861	0.670	0.936	−0.015	0.717	13	13	13	13	13
D1DBT	0.812	0.911	0.777	0.985	0.191	0.801	0.976	17	17	17	17
C2DBT	0.930	0.827	0.884	0.890	0.123	0.640	0.806	0.861	17	17	17
PRIS	−0.173	−0.317	−0.151	−0.208	−0.025	−0.066	−0.023	−0.187	−0.061	19	19
PHY	0.380	0.251	0.506	0.268	0.526	0.726	0.150	0.244	0.248	0.296	19

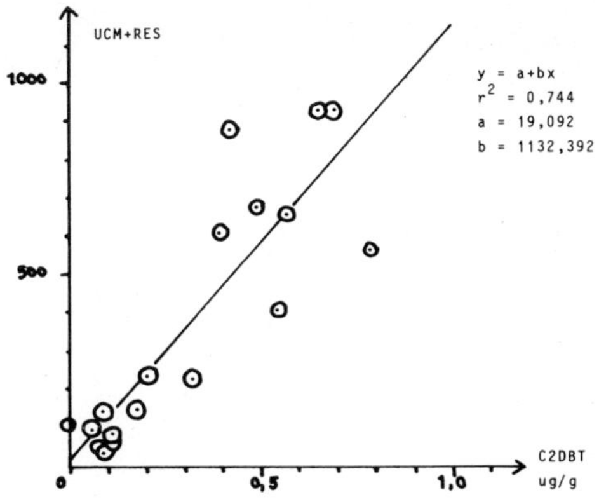

FIGURE 7 Linear regression between the analysis results UCM+RES and dimethyl dibenzothiophene. (Mytilus).

References

1. J. Paasivirta, R. Herzschuh, M. Lahtiperä, J. Pellinen and S. Sinkkonen, *Chemosphere* **10,** 919 (1981).
2. C. Grahl-Nielsen, T. Neppelberg, K. H. Palmork, K. Westerheim and S. Wilhelmsen, *Organiska miljögifter i vatten NORDFORSK Miljövårdsekreteriatet publ. 1976*: **2,** 221 (1976).
3. D. C. Malins, T. K. Collier, L. C. Thomas and W. T. Roubal, *Intern. J. Environ. Anal. Chem.* **6,** 55 (1979).
4. M. Ogata and Y. Miyake, *J. Chromatogr. Sci.* **18,** 594 (1980).
5. A. M. Krstulovic, D. M. Rosie and P. R. Brown, *Anal. Chem.* **48,** 1383 (1976).

Investigation of the Sources of Aliphatic Hydrocarbons in the Mussel *Mytilus edulis* from North Sea Oil Production Platforms by Capillary glc and CGCMS†

P. J. C. TIBBETTS, S. J. ROWLAND‡, L. L. TOVEY and R. LARGE

Masspec Analytical, Wallbridge, Stroud, Glos. GL5 3JA, U.K.
‡Organic Geochemistry Unit, School of Chemistry, University of Bristol, Cantocks Close, Bristol, BS8 1TS, U.K.

(*Received November* 27, 1981)

Aliphatic hydrocarbons isolated from mussels collected over a 20 month period from three North Sea Forties field oil production platforms have been examined by capillary gas chromatography and gas chromatography-mass spectrometry.

The biological hydrocarbons consist predominantly (300–700 $\mu g\,g^{-1}$ lipid) of C_{31} and C_{33} *n*-alkenes with 2, 3 and 4 double bonds probably derived from a dietary intake of *Emiliania huxleyi*, a microscopic coccolithophorid alga. In some of the mussels C_{18}, C_{20} and C_{22} *n*-alkanes are present in unusually high abundance compared to their odd carbon number homologues. Possible reasons for this distribution are discussed.

The concentrations of fossil fuel hydrocarbons (mainly unresolved complex mixtures) in the mussels ranged from 330 to 5,298 $\mu g\,g^{-1}$ of lipid. No significant increase in values was detected two months after the start of discharge of treated Forties production water from the Forties D platform.

To determine the pollutant hydrocarbon sources, the sterane and terpane distributions of hydrocarbon fractions isolated from the mussels and from a number of fossil fuels, including Forties crude oil, were examined by mass fragmentography. The steranes in the mussels from the Forties C and D platforms contained higher proportions of regular 14α(H), 17α(H) components than Forties oil in which the steranes were mainly 13β(H), 17α(H) diasteranes. In

---

†Presented at the Workshop on the Chemistry and Analysis of Hydrocarbons in the Environment, Barcelona, November 1981.

addition, the terpane distributions of the mussel fractions differed markedly from that of Forties crude oil, particularly in the relative abundance of diterpanes to triterpanes, which was higher in the mussel fractions than in the oil, and in the absence of 17α(H), 18α(H), 21β(H)-18, 30-bisnorhopane, a known component of Forties crude. Furthermore, the ratio of C_{29} to C_{30} triterpanes was greater than unity in the mussels but much less than unity in the Forties oil. These data preclude Forties or other similar North Sea crudes as the major source of pollutant aliphatic hydrocarbons in the mussels. The relative concentrations and distributions of steranes and triterpanes suggest that the most likely source of pollutants is a Middle Eastern based oil derived either from rig activities, or from background pollution in the North Sea. The low concentration of $\geq C_{27}$ steranes in the more polluted mussels (e.g. Forties B; UCM > 200 ppm dry wt.) suggests that gas oils used on the platforms may be the major source of petrogenic hydrocarbons in these samples.

KEY WORDS: Hydrocarbons, North Sea, hopanes, steranes, mass fragmentography, mussels.

1. INTRODUCTION

The mussel, *Mytilus edulis* is common around certain parts of the coastlines of the United States and Great Britain and has been used as a sentinel organism to monitor marine pollution.[1,2,3]

A considerable amount of data on the fossil fuel hydrocarbon content of coastal mussels is available but since the molluscs are sessile in their adult form and require a permanent structure to adhere to, no previous studies of mussels growing offshore have appeared in the literature. The installation of offshore oil production platforms in the North Sea has provided ideal sites for mussel growth and *Mytilus edulis* is the dominant organism on the top one hundred feet of the four Forties platforms in the U.K. Sector.[4]

The mussel population has become established on the platform legs by natural recruitment.

The Forties field was discovered in November 1970, the platforms (A–D; Figure 1) were installed in 1974/5 and oil production (500–550 thousand barrels/day;[5]) commenced in November 1975.

We report here the results of a study established to determine the typical concentration and distributions of fossil fuel and biological hydrocarbons in mussels from the Forties oilfield. We have examined specimens from the Forties D platform before and after the discharge of treated oil production water to monitor any resulting changes in their hydrocarbon composition and we have attempted to interpret the hydrocarbon distributions of the mussels from all the platforms in terms of their likely sources.

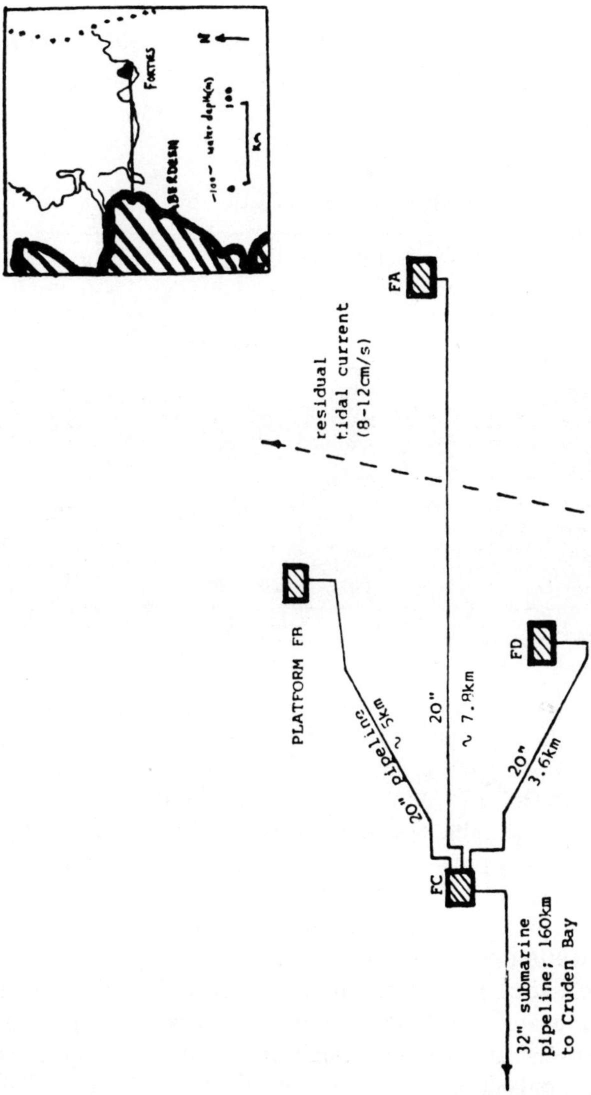

FIGURE 1 Location of oil production platforms within the "Forties" field, North Sea (scale only approximate).

2. EXPERIMENTAL

2.1 Samples

Details of the samples of *Mytilus edulis* collected by divers from the Forties oil platforms are shown in Table I. A sample of production water from the Forties field was supplied by B.P.

TABLE I
Details of samples examined

Date		Platform	Platform leg	Depth (ft)	No. of mussels extracted
October	1978	B		Surface	
June	1980	B	SW	Surface	8
July	1979	C	NW	Surface	4[a]
July	1979	C	NW	50	4
July	1979	C	NW	100	4
June	1980	C	SW	Surface	8
October	1979	D	NE	Surface	9
June	1980	D	SW	19	5
June	1980	D	SW	80	5
June	1980	D	SW	136	5

[a]4 sets of 4 mussels used in replicate analyses.

2.2 Methods

Mussels were transported live to the laboratory where they were shucked. The total body mass of the shucked mussels from each location was macerated by fine chopping with a scalpel. Each sample was then examined by the method of Dole and Meinhertz.[6] Specifically, propan-2-ol:light petroleum (4:1; 100 ml) was added prior to extraction by ultrasonication (10 minutes). After centrifugation the liquid was decanted and partitioned between light petroleum (72 ml) and water (48 ml). The procedure was repeated and the light petroleum layers combined. Evaporation of solvent at ambient temperature (rotary evaporator) left a yellowish oil (total lipids).

Anthracene (3 μg), 2,21-dimethyldocosane (3 μg) and squalane (12 μg) were added as solutions in hexane to each extract as internal standards to allow quantitation of hydrocarbon levels. The extracts were treated with ethanolic potassium hydroxide (5%) at *ca.* 30°C overnight. After saponification, water (10 ml) was added and the aqueous solution extracted twice (with light petroleum). The light petroleum extract was evaporated to dryness and fractionated by thin-layer chromatography (Silica Gel G, 95+5 light petroleum:diethyl ether). Fractions of aliphatic hydrocarbons (Rf = *n*-heptadecane), and aromatic hydrocarbons (Rf = phenanthrene) were scraped off and eluted with diethyl ether. Forties production water (200 ml) was extracted with hexane (75 ml) followed by hexane:ethyl acetate (1:1; 75 ml) and the extracts combined. After evaporation of the solvent the extract was fractionated by tlc as above.

Capillary gas–liquid chromatography (glc)

A Finnigan 9500 gas chromatograph equipped with flame ionisation detector was used with 20 m × 0.2 m (i.d.) OV-1 WCOT glass columns. The oven temperature was programmed from 60–260°C at 6°/minute. The carrier gas was helium (*ca.* 2 ml/minute flow rate). Injector and detector temperatures were 300°C.

Computerised gas chromatography-mass spectrometry (CGCMS)

A Finnigan 3200 gcms system incorporating a 6100 data system was used to scan a mass range of 50–560 a.m.u. every 2 seconds. The emission current was 300 μA at 70 eV ionisation energy. The glc conditions were as above.

3. RESULTS

The concentrations of aliphatic biological and petrogenic hydrocarbons found in the Forties mussels are given in Table II. All concentrations were calculated by measurement of peak heights or areas (UCM) and comparison with known amounts of internal standards. The values for the UCM areas are not estimated as accurately as those of the resolved components because of the small areas of the internal standard peaks compared to their heights. Examination of procedural and reagent blanks indicated minimal contamination from the analytical procedures. Values are expressed as μg per g of dry weight and as μg per g of lipid, the latter to allow for seasonal variations in mussel lipid concentrations,[7] since the amount of lipid per mussel varied from 13–90 mg over the period of the study.

TABLE II

Concentrations of aliphatic hydrocarbons in *Mytilus edulis* from "Forties" production platforms in $\mu g\,g^{-1}$ dry weight and $\mu g\,g^{-1}$ per lipid (in parenthesis)

Sample depth	Date		Biological hydrocarbons: Squalene	Biological hydrocarbons: Long-chain alkenes	Pollutant hydrocarbons: Σ_{13}^{27} *n*-alkanes[c]	Pollutant hydrocarbons: Aliphatic UCM[b]	Total pollutant hydrocarbons	Dry weight per mussel	Sterane distribution	Triterpane distribution
FB 0′	October	1978	0.4 (7.2)	n.m. (n.m.)	5.8 (104.2)	289 (5194)	295 (5298)	n.m.	Sterane content too low to be discernible	Norhopane/hopane ratio >1. Dissimilar to Forties crude
FB 0′	June	1980	n.d. (n.d.)	n.d. (n.d.)	7.4 (46.5)	380 (2386)	387 (2433)	0.14	Sterane content too low to be discernible	Norhopane/hopane ratio >1. Diterpanes abundant. Dissimilar to Forties crude
FC 0′	June	1979[a]	0.9 (17.8)	15.0 (301)	4.6 (93.3)	59 (1165)	64 (1258)	0.53	C_{27}-C_{29} regular and rearranged steranes present in approx. equal abundance. Dissimilar to Forties crude	Norhopane/hopane ratio >1. Diterpanes abundant. Biological triterpanes also present. Dissimilar to Forties crude
FC 0′	June	1980	n.d. (n.d.)	>15.0 (>726)	22.3 (155.1)	218 (1516)	240 (1671)	0.32	Sterane content too low to be to be discernible	Norhopane/hopane ratio >1. Diterpanes abundant. Dissimilar to Forties crude
FC 50′	July	1979	0.8 (24.0)	n.m.	1.1 (12.7)	10 (297)	11 (330)	0.85	Sterane content too low to be discernible	Triterpane content too low to be discernible
FC 100′	July	1979	0.8 (30.0)	n.m.	1.9 (70.5)	12 (445)	14 (516)	0.48	Sterane content too low to be discernible	Triterpane content too low to be discernible
FD 0′	October	1979	n.m. (n.m.)	1.4 (86)	1.9 (116.3)	42 (2570)	44 (2686)	1.40	C_{27}–C_{29} regular and rearranged steranes present in approx. equal abundance. Dissimilar to Forties crude	Norhopane/hopane ratio >1. Diterpanes abundant. Dissimiler to Forties crude
FD 19′	June	1980	n.d. (n.d.)	n.d. (n.d.)	2.4 (38.9)	41 (666)	43 (705)	1.50	C_{27}–C_{29} regular and rearranged steranes present in approx. equal abundance. Dissimiler to Forties crude	Norhopane/hopane ratio >1. Diterpanes abundant. Dissimilar to Forties crude
FD 80′	June	1980	n.d. (n.d.)	n.m.	0.8 (11.0)	34 (466)	35 (477)	0.84	C_{27}–C_{29} regular and rearranged steranes present in approx. equal abundance. Dissimilar to Forties crude	Norhopane/hopane ratio >1. Diterpanes abundant. Dissimilar to Forties crude
FD 136′	June	1980	n.d. (n.d.)	n.d. (n.d.)	1.3 (29.2)	40 (900)	41 (929)	2.11	C_{27}–C_{29} regular and rearranged steranes present in approx. equal abundance. Dissimilar to Forties crude	Norhopane/hopane ratio >1. Diterpanes abundant. Dissimilar to Forties crude

Notes
[a]Average of 4 analyses.
[b]All UCM values ±10%.
[c]The validity of using *n*-alkanes as pollutant indicators has been questioned because of the high even C number predominance (see text).
n.m. not measured.
n.d. not detected.

4. DISCUSSION

4.1 Biogenic hydrocarbons

The principal biogenic hydrocarbons identified in the mussels were alkenes (Figure 2). In addition to small amounts of squalene and n-$C_{21:6}$ (0–30 $\mu g\,g^{-1}$ lipid), C_{31} and C_{33} n-alkenes with 2,3 and 4 double bonds predominated (300–700 $\mu g\,g^{-1}$ lipid, Figure 2). The latter have been attributed to dietary intake of a marine coccolithophorid alga, *Emiliania huxleyi*.[8]

As expected from the known seasonal variations in the distribution of this alga in the North Sea area, the abundance of these alkenes shows some seasonal variation. More data from samples taken throughout the year is needed in order to fully monitor these variations, but collection of mussels by diving during winter months is often restricted due to adverse weather conditions.

All of the mussels examined contain a series of n-alkanes (*ca.* C_{16-31}, Figure 2) often maximising at n-octadecane. Such a distribution is normally attributed to fairly recent pollution from fossil fuel. Whist this may indeed be the case, it is interesting to note that the n-C_{16}–n-C_{22} alkanes in the mussels are frequently characterised by a predominance of the even carbon number alkanes over the odd carbon number alkanes. This is particularly pronounced in samples collected in 1980 (Figure 3). The distribution is unusual; more commonly n-alkanes show no even/odd predominance (e.g. crude oils) or a distinct odd/even predominance (e.g. most higher plants[9,10] and algae[11]). However, n-alkanes showing a pronounced even/odd predominance have been observed in a number of bacteria and a yeast.[12,13,14]

In other cases an enhanced proportion of even carbon number n-alkanes has been attributed to bacterial action, such as oxidation and decarboxylation of existing, predominantly odd carbon number n-alkanes or reduction of even carbon number fatty acids.[15,16,17]

Cranwell[18] noted a significant predominance of n-C_{16}, n-C_{18} and n-C_{20} alkanes in the bound lipids of *Oscillatoria agardhii*, a blue-green alga, after bacterial degradation (under anaerobic conditions). The observed pattern of n-alkanes in the Forties mussels may be the result of ingestion of algal detritus, perhaps bacterially degraded in the water column, or a direct contribution of bacterial lipids. We intend to investigate these hypotheses by future experiments involving the feeding of radio labelled substrates to mussels.

Similar n-alkane distributions to that observed in the Forties mussels were found by Yasuda and Fukamiya[19] in short-necked clams from Hiroshima Bay, Japan; by Douglas *et al.*,[20] in mussels from Sullom Voe,

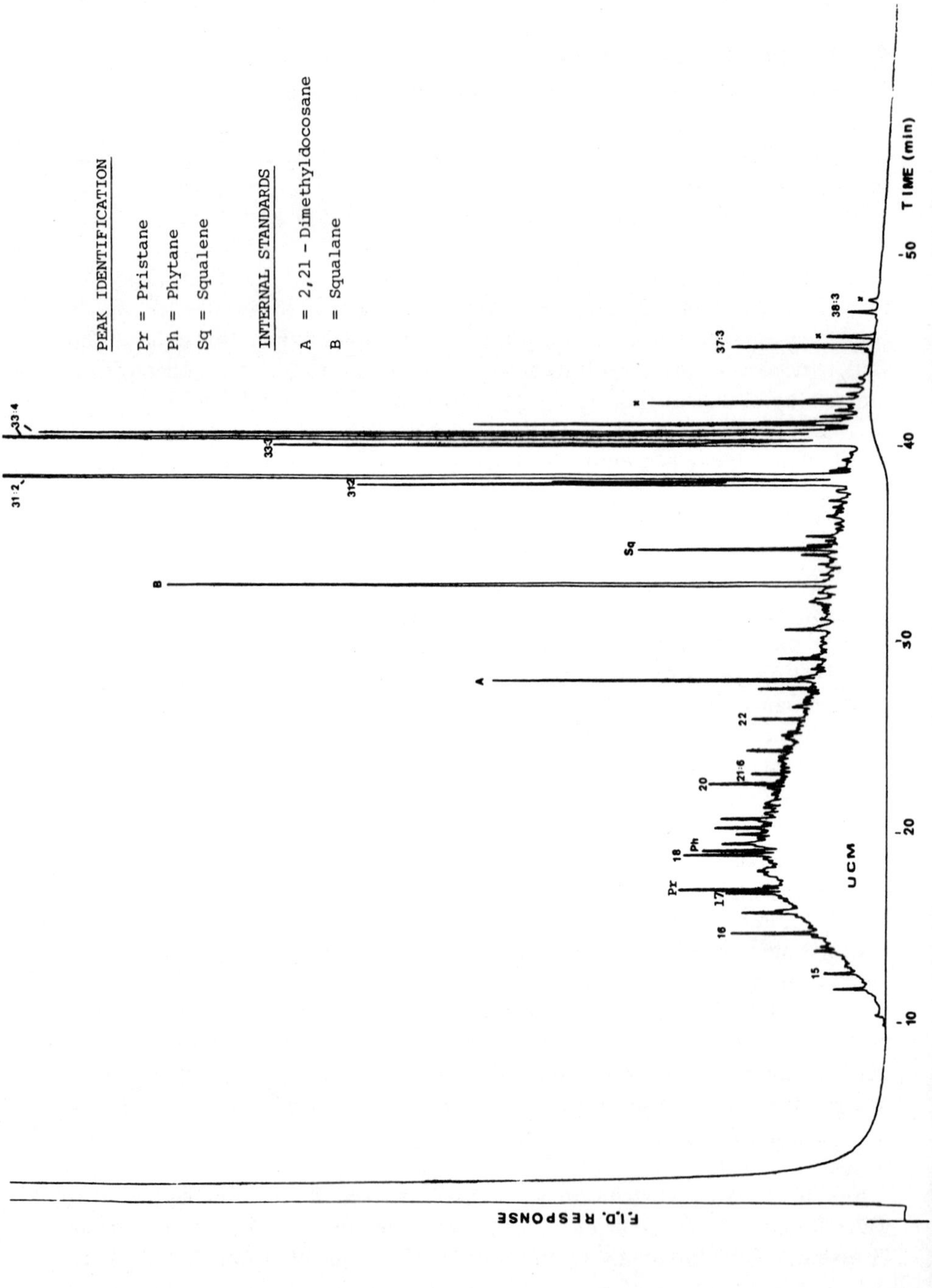
PEAK IDENTIFICATION
Pr = Pristane
Ph = Phytane
Sq = Squalene
INTERNAL STANDARDS
A = 2,21 - Dimethyldocosane
B = Squalane
F.I.D. RESPONSE
TIME (min)
10
20
30
40
50
UCM
15
16
17
18
20
21:6
22
Pr
Ph
A
B
Sq
31:2
33:3
33:4
37:3
38:3
x

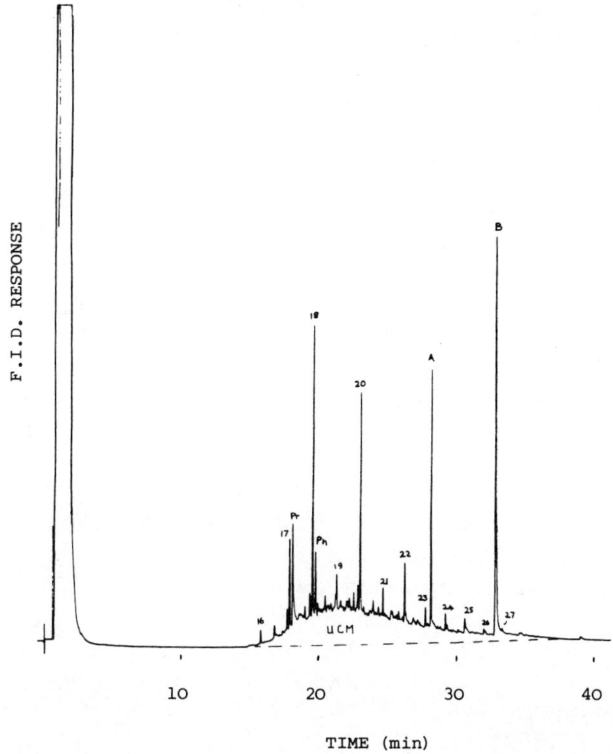

FIGURE 3 GLC trace of aliphatic hydrocarbons in *Mytilus edulis* from Forties D, 1980 (19 feet depth).

Scotland, and by Mackie *et al.*,[3] in unpolluted mussels from Inveraray on the western coast of Scotland. The latter authors suggest sedimentary alkanes as the source of the distribution.

Although some crude oils and ancient sediments also contain enhanced even numbered *n*-alkanes[21,22,23] the distributions are different (often n-C_{22-32} *n*-alkanes predominate) from that observed in the mussels. Forties oil does not show a distinct even/odd *n*-alkane predominance, although some of the surficial sediments in the Forties field contain very similar *n*-alkane distributions to the mussels.

4.2 Petrogenic hydrocarbons

The petrogenic hydrocarbons of the mussels are mainly present as the unresolved complex mixture (UCM), quite commonly observed in polluted mussels.[2,24] Most of the published data for UCM hydrocarbon concentrations in mussels are for specimens taken from coastal areas. Only one literature example of offshore mussel studies is known to the authors, (Rowland and Volkman,[8] Forties C mussels collected in 1979). It is interesting to compare the UCM concentrations of coastal and offshore mussels. Since most literature values are quoted as per dry weight of mussel we will also use this unit for discussion here. Most of the Forties mussels have UCM concentrations (*ca.* 30–60 $\mu g\,g^{-1}$ per dry weight) comparable to those from areas near high urban populations (20–40 $\mu g\,g^{-1}$;[24]). Samples collected at depth from Forties C and D in 1979 have lower values more similar to mussels from pristine coastal areas;[2,3,25] (Masspec Analytical, unpublished results). However, certain of the mussels growing at the surface, notably those from the Forties B platform, have UCM concentrations of 200–300 $\mu g\,g^{-1}$ (Table II), which approach those found in mussels from areas of known oil contamination.[24,26,27].

A few other resolved components of probable fossil fuel origin (e.g. pristane and phytane) are also present at low concentrations in the mussels (Table II).

One of the aims of the present study was to monitor the accumulation of petrogenic hydrocarbons in the mussels with relation to the discharge of treated production water from the Forties D platform. No discharge of production water was made from the Forties B and Forties C platforms prior to sample collection.

The October 1979 Forties D mussels were sampled prior to any discharge of production water from the platform and discharge of production water containing less than 40 $\mu g\,g^{-1}$ total oil (measured by infrared spectroscopy), was started in April 1980. By the time that the mussels were sampled in June 1980 there had been two months of discharge (total amount *ca.* 5 tonnes of oil; B.P.). Perhaps surprisingly, the values of aliphatic hydrocarbons obtained (Table II) for the latest samples show that there had been no further accumulation petrogenic hydrocarbons, indeed, the concentrations have decreased when compared to the previous surface sample. However, no surface sample was taken from Forties D platform after discharge commenced, and the effects of the production water discharge (which takes place above the water surface) may not be expected to be reflected in the lower samples. It should be noted in this respect that the depth profile on Forties C showed that the levels of petrogenic hydrocarbons in surface samples were several times higher than those at 50′ and 100′.

Highly specific and sensitive analytical techniques are necessary both to differentiate biogenic from pollutant hydrocarbons in mussels[24] and to distinguish between the variety of sources of hydrocarbon pollution that exist in the marine environment. Computerised gas chromatography–mass spectrometry (CGCMS) with capillary columns is becoming increasingly used for such studies.[24,28]

The distribution of steranes (I) and triterpanes (II) (Figure 6), which are minor but ubiquitous components of crude oils and most fossil fuels, has been proposed to distinguish sources of crude oil pollution in the marine environment.[28,29,30,31]

For instance, Risebrough *et al.*,[24] compared the distribution of dimethyl monoaromatic steranes in *Mytilus* specimens from the U.S. Mussel Watch program with that of the filtered particulate matter from the surrounding seawater and noted some similarities between the samples. Albaiges[32] was able to identify Iranian crude oils as the source of pelagic tar balls in the Western Mediterranean by comparing the distribution of steranes, methyl steranes, triterpanes and diterpanes to those of the oil.

In all of the Forties mussels, except those from the Forties B platform (Table II), the distribution of steranes and triterpanes, where present, were essentially identical to each other but different from that of Forties production water (Figures 4 and 5). A major difference in the mussels compared to the production water is the relative proportions of norhopane (II, $R=C_2H_5$) (peak 3, Figure 4) and hopane (II, $R=CH(CH_3)_2$; peak 4, Figure 4). This ratio is >1 in the mussels, whereas it is only *ca.* 0.5 in the production water.

Also Forties oil, like the majority of North Sea crudes contains a C_{28} triterpane, 17α(H), 18α(H), 21β(H)-28, 30-bisnorhopane (III, Figures 4 and 6[33]). This compound is not present in the mussels (Figure 4). Other differences exist; for example the relative abundance of the diterpanes (Figure 4, peak 1) is greater in the mussel fractions than in the Forties production water. These observations suggest that the oil in the Forties C and D mussels is not Forties and is probably not a North Sea crude. Further evidence for this is obtained by examination of the distribution of steranes in the production water and in the mussels. The dominant steranes in the Forties production water (and most North Sea oils) are diasteranes (so-called "rearranged" steranes; IV). The mussel samples contain greater proportions of regular relative to diasteranes, than the production water. These differences are reflected in the m/z 217 fragmentograms of the samples (Figure 5). Although some changes in sterane distributions can be brought about by very extensive microbial action,[34,35] the differences in patterns observed in the mussels are considered too large to be accounted for by bacterial modification. Also, such an effect appears to result in a decrease in relative abundance of steranes to diasteranes, the opposite of the observations here.

Since the distributions of steranes and triterpanes from the mussels were

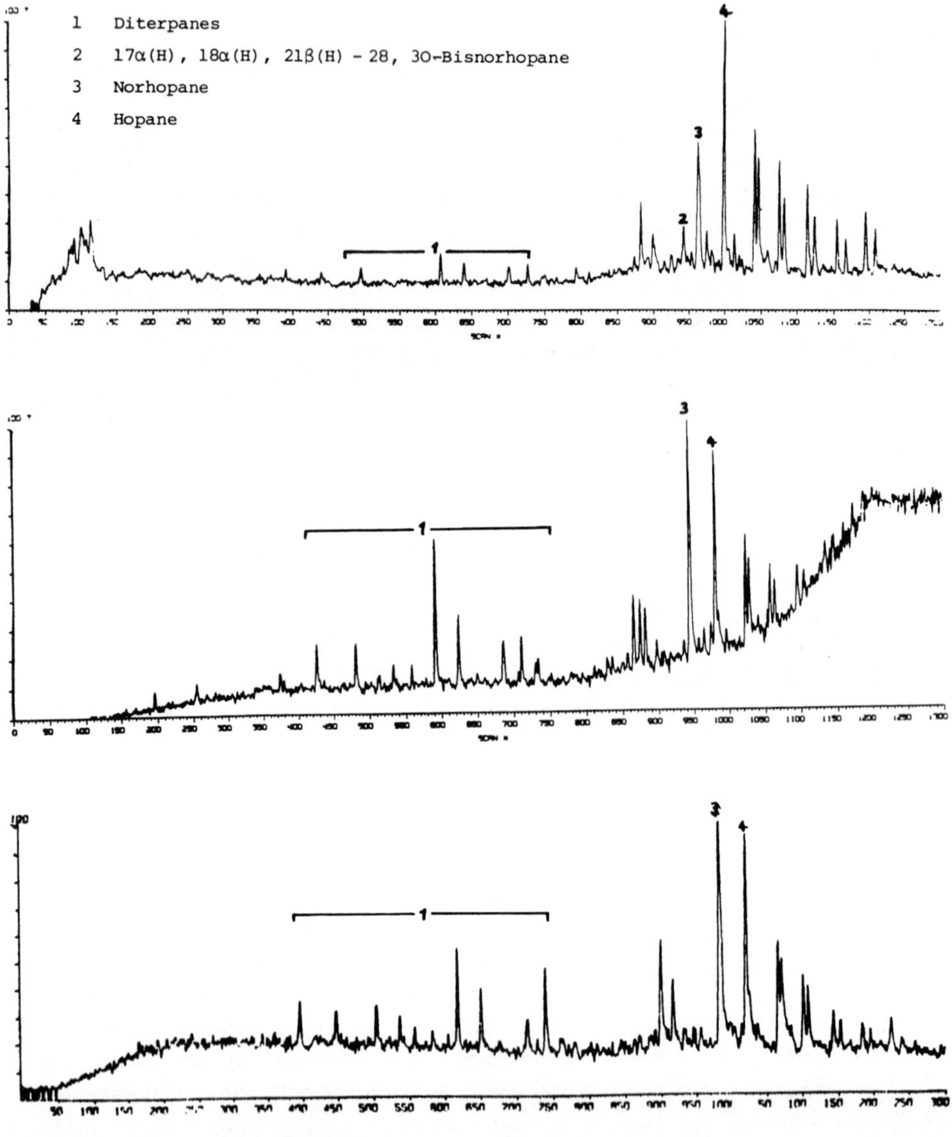

FIGURE 4 Mass fragmentograms (m/z 191) of aliphatic hydrocarbons from (a) Forties production water, (b) *Mytilus edulis*, Forties D, 1980 (19 feet depth) and (c) 50:50 blend of Murban and Iranian light crude oils.

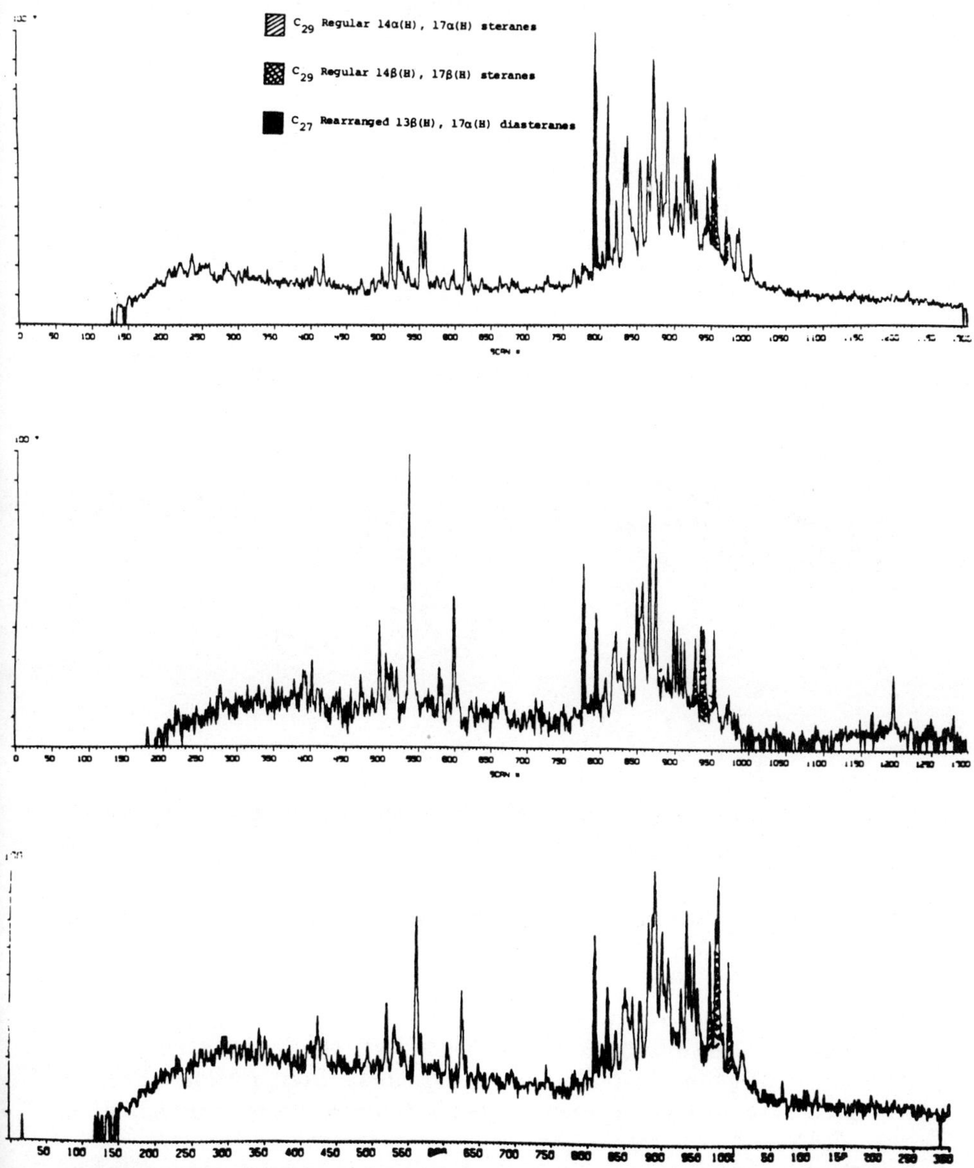

FIGURE 5 Mass fragmentograms (m/z 217) of aliphatic hydrocarbons from (a) Forties production water, (b) *Mytilus edulis*, Forties D, 1980 (19 feet depth) and (c) 50:50 blend of Murban and Iranian light crude oils.

STRUCTURES

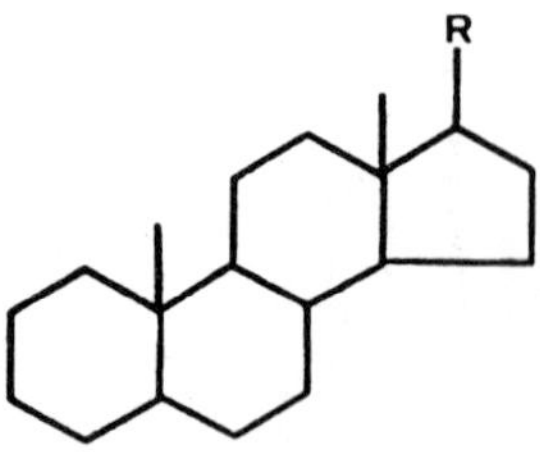

(I)

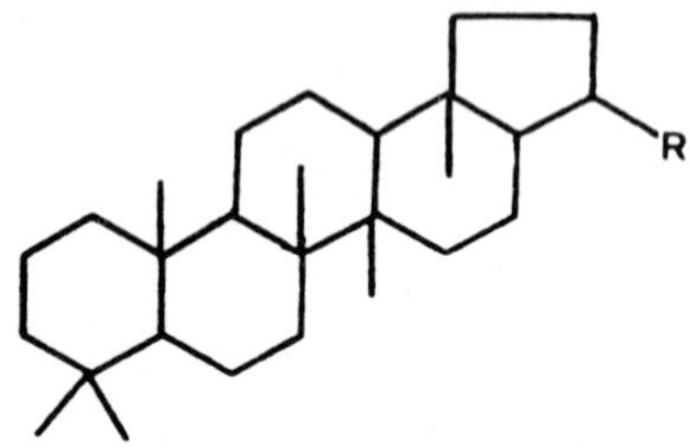

(II)

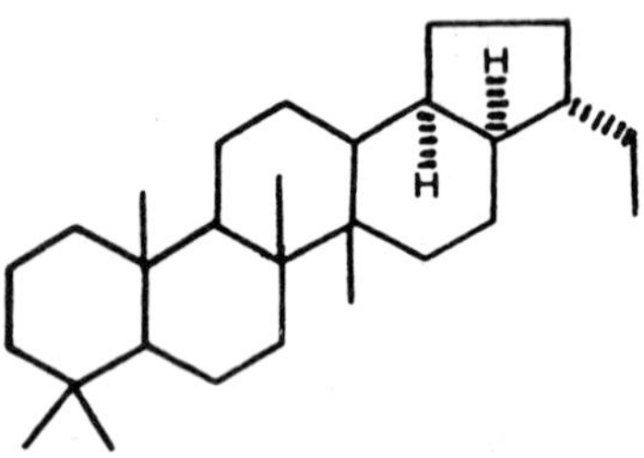

(III)

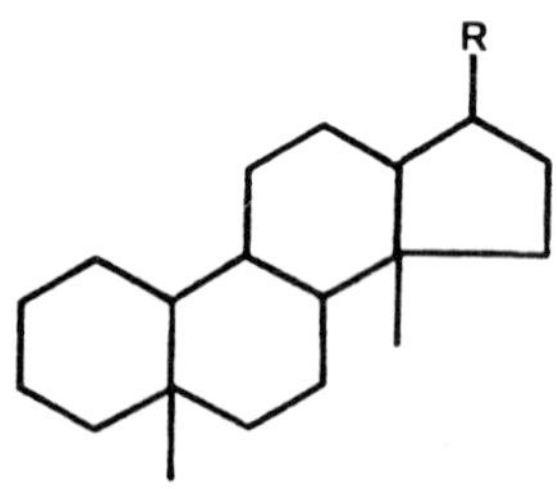

(IV)

FIGURE 6 Hydrocarbon structures: Steranes (I), hopanes (II, III) and rearranged steranes (IV).

not similar to production water contaminated with Forties oil, several other fossil fuels were examined and compared with the mussels. Of the oils examined, the distribution of steranes and triterpanes in a blend of Murban and Iranian crude oils was found to be most similar to that of the mussels. The distributions were not identical even in this example, but this does suggest that the pollutants may originate from a Middle Eastern fossil fuel. Several possible sources of this Middle Eastern fuel pollution exist.

At the time of the study, gas oil used as fuel for turbines, was stored in hollow platform legs. During filling seawater is displaced from the legs and discharged to the sea via a sea sump, a simple vertical oily-water separator, which relies upon slow flow and which also accepts miscellaneous surface drainage from the platform. At fast flow rates, gas oil can be entrained in the discharge water, and this could be the origin of some of the hydrocarbons in the mussels. A direct leakage of stored gas oil is thought unlikely but some spillage could occur during tank filling. To test this hypothesis, a gas oil cut from a Middle Eastern crude was examined. CGCMS analysis of the oil showed no traces of steranes $\geqq C_{27}$ or triterpanes $\geqq C_{27}$ and negligible *n*-alkanes $> C_{26}$. It therefore seems unlikely that the steranes and triterpanes in the mussels originate from this source. However, the most polluted samples (e.g. Forties B) are contaminated mainly by a low molecular weight UCM, and only contain very low levels of $\geqq C_{27}$ steranes. The distribution of petrogenic hydrocarbons in these samples, shows similar characteristics to that of the gas oil, and may therefore have originated from this source.

Other possible sources of petrogenic pollution include; general background contamination from shipping or spillage of lubricating oils used on the platforms. To be consistent with the sterane and triterpane data, polluting oils from these sources must be of Middle Eastern origin or similar, and of the correct molecular weight range.

In the case of general background pollution, adequate data from control mussels growing offshore but away from oilfields, are not yet available to test the hypothesis.

Workboats and support vessels visiting the platform are diesel-powered, so localised inputs of fuel steranes and triterpanes from this source can probably be discounted.

Lubricating oils, which are sometimes based on Middle Eastern crudes[36] might be expected to contain the sterane and triterpane distributions observed. The mussel-accumulated petrogenic hydrocarbons could derive principally from this source. Since lube oils are invariably dewaxed the relatively low concentrations of *n*-alkanes present must then derive from other sources; for example from gas oil (low carbon number) and/or biological sources such as bacteria and algae. The input of petroleum hydrocarbons from lube oil sources might be expected to vary with platform activity.

To summarise, the input of steranes and triterpanes to the Forties mussels appears to arise from background pollution, or lubricating oil, each of the Middle Eastern type and not from North Sea oils.

The distribution of petrogenic hydrocarbons observed in the more polluted mussels (e.g. Forties B) suggests that gas oil may be the major source of pollution. Further differentiation between these sources of pollution requires the collection of offshore mussels from control areas remote from oil production activities and examination of lubricating oils. This work is under progress as part of a continuing mussel monitoring program.

Acknowledgements

This study was sponsored by B.P. Petroleum Development Ltd., to whom we are grateful for permission to publish, and was carried out in conjunction with Dr. R. Ralph of the Offshore Marine Studies Unit (University of Aberdeen). S. J. Rowland would like to thank S.R.C. for a studentship (79101023). We are grateful to Dr. N. J. Haskins for help with CGCMS.

References

1. E. D. Goldberg, V. T. Bowden, J. W. Farrington, G. Harvey, J. H. Martin, P. L. Parker, R. W. Risebrough, W. Robertson, E. Schneider and E. Gamble, *Environ. Conserv.* **5,** 101 (1978).
2. National Academy of Sciences, *The International Mussel Watch*, National Academy of Sciences, Washington D.C. (1980).
3. P. R. Mackie, R. Hardy, K. J. Whittle, C. Bruce and A. S. McGill, *Proceedings 4th International Symposium on Chemistry, Biology and Effects of Polynuclear Aromatic Hydrocarbons, 1979*, 379 (1981).
4. R. Ralph and K. Goodman, *New Scientist*, **82,** 1018 (1979).
5. J. K. Klitz, *North Sea Oil. Resource Requirements for Development in the U.K. Sector*, Pergamon, 159 (1980).
6. V. P. Dole and H. Meinhertz, *J. Biol. Chem.* **235,** 2395 (1960).
7. J. J. Stegeman and J. M. Teal, *Mar. Biol.* **23,** 37 (1973).
8. S. J. Rowland and J. K. Volkman, *Sea, Mar. Env. Res.*, (in press).
9. A. G. Douglas and G. Eglinton, *Comparative Phytochemistry*, (Ed. T. Swain, Academic Press) pp. 57–77 (1966).
10. G. Eglinton, R. J. Hamilton, R. A. Raphael and A. G. Gonzalez, *Nature* **193,** 739 (1962).
11. C. W. Bird and J. M. Lynch, *Chem. Soc. Rev.* **3,** 309 (1974).
12. P. W. Albro and C. K. Huston, *J. Bacteriol* **88,** 981 (1964).
13. J. Han, E. D. McCarthy, W. Van Hoeven, M. Calvin and W. H. Bradley, *Proc. Natl. Acad. Sci.* **59,** 29 (1968).
14. J. Han and M. Calvin, *Proc. Natl. Acad. Sci.* **64,** 436 (1969).
15. D. W. Nooner and J. Oró, *Lipids* **8,** 489 (1973).
16. P. G. Simmonds, D. W. Nooner, A. Zlatkis and J. Oró, *J. Am. Oil. Chem. Soc.* **45,** 34 (1968).
17. D. H. Welte and G. Ebhardt, *Geochim. Cosmochim. Acta.* **32,** 465 (1968).
18. P. Cranwell, *Freshwater Biol.* **9,** 305 (1979).
19. S. Yasuda and N. Fukamiya, *Nippon Suisan Gakkaishi* **44,** 521 (1978).
20. A. G. Douglas, P. B. Hall, B. Bowler and P. F. V. Williams, *Proc. Roy. Soc.* (*Edinburgh*) **80B,** 113 (1981).
21. J. Albaiges and J. M. Torradas, *Nature* **250,** 567 (1974).
22. A. G. Douglas and P. J. Grantham, in *Advances in Organic Geochemistry 1973* (B. Tissot and F. Bienner, eds.), 261–276. Editions Technip (1974).
23. H. Dembicki Jr., W. G. Meinschein and D. E. Hattin, *Geochim. Cosmochim. Acta.* **40,** 203 (1976).
24. R. W. Risebrough, B. W. De Lappe, W. Walker, A. M. Springer, M. Firestone-Gillis, J. Lane, W. Sistek, E. F. Letterman, J. C. Shropshire, R. Vick and A. S. Newton, *Analytical Techniques in Environmental Chemistry* (J. Albaiges, ed.) Pergamon, Oxford, 33 (1980).

25. S. A. Wise, S. N. Chester, H. S. Hertz, W. E. May, F. R. Guenther and L. R. Hilport, *Analytical Techniques in Environmental Chemistry* (J. Albaiges, ed.) Pergamon, Oxford, 41 (1980).
26. D. Straughan, in *Proceedings of Symposium on Physiological Response of Marine Biota to Pollutants, 1975*, p. 289 (1977).
27. J. L. Smith and K. Burns, *Hydrocarbons in Westernport Bay Mussels.* Final Report, Hydrocarbons in Victorian Ecosystems, Task 701–702. Westernport Regional Laboratory and Marine Chemistry Unit, Ministry for Conservation, Australia (1978).
28. J. Albaiges and P. Albrecht, *Int. J. Environ. Anal, Chem.* **6,** 171 (1979).
29. J. G. Pym, J. E. Ray, G. W. Smith and E. V. Whitehead, *Anal. Chem.* **47** (9), 1617 (1975).
30. M. Dastillung and P. Albrecht, *Mar. Pol. Bull.* **7,** 13 (1976).
31. B. R. T. Simoneit and I. R. Kaplan, *Mar. Env. Res.* **3,** 113 (1980).
32. J. Albaiges, *Analytical Techniques in Environmental Chemistry* (J. Albaiges, ed.) Pergamon, Oxford, 69 (1980).
33. P. J. Grantham, J. Posthuma and K. De Groot, *Advances in Organic Geochemistry 1979* (A. Douglas and J. Maxwell, eds.) Pergamon Press, 29 (1980).
34. W. K. Seifert and J. M. Moldowan, *Geochim. Cosmochim. Acta.* **43,** 111 (1979).
35. N. S. Goodwin, P. J. D. Park and T. Rawlinson, *Advances in Organic Geochemistry 1981*, (in press).
36. J. M. Hunt, *Petroleum Geochemistry and Geology*, Freeman and Co, San Francisco, U.S.A. (1979).

Hydrocarbons in the Western Mediterranean Sea, 1981†

ROZENN HÔ, JEAN-CLAUDE MARTY and ALAIN SALIOT

Laboratoire de Physique et Chimie Marines de l'Université Pierre et Marie Curie, ERA CNRS, Tour 24, 4 place Jussieu, 75230 Paris, Cedex 05, France

(*Received December 10, 1981*)

Hydrocarbons in the western Mediterranean Sea were analysed for surface microlayer and subsurface waters during April 1981. Three methods were used for determining both concentrations and markers distribution patterns: U.V. spectrofluorimetry, an original HPLC/U.V. spectrophotometry method and capillary gas liquid chromatography.

No systematic relationship was found between concentrations determined by U.V. spectrofluorimetry and HPLC/U.V. spectrophotometry. Concentrations variation ranges are: 0.01–0.20 $\mu g\,l^{-1}$ for dissolved aromatic hydrocarbons; 0.02–0.13 $\mu g\,l^{-1}$ for particulate aromatic hydrocarbons; 0.02–20.57 $\mu g\,l^{-1}$ for dissolved *n*-alkanes; 0.13–0.83 $\mu g\,l^{-1}$ for particulate *n*-alkanes. Surficial waters appear as very slightly polluted with an exception, the Ligurian Sea for which highest concentrations were encountered.

There is a slight enrichment of particulate hydrocarbons in the surface microlayer in the range 1–3. But, dissolved hydrocarbons can be either depleted or enriched up to 270 times in the surface microlayer, indicating the extreme efficiency of sea-air exchange processes.

Aromatic fingerprints and *n*-alkane distribution patterns are discussed both for pollution and fractionation processes indicators at the sea-air interface.

KEY WORDS: Mediterranean Sea, dissolved hydrocarbons, particulate hydrocarbons, sea water, sea-air interface.

INTRODUCTION

The western Mediterranean Sea is exposed to serious pollutions especially by hydrocarbons, due to an increasing ship traffic and the industrialization of coastal states. Annual inputs of petroleum hydrocarbons to the Mediterranean are estimated in the range 0.5–1.0 million tons.[1]

---

†Contribution PHYCEMED no. 81 02. Presented at the Workshop on the Chemistry and Analysis of Hydrocarbons in the Environment, Barcelona, November 1981.

This context points out a critical situation for the semiclosed Mediterranean Sea owing to geographic and hydrologic characteristics.

Although petroleum monitoring programmes have been developed,[2] there is a lack concerning Mediterranean open sea waters, except a few data by Horn *et al.*,[3] Morris *et al.*[4] and Albaiges *et al.*[5] dealing with surface pelagic tar evaluations, Faraco and Ros,[6] Brown and Huffman[7] and Zsolnay[8] about hydrocarbons in open sea waters. It has appeared necessary to dispose of a new reference state for the open Mediterranean Sea. This has been realised during the PHYCEMED cruise on the R/V *Le Suroit* in April 1981. During this cruise, surficial water (0.2 m depth) and surface microlayer sampled by a metallic screen, have been collected taking all precautions against contamination.

Analyses of hydrocarbons and other geochemical markers have been performed separately on dissolved and particulate matter. Various global and precise techniques have been used. U.V. spectrofluorimetry on sea water hexane extracts gives total aromatic hydrocarbon concentrations. High pressure liquid chromatography coupled to U.V. spectrophotometry enables to obtain a fingerprint of aromatic hydrocarbons separated upon their degree of aromaticity. Finally, gas liquid chromatography on capillary column allows to analyse *n*-alkanes, isoprenoids and saturated unresolved compounds and evaluate respective biogenic and anthropogenic contributions. Data are discussed attempting to correlate aromatic hydrocarbon concentrations obtained through different methods and focusing on the accumulation or depletion of hydrocarbons in the surface microlayer and associated sea-air exchange processes.

EXPERIMENTAL

Sampling sites (Figure 1)

Samples have been collected during the PHYCEMED cruise of the R/V *Le Suroit* in the western Mediterranean Sea, in April 1981. Stations have been chosen corresponding to different hydrologic systems. Dominant characteristics of the five sites are indicated in Table I.

Sampling

Water samples were collected from a rubber Zodiac boat, at about 1–2 km upwind of the main ship. Surface microlayer samples (20 litres) were obtained using a stainless steel screen (1.25 mm square openings in the mesh) collecting a water film approximately 0.44 mm thick. Subsurface water samples (20 litres) were obtained by immersing a 25 litres glass

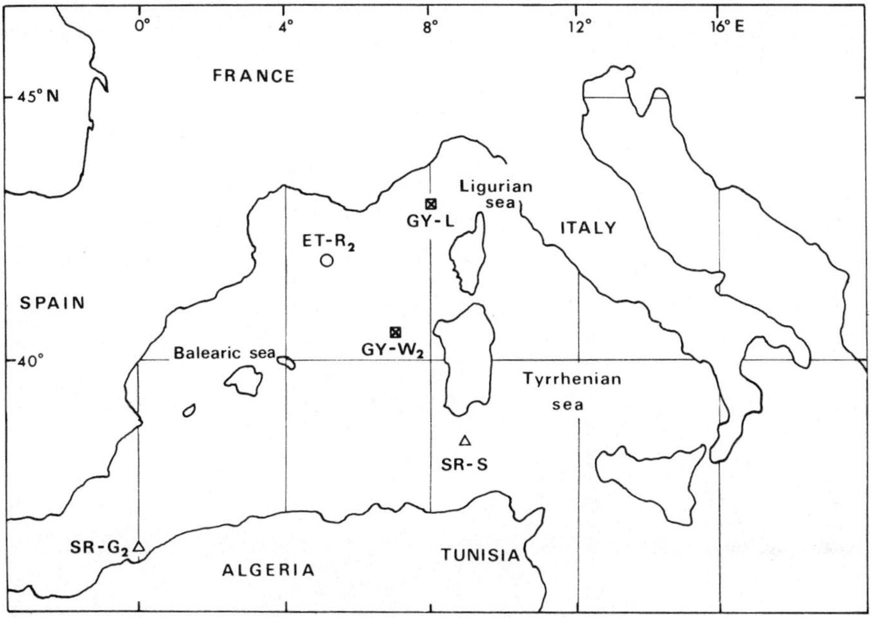

FIGURE 1 Sampling sites, PHYCEMED cruise, April 1981. ○ estuary station: ET-R_2, △ strait stations: SR-G_2 and SR-S, ⊠ gyre circulation stations: GY-L and GY-W_2.

bottle about 0.2 m below the water surface. Thus, two samples were collected for each station and treated immediately on board ship.

Treatment and analysis of the samples

The analytical flow diagram is shown in Figure 2.

Analysis of total aromatic hydrocarbons by U.V. spectrofluorimetry On board ship, one litre of sea water was extracted three times with spectro U.V. grade hexane (2 × 20 ml; 1 × 10 ml). In the laboratory, the joined extracts were dried over Na_2SO_4 and evaporated under nitrogen. They were stored at −25°C. The extract was dissolved in 5 ml of hexane and then analysed using a Perkin Elmer MPF-3L spectrofluorimeter according to the method of Keizer and Gordon.[9] Measurement conditions were: cell, 1 cm; λ excitation 310 nm, slit width 3 nm; λ emission 320–420 nm, slit width 4 nm; λ measurement 360 nm. Fluorescence response was calibrated with solutions of chrysene in the concentration range comparable to the samples. Results are expressed in $\mu g\ l^{-1}$ of chrysene equivalent. It must be

TABLE I

Stations sampled in the Mediterranean, PHYCEMED cruise, April 1981

Station code number	Type	Date	Lat. °N	Long. °E	Sea conditions
GY-L	Gyre	7 April	43°05,4	08°00,5	rough
ET-R_2(1)	Estuary	10 April	42°02,3	05°05,4	rough
ET-R_2(2)	Estuary	11 April	42°02,3	05°05,4	slight
SR-G_2	Strait	18 April	36°16,9	00°06,3	slight
SR-S	Strait	20 April	38°22,9	09°24,8	smooth
GY-W_2	Gyre	23 April	40°36,6	07°10,7	moderate (rain)

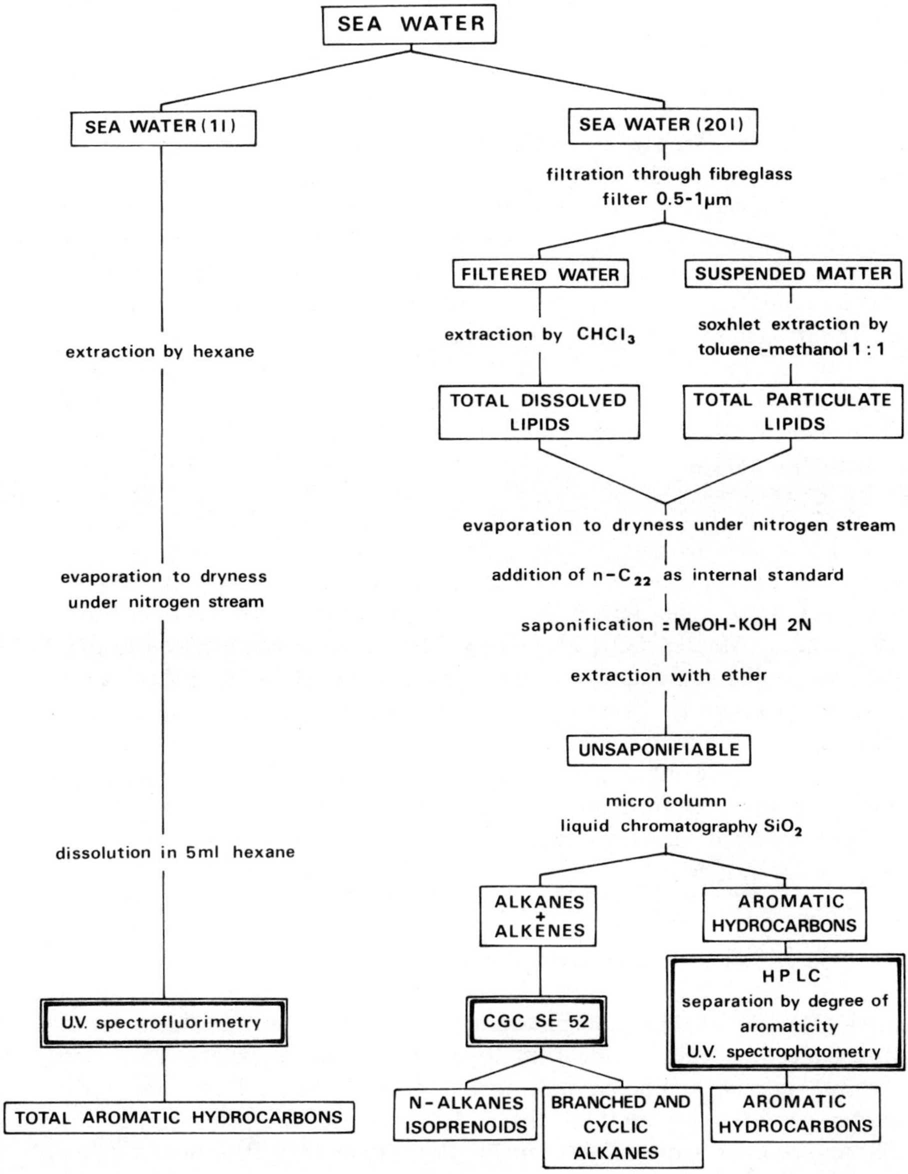

FIGURE 2 Analytical flow diagram.

emphasized that this method is not specific for petroleum hydrocarbons, but detects any fluorescing compounds that are not lost during the evaporation step.

Isolation of dissolved and particulate hydrocarbons Particulate matter was obtained from sea water by filtration under reduced pressure on Whatman GF/F glass fibre filters (15 cm diameter), collecting particles with a diameter greater than 1 μm. Filters were then kept frozen until extraction for 36 hours (solvent being changed after 24 hours) in a soxhlet apparatus using a 1:1 mixture of toluene and methanol. The extract was dried over $CaCl_2$, evaporated under vacuum at reduced temperature ($t < 40$°C), dried under nitrogen stream, then stored frozen until analysis.

Dissolved lipids were extracted on board three times by chloroform (twice at pH 8, once at pH 2). Chloroform extracts were added together, dried over $CaCl_2$ and evaporated in a Büchi rotavapor. They were stored frozen until analysis.

Column chromatography on total lipids Internal standard, $n-C_{22}$ was added to each extract in order to evaluate the efficiency of following analytical steps. The extract was saponified under nitrogen in a Sovirel SVL tube by MeOH-KOH 2N during 2 hours. After acidification to pH 2, the unsaponifiable fraction was extracted by ether (3×1 ml). Ether extracts were evaporated to dryness under nitrogen stream and separated by adsorption chromatography on a microcolumn filled with SiO_2. The first fraction eluting (6 ml hexane) contained alkanes and alkenes. The second fraction containing aromatic hydrocarbons was eluted successively by 3 ml of hexane + 60 μl of ethyl acetate, 3 ml of hexane + 15 μl of ethyl acetate and 2 ml of hexane.

Analysis of saturated hydrocarbons by capillary gas chromatography The fraction containing saturated and olefinic hydrocarbons was analysed using a Girdel 3000 gas chromatograph with flame ionization detector and Ross injector. The column employed was glass capillary 25 m long, 0.25 mm internal diameter, wall coated with SE 52. Helium was used as carrier gas at a flow of 2 ml min^{-1}. Temperature was programmed from 120 to 270°C at a rate of 4°C min^{-1}. Peak areas were determined using a Varian Vista integrator. *N*-alkanes and isoprenoids were identified by comparison of their retention times with those of known standards from n-C_{11} to n-C_{36} injected in the same conditions, and quantified in respect to the internal standard, n-C_{22}. The same calibration factor was used for the unresolved envelope.

Analysis of aromatic hydrocarbons by high pressure liquid chromatography/U.V. spectrophotometry Aromatic hydrocarbons were analysed by HPLC according to the method of Moinet,[10] described in Moinet *et al.*[11] This method allows to obtain a HPLC fingerprint corresponding to the separation of hydrocarbons by increasing degree of aromaticity. Briefly, aromatic hydrocarbons were fractionated on an analytical column 20 cm long, 4·7 mm internal diameter filled with 5 μm SiO_2 bonded-NH_2 stationary phase. Sample injection was realised using a Rheodyne 7120 valve, with an injection loop of 20 μl. The pump, Chromatem Altex 380, was used with a pulsation damper at a flow of 1.8 ml min^{-1}. Aromatic hydrocarbons were detected by ultraviolet absorption at 254 nm with a Schoeffel spectromonitor. In these conditions aromatic hydrocarbons are separated essentially by degree of aromaticity, with little effect of alkyl substituents.[12] On HPLC chromatograms presented, retention times of known standards from benzene (Be) to perylene (Pe) are indicated. Peaks emerging from the envelope correspond to mono- to penta-aromatic series.

Blanks, limits of detection Great care was taken against contamination during on board ship and laboratory work up. Solvents were distilled before the cruise and a second time on board just before use. Filters, Na_2SO_4, $CaCl_2$ were extracted in a soxhlet with distilled chloroform and blanks showed that only traces of interfering compounds were present in our samples.

Data were corrected for losses during laboratory work up, using recovery of the internal standard (n-C_{22}). Although drastic precautions have been taken during evaporation steps, losses of the more volatile compounds occur. Consequently, data given for $<C_{14}$ alkanes and <3 rings aromatic hydrocarbons are only semi-quantitative.

The analytical precision, taking into account extraction, chemical steps during sample work up, and gas liquid or high pressure chromatography analysis has been estimated as 25% for saturated and aromatic hydrocarbons. The corresponding limits of detection are as follows: 0.02 $\mu g\,l^{-1}$ for total aromatic hydrocarbons (U.V. spectrofluorimetry), 0.005 $\mu g\,l^{-1}$ for individual n-alkanes (CGC), and 0.001 $\mu g\,l^{-1}$ for aromatic fractions (HPLC/U.V. spectrophotometry).

RESULTS AND DISCUSSION

Concentrations

Aromatic hydrocarbons Concentrations of aromatic hydrocarbons evaluated using U.V. spectrofluorimetry and HPLC/U.V. spectrophotometry are given in Table II.

TABLE II

Western Mediterranean sea; PHYCEMED cruise, April 1981. Concentrations of aromatic hydrocarbons determined by U.V. spectrofluorimetry and HPLC/U.V. spectrophotometry for dissolved and particulate matter, expressed in $\mu g\ l^{-1}$ as chrysene equivalent

Station		Aromatic hydrocarbon concentrations		
			HPLC/U.V. spectrophotometry	
		U.V. spectrofluorimetry	Dissolved	Particulate
GY-L	microlayer	0.35	0.049	0.131
	subsurface	0.47	0.111	0.081
ET-R_2(1)	microlayer	0.26	0.165	0.061
	subsurface	0.84	0.046	0.028
ET-R_2(2)	microlayer	0.28	0.229	0.102
	subsurface	0.49	0.127	0.046
SR-G_2	microlayer	5.15	0.084	0.029
	subsurface	0.20	0.077	0.022
SR-S	microlayer	0.25	0.105	0.077
	subsurface	0.07	0.009	0.026
GY-W_2	microlayer	0.70	0.197	0.122
	subsurface	0.33	0.061	0.039

Concentrations in total aromatic hydrocarbons (U.V. spectrofluorimetry) vary in the range 0.07–5.15 $\mu g\,l^{-1}$. This concentration range is less than the range reported for coastal waters of the Mediterranean, using also U.V. spectrofluorimetry,[2] indicating that open sea waters are not heavily polluted. It must be mentioned here that after Keizer and Gordon,[9] concentrations in excess of 2 $\mu g\,l^{-1}$ indicate either the presence of petroleum residues or sample contamination.

Figure 3 shows that there is no relationship between concentrations determined by the two methods: U.V. spectrofluorimetry and HPLC/U.V. spectrophotometry. Some of the values obtained by U.V. spectrofluorimetry are highly overestimated *versus* concentrations by U.V. spectrophotometry, as for example samples SR-G_2 microlayer and ET-R_2(1) subsurface. Such discrepancies have been previously related and can be interpreted by the interference of fluorescing compounds which are not aromatic hydrocarbons. This point underlines difficulties of the interpretation of aromatic hydrocarbon concentrations issued from spectrophotometric methods without a purification step of the initial extract. As a consequence, we have selected HPLC/U.V. spectrophotometry

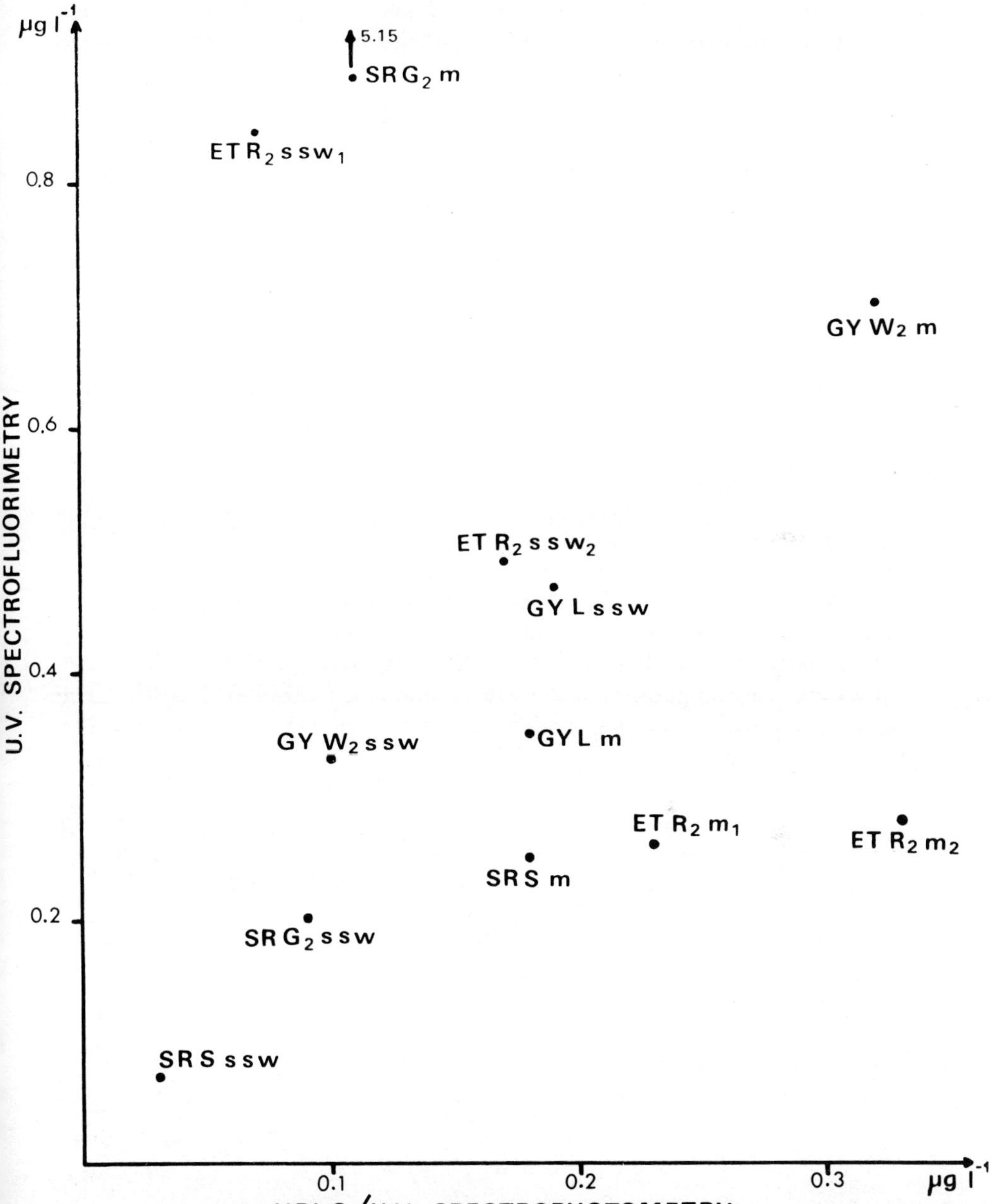

FIGURE 3 Total hydrocarbon concentrations determined by U.V. spectrofluorimetry *versus* dissolved + particulate hydrocarbon concentrations determined by HPLC/U.V. spectrophotometry. m: microlayer, ssw: subsurface water.

concentrations because in this method aromatic hydrocarbons are purified on a SiO_2 microcolumn. This method enables also to distinguish between dissolved and particulate compounds.

Dissolved aromatic hydrocarbon concentrations vary from 0.01 to 0.20 $\mu g\,l^{-1}$, particulate in the range 0.02–0.13 $\mu g\,l^{-1}$. There is about 2 to 3 times less of particulate hydrocarbons in the subsurface water. Particulate hydrocarbons are accumulated in the surface microlayer with an enrichment factor varying from 1 to 3. The enrichment factor (E.F.) is defined as the ratio between the concentration of a compound in microlayer and in subsurface water. Such values have been reported for both coastal and open ocean locations.[13–15]

Slight enrichments of up to 3 or 4 are typical of the microlayer under ambient surface conditions. In areas of high pollution by petroleum, the surface film can be enriched many thousands of times with hydrocarbons as a result of multilayer oil films.[16]

Contrarily drastic changes are observed for dissolved hydrocarbons: depletion for sample GY-L (E.F. = 0.43), high accumulation for sample SR-S (E.F. = 12).

Saturated hydrocarbons Dissolved and particulate *n*-alkane and total saturated hydrocarbon concentrations are listed in Table III.

Concentrations in dissolved *n*-alkanes vary from 0.02 to 20.57 $\mu g\,l^{-1}$. The variation range for particulate *n*-alkanes is 0.13–0.83 $\mu g\,l^{-1}$. The relative importance of *n*-alkanes *versus* total saturated hydrocarbons is highly variable, from 9 to 93% for dissolved hydrocarbons. This percentage is less for hydrocarbons associated with suspended matter: 5–22%. The fractionation of hydrocarbons between the dissolved and particulate states which is the result of various physico-chemical processes is largely dependent on local conditions; the ratio dissolved/particulate vary from 10 times less to 40 times higher.

As for aromatic compounds, if there is a slight but constant enrichment of particulate hydrocarbons in the surface microlayer, an inverse situation characterizes the dissolved pool of hydrocarbons: high depletion for sample GY-L (E.F. = 8×10^{-3}), important accumulation for ET-R2(1) (E.F. = 268). Another observation must be mentioned: the extreme local variability of both dissolved and particulate concentrations as shown for the same station ET-R2 sampled at 24 hours interval; the enrichment factor for dissolved *n*-alkanes decreases from 268 to 2.

Total hydrocarbons In order to compare the PHYCEMED data with those reported earlier, we have listed in Table IV total hydrocarbon concentrations by adding both aromatic (U.V. spectrophotometry) and saturated (CGC), dissolved and particulate concentrations.

TABLE III

Western Mediterranean Sea: PHYCEMED cruise, April 1981. Concentrations of dissolved and particulate *n*-alkanes and saturated hydrocarbons determined by CGC, expressed in $\mu g\,l^{-1}$. a: non determined.

Station		Dissolved hydrocarbons			Particulate hydrocarbons		
		n-alkanes	Total saturated hydrocarbons	*n*-alkanes / Saturated hydrocarbons %	*n*-alkanes	Total saturated hydrocarbons	*n*-alkanes / Saturated hydrocarbons %
GY-L	microlayer	0.17	1.88	9	0.83	7.23	11
	subsurface	20.57	22.16	93	0.52	3.52	15
ET-R_2(1)	microlayer	13.30	34.20	39	0.18	1.37	13
	subsurface	0.05	—[a]	—	0.17	1.61	11
ET-R_2(2)	microlayer	2.59	—	—	0.27	1.27	22
	subsurface	1.22	2.67	46	0.28	2.32	12
SR-G_2	microlayer	—	—	—	0.18	1.45	12
	subsurface	0.15	2.14	7	0.13	2.65	5
SR-S	microlayer	0.09	0.79	12	0.48	—	—
	subsurface	0.02	0.08	34	0.21	0.98	21
GY-W_2	microlayer	0.31	3.52	9	0.38	4.02	9
	subsurface	0.26	2.56	10	0.13	0.79	16

J

TABLE IV

Western Mediterranean Sea: PHYCEMED cruise, April 1981. Concentrations of total (dissolved + particulate) hydrocarbons

Station		Total D + P hydrocarbons	Aromatic hydrocarbons / Total hydrocarbons %
GY-L	microlayer	9.30	1.9
	subsurface	25.87	0.8
$ET\text{-}R_2(1)$	microlayer	35.80	0.6
	subsurface	1.73[a]	4.0
$ET\text{-}R_2(2)$	microlayer	4.18[a]	7.9
	subsurface	5.16	3.3
$SR\text{-}G_2$	microlayer	—	—
	subsurface	4.89	2.0
SR-S	microlayer	1.45[a]	12.4
	subsurface	1.08	2.8
$GY\text{-}W_2$	microlayer	8.05	4.0
	subsurface	3.45	3.0

[a]Non resolved saturated components are not included.

The Mediterranean has been the subject of several investigations for non volatile hydrocarbons as by Horn *et al.*,[3] Morris *et al.*[4] and Albaiges *et al.*[5] for surface pelagic tars; by Monaghan *et al.*,[17] Faraco and Ros,[6] and Zsolnay[8] for hydrocarbons in surficial water. But a few data is available for the western Mediterranean. The Ligurian Sea has been studied between Nice and Calvi in 1972[14] and off Villefranche bay in 1976–1977.[18]

Concentrations reported were in the range 4–40 $\mu g\,l^{-1}$, which indicates in respect with relatively high values encountered in 1981 that the central part of the Ligurian Sea is effectively polluted. This can be correlated with the gyre type superficial water circulation, accumulating different urban run off and refinery wastes.

Algerian coastal waters have been sampled in 1974–1975 and Zsolnay in 1979 reported a mean concentration of 20 $\mu g\,l^{-1}$. In 1981, the concentration found in SR-G2 is about 5 $\mu g\,l^{-1}$.

On a global scale, total concentrations obtained in 1981 vary in the range from a few $\mu g\,l^{-1}$ to 35 $\mu g\,l^{-1}$. Percentages of aromatic vs total hydrocarbons are low, below some per cent. Regarding concentrations, the western Mediterranean appears as non-heavily polluted with an exception for the Ligurian Sea (station GY-L).

Aromatic fingerprints and *n*-alkane distribution patterns; utilization as pollution and fractionation processes indicators at the sea-air interface.

To illustrate the potential use of the precise analysis of hydrocarbons we have selected two stations representative of two opposite conditions: first, SR-S, which could be considered obviously as a reference station with a very low hydrocarbon background level and GY-L which appears to be slightly polluted.

Station SR-S HPLC chromatograms of dissolved and particulate hydrocarbons, both for microlayer and subsurface water are presented in Figure 4.

Two series, triaromatic (phenanthrene) and tetraaromatic (pyrene) predominate in subsurface water. Very different fingerprints are observed for the microlayer. They are characterized by the predominance of triaromatic structures (phenanthrene) for particles and lower aromaticity degree structures for dissolved hydrocarbons.

The high value of the enrichment factor, 12, is related to the accumulation of volatile compounds in the microlayer, a preliminary step before evaporation into the atmosphere. According to this explanation the enrichment factor for $<C_{20}$ *n*-alkanes which are less volatile than phenanthrene has a lower value, 3. Smooth sea conditions encountered at this station (Table I) allows this steady-state accumulation vs evaporation.

Station GY-L The subsurface is characterized by two different types of aromatic fingerprints, with a marked predominance of low aromaticity compounds: monoaromatic (benzene) in the particulate fraction, diaromatic (fluorene) in the dissolved (Figure 5).

Note that the evaluation of these compounds is only semi-quantitative, due to losses during evaporation steps but particularly important in connection with toxicity problems. It is well known that benzene-type monoaromatic hydrocarbons, are highly toxic for the marine life.[19, 20]

Particulate hydrocarbons are accumulated in the surface microlayer (mean E.F. = 2), the monoaromatic fraction being evaporated. In the dissolved pool, the more volatile compounds have disappeared during the microlayer formation.

The examination of *n*-alkane distribution patterns (Figure 6) and saturated hydrocarbon enrichment factors lead to the same figure. Dissolved *n*-alkanes are not accumulated (E.F. $= 8 \times 10^{-3}$). Particulate *n*-alkane distributions suggest that there is a more pronounced biological activity in the surface film than for other samples. This is

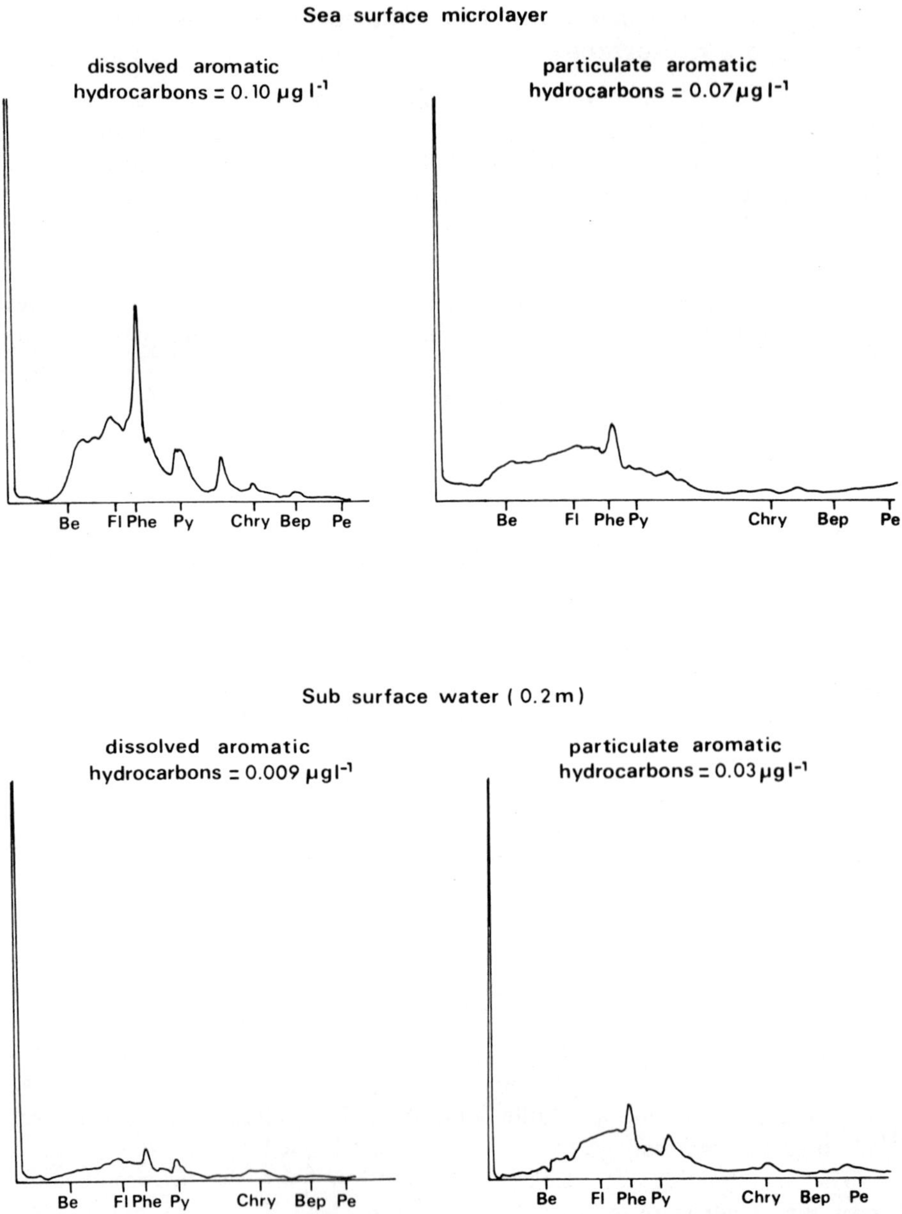

FIGURE 4 Aromatic hydrocarbon fingerprints obtained by HPLC/U.V. spectrophotometry, station SR-S.

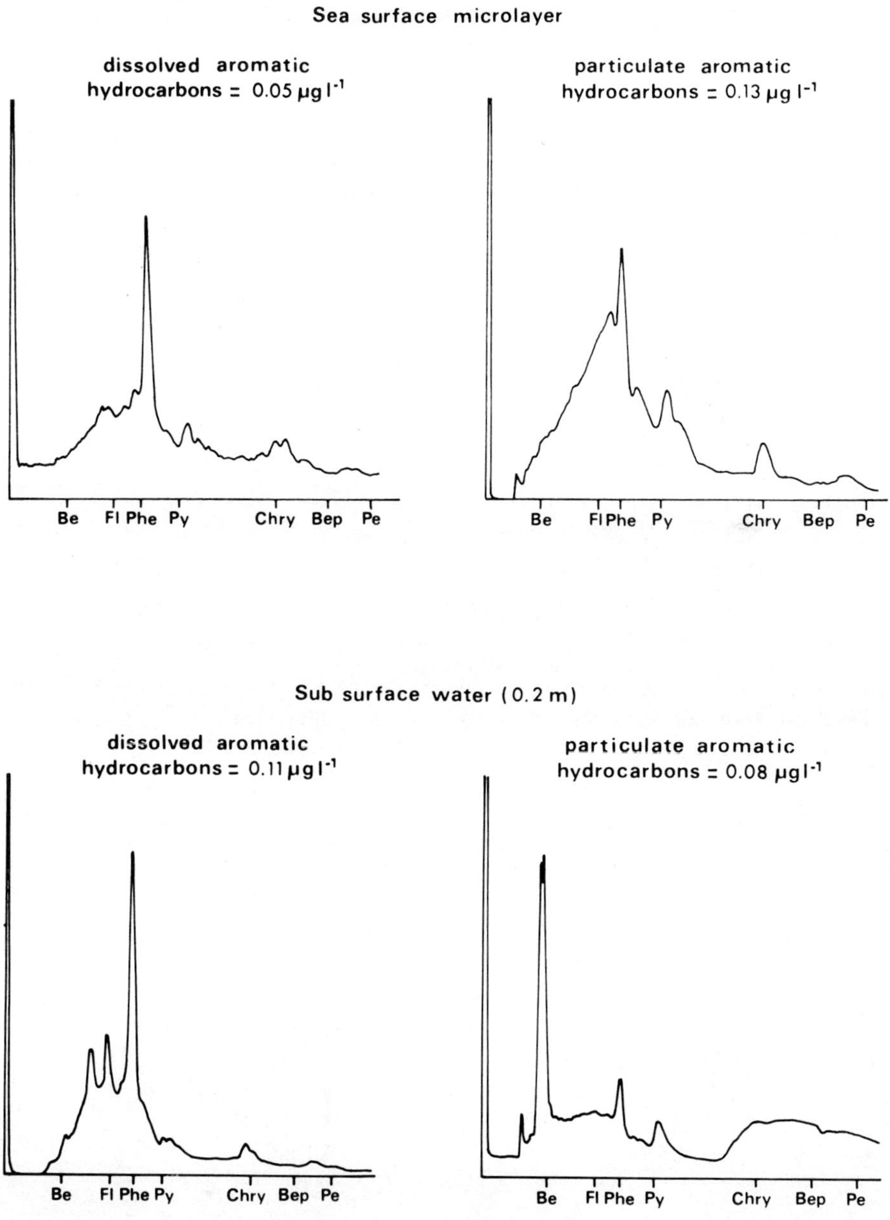

FIGURE 5 Aromatic hydrocarbon fingerprints obtained by HPLC/U.V. spectrophotometry station GY-L.

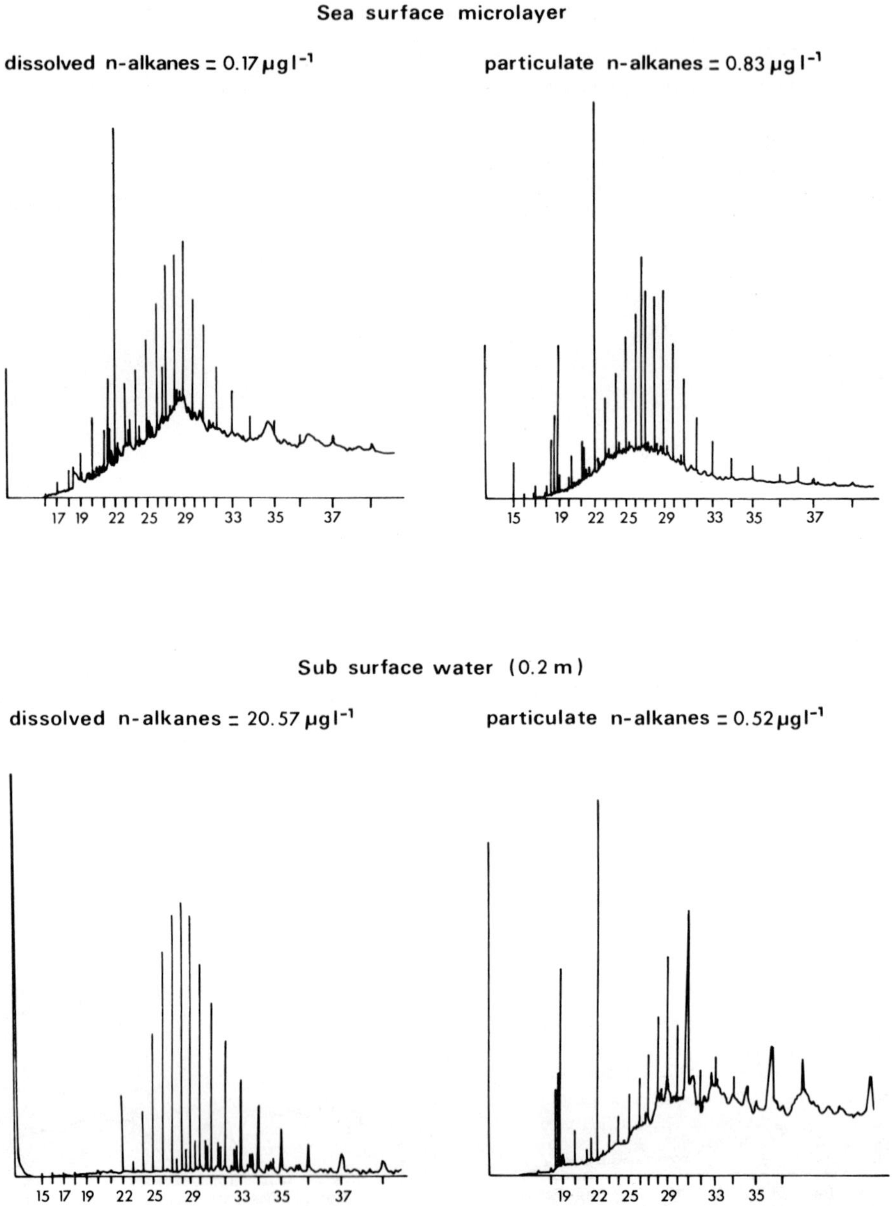

FIGURE 6 Saturated hydrocarbon chromatograms showing *n*-alkane distribution patterns, station GY-L.

indicated by the importance of n-C_{15} and n-C_{17}, associated with planktonic activity.[21] Owing to the more degradability of n-alkanes vs other branched and cyclic compounds, this could explain the decrease of n-alkane importance *versus* total saturated hydrocarbons in the microlayer (9%) in respect to subsurface water (27%).

Of note is a terrestrial contribution in particulates of subsurface water, shown by the predominance of an odd carbon number series in the range C_{25}–C_{31}. This illustrates the decoupling between the two organic matter pools, dissolved and particulate, recently clearly demonstrated by Boehm[22] and Barbier *et al.*[23]

All these observations show that the microlayer is a site of specific biological and physico-chemical processes and active exchanges between the surficial water and the atmosphere. In fact, for polluted stations, the examination of hydrocarbons budget in the microlayer allows to precise the efficiency of the degradation and transfer from the sea to the atmosphere of hydrocarbons, both biogenic and petrogenic.

Acknowledgements

We thank Miss J. Fournier for assistance for U.V. spectrofluorimetry and Miss A. Lorre for providing HPLC/U.V. spectrophotometry measurements.

References

1. P. Le Lourd, *Ambio.* **6,** 317 (1977).
2. UNEP, *Preliminary Report on the State of Pollution of the Mediterranean Sea*, UNEP/IG.11/INF. 4, United Nations Environment Programme, Geneva, Switzerland (1977).
3. M. H. Horn, J. M. Teal and R. H. Backus, *Science* **168,** 245 (1970).
4. R. J. Morris, J. N. Butler and A. Zsolnay, *Environ. Conserv.* **2,** 275 (1975).
5. J. Albaiges, J. Borbon and J. Ros. *IV Journées Etud. Pollutions*, C.I.E.S.M. (Antalya) pp. 103–109 (1978).
6. F. Faraco and J. Ros, *IV Journées Etud. Pollutions,* C.I.E.S.M. (Antalya) pp. 117–121 (1978).
7. R. A. Brown and H. L. Huffman Jr., *Science* **191,** 847 (1976).
8. A. Zsolnay, *Mar. Chem.* **7,** 343 (1979).
9. P. D. Keizer and D. C. Gordon, Jr., *J. Fish. Res. Board Can.* **30,** 1039 (1973).
10. A. Moinet, *Etude de Marqueurs Géochimiques Extraits de Sédiments Marins Récents: les Hydrocarbures Aromatiques Polycycliques.* Thèse 3ème cycle. Université P. et M. Curie, Paris, 75 p.
11. A. Moinet, M. Ewald, P. Albrecht and A. Saliot (in prep.).
12. S. A. Wise, S. N. Chesler, H. S. Hertz, L. R. Himpert and W. E. May, *Anal. Chem.* **49,** 2506 (1977).
13. T. L. Wade and J. G. Quinn, *Mar. Pollut. Bull.* **6,** 54 (1975).
14. J. C. Marty and A. Saliot, *Deep-Sea Res.* **23,** 863 (1976).

15. J. C. Marty, A. Saliot, P. Buat-Menard, R. Chesselet and K. A. Hunter, *J. Geophys. Res.* **84,** 5707 (1979).
16. R. J. Morris, *Mar. Pollut. Bull.* **5,** 105 (1974).
17. P. H. Monaghan, D. E. Brandon, R. A. Brown, T. D. Searl and J. J. Elliott, *Measurement and interpretation of nonvolatile hydrocarbons in the ocean. I. Measurements in Atlantic, Mediterranean, Gulf of Mexico, and Persian Gulf.* AID. 1DJB 74, EPR.4EX.74. U.S. Department of Commerce Maritime Administration, Washington, D.C. (1974).
18. M. Goutx and A. Saliot, *Mar. Chem.* **8,** 299 (1980).
19. National Academy of Sciences, *Petroleum in the Marine Environment*. National Academy of Sciences, Washington, D.C. (1975).
20 J. M. Neff, *Polycyclic Aromatic Hydrocarbons in the Aquatic Environment. Sources, Fates and Biological Effects.* Applied Science Publishers. London. 262 pp. (1979).
21. A. Saliot, *Marine Organic Chemistry*. (E.K. Duursma and R. Dawson Eds.) Elsevier Scientific Publishing Company, Amsterdam, pp. 327–374 (1981).
22. P. D. Boehm, *Mar. Chem.* **9,** 255 (1980).
23. M. Barbier, D. Tusseau, J. C. Marty and A. Saliot, *Oceanologica Acta* **4,** 77 (1981).

Polycyclic Aromatic Hydrocarbons in Mediterranean Coastal Sediments†

GILBERT MILLE, JIAN YU CHEN and HENRI J. M. DOU

Centre de Spectroscopie Moléculaire et L.A. 126 de Chimie Moléculaire et Pétroléochimie, Faculté des Sciences et Techniques, Rue Henri Poincaré, 13397 Marseille, Cedex 4, France

(*Received November 27, 1981*)

Several Mediterranean coastal sediment samples from polluted and non polluted areas of the French coast have been analysed for polycyclic aromatic hydrocarbons. The amounts of selected PAH have been determined in order to identify possible sources of these compounds. Seasonal variations have been observed and are discussed.

KEY WORDS: PAH, Mediterranean sediments.

INTRODUCTION

The polycyclic aromatic hydrocarbons are the largest class of chemical carcinogens known today and they are found at various concentration levels throughout the environment. No data are available for polycyclic aromatic hydrocarbons in Mediterranean coastal sediments. Following studies on the whole hydrocarbon content of Mediterranean sediments,[1] we have undertaken more detailed analysis focusing on the polycyclic aromatic hydrocarbon fraction in order to:

1) determine the amount and the structures of the major components, the carcinogenic activity of which is directly linked to their structures.

2) know the possible sources of these compounds.

†Presented at the Workshop on the Chemistry and Analysis of Hydrocarbons in the Environment, Barcelona, November 1981.

EXPERIMENTAL

Sampling sites: The location of the sampling stations are given in Figure 1.

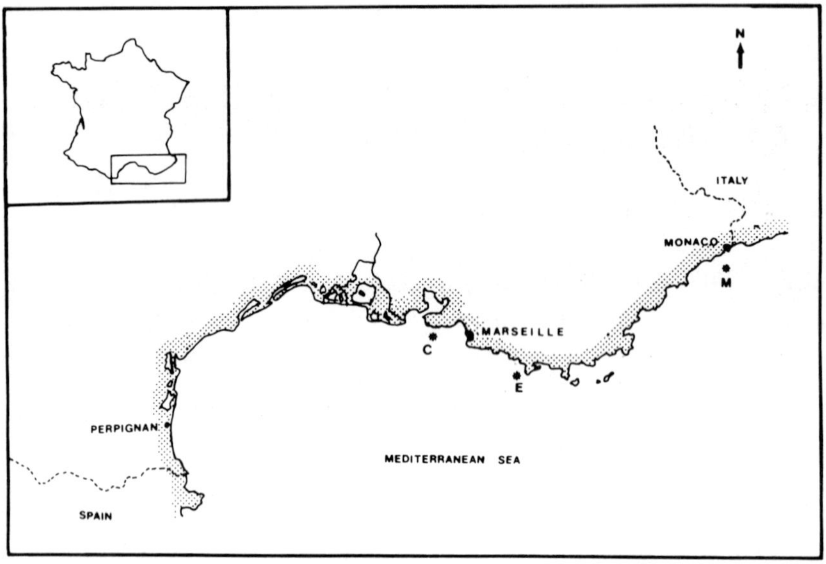

FIGURE 1 Station locations for studies of PAH in sediments.

All the samples were taken between 3 and 10 metres deep. They were taken by divers and are representative of the upper 0–5 cm of surface sediments. Samples taken include:

— Zone C (Côte Bleue): This zone is located between Marseille and the Rhone Delta. Several samples were taken at different places, representing non-polluted areas (H-4, H-5) and possibly polluted areas near a refinery (L). Several samples were also taken at Station 4, one at each season during 1979.

— Zone E (Les Embiez): This zone is located 50 km east from Marseille. It is a big resort area densely populated in summer with an important coastal navigation and boating activities (1l, 1n).

— Zone M (Monaco): This non-polluted area also supports substantial coastal navigation in summer. (F, G).

Extraction and analysis

Hydrocarbons were extracted from wet sediments (~50 g) by direct saponification via alkaline hydrolysis. Samples were heated under reflux for 4 hr in a mixture of 0.5 N KOH in 95% methanol/toluene 2/1 and then elemental sulfur was removed from the extract by percolating through a column of activated Cu/Al_2O_3. Fractions containing hydrocarbons were isolated from column chromatography on a silica gel (5% deactivated with H_2O) column.[2]

PAH analysis was performed by synchronous UV fluorescence spectroscopy, capillary gas chromatography and combined capillary gas chromatography/mass spectrometry.[3]

RESULTS AND DISCUSSION

The gravimetric results of the hydrocarbon analysis of the Mediterranean coastal sediments are listed in Table I and concentrations of some selected PAH are given in Table II. Gas chromatograms of aromatic fractions for Stations H4 (Côte Bleue) and F (Monaco) are presented in Figures 2 and 3.

TABLE I

Hydrocarbon concentrations in Mediterranean coastal sediments (μg/g dry weight)

	Côte Bleue(C)			Les Embiez (E)		Monaco (M)	
	H4	H5	L	11	1n	F	G
Total extract	141	134	1610	900	990	203	70
Aliphatic hydrocarbons	12	16	720	78	56	65	46
Aromatic hydrocarbons	1.2	1.3	232	15	13	12.1	5.2
%(aromatic/aliph.)	10	8	31	19	23	19	11
%(aromatic/hydroc)	9	7.5	24	16	19	16	10

Seasonal variations of the PAH content for sediment from Station 4 are indicated by the ratios of concentrations of some selected PAH (Table III) and by the UV fluorescence spectra (Figure 5).

The amounts of hydrocarbons found in Mediterranean coastal sediments are generally low (<100 μg/g dry weight) except for Station L (952 μg/g) which is located near an oil refinery. The aromatic fraction represents from 7 to 23% of the total hydrocarbons except for the more contaminated station L which was 31%. The values found (1–33 μg/g and 232 μg/g for station L) are close to those found in North Atlantic,[4] lake or river sediments.[5, 6]

The relative concentrations of selected PAH are not the same for all sediments (Table II) and two different patterns can be seen (Figure 4)

TABLE II

Selected aromatic hydrocarbon concentrations in Mediterranean coastal sediments (ng/g dry weight).

Peak number	m/e	Compounds	Côte Bleue (C)			Monaco (M)		Les Embiez (E)	
			H4	H5	L	F	G	11	1n
4	178	Phenanthrene	37.5	11.6	120	8.6	3.8	4.2	5.2
5	178	Anthracene	2	0.8	15	—	—	—	—
6	192	C_1 Phenanthrenes	20	10.7	280	7	5	7.1	12
11	206	C_2 Phenanthrenes	12	10.9	451	3.5	3	6.4	10.6
12	202	Fluoranthene	46	26.9	318	46.5	15	8	13
13	202	Pyrene	28.5	16.8	273	50.4	7.2	6	10
15	216	C_1 Fluoranthene or pyrene	26	15	N.D.	15	4	N.D.	N.D.
	PAH $C_{18}H_{12}$		48	21.5	N.D.	32	21.3	N.D.	N.D.
16	228	Benz(a)anthracene	12.5	5	N.D.	16.3	8.1	N.D.	N.D.
17	228	Chrysene	22	11	N.D.	14.7	9.2	N.D.	N.D.
	PAH $C_{20}H_{12}$		52.5	33.5	N.D.	45	19.6	N.D.	N.D.
18	252	Benzofluoranthenes	20	14	N.D.	20	5.2	N.D.	N.D.
19	252	Benz(e)pyrene	8	4	N.D.	} 10.5	} 4.3	N.D.	N.D.
20	252	Benz(a)pyrene	9	5	N.D.			N.D.	N.D.
21	252	Perylene	3	2	N.D.	12	7	N.D.	N.D.
22	278	Dibenzanthracene	—	—	N.D.	6	—	N.D.	N.D.
23	276	Benzo(g, h, i)perylene	—	—	N.D.	7.5	—	N.D.	N.D.

N.D. = Not Determined.

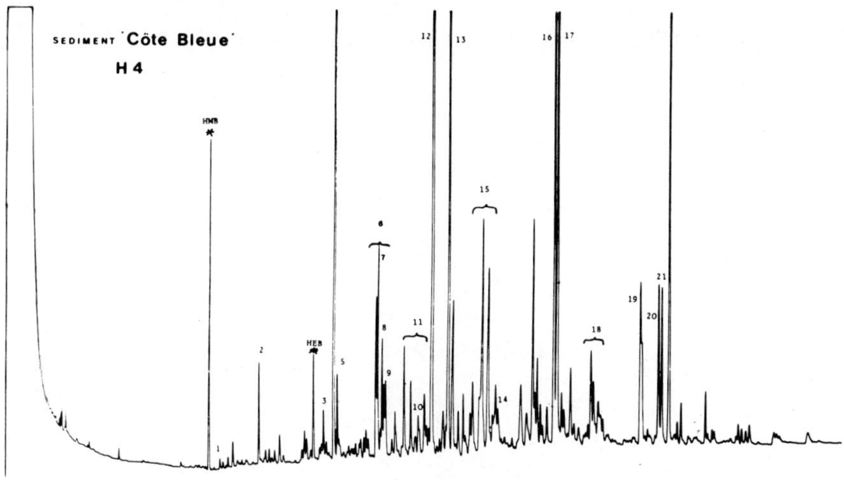

FIGURE 2 Glass capillary gas chromatogram of aromatic hydrocarbons in Mediterranean surface sediments Côte Bleue station H4. S.E. 52 column (15 m × 0.32 mm) installed on a Carlo Arba Model 2150 gas chromatograph equipped with split/splitless injection; Helium carrier gas at 0.55 kg/cm^2. Program 70—270°C at 3°C/mn.

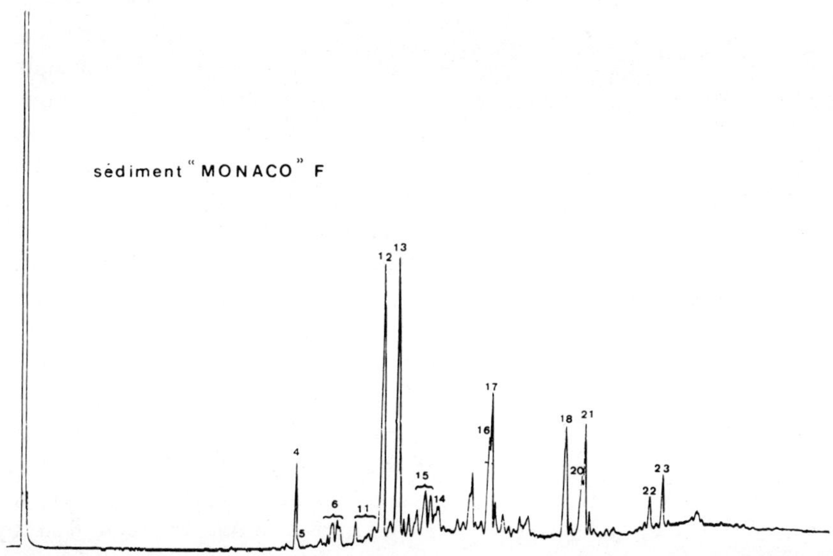

FIGURE 3 Glass capillary gas chromatogram of aromatic hydrocarbons in Mediterranean surface sediments Monaco. Station F. S.E. 52 column (25 m × 0.32 mm) installed in a Carlo Erba Model 4160 gas chromatograph equipped with on column injection. Helium gas carrier gas at 0.55 kg/cm^2. Program 70—300°C at 5°C/mn.

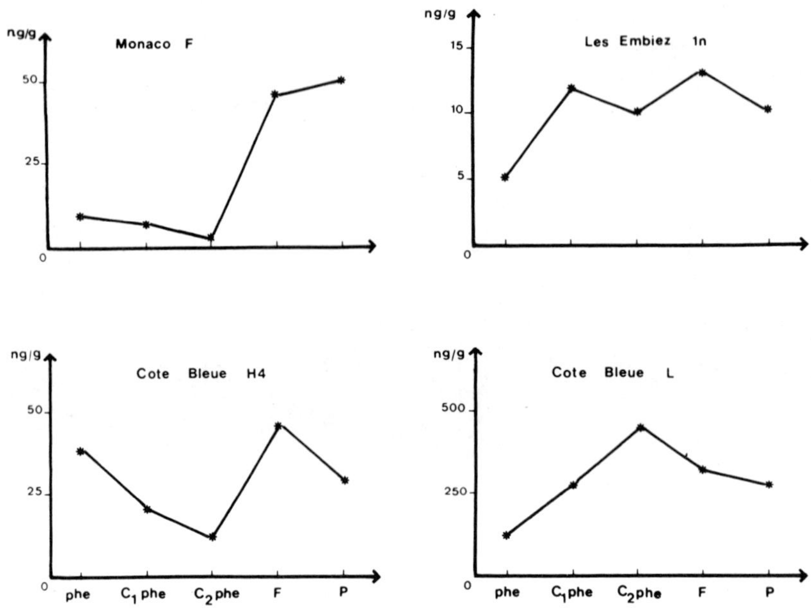

FIGURE 4 Distributions of some selected PAH in Mediterranean coastal sediments.

which indicate different possible sources for polycyclic aromatic hydrocarbons in surface sediments. These sources of polycyclic aromatic hydrocarbons in the marine environment are numerous and are discussed in detail by Giger and Blumer,[7] Youngblood and Blumer,[8] Laflamme and Hites,[9] Hites *et al.*[10] To summarize these discussions, polycyclic aromatic hydrocarbons found in the environment may be derived from:

— Direct deposition of petroleum (ballasting/deballasting operations of tankers, discharge of oily bilge-water, tank washing, refiner effluents, discarded lubricants and other oils, accidents, natural seepage...). In this case PAH are generally present as complex mixtures containing both alkylated and non-alkylated species but there is a deficiency in the unsubstituted compounds.

— Combustion is often an important source of PAH. Natural combustion (forest fire, burning land plants) and anthropogenic combustion (fossil fuels) leads to a mixture of PAH characterized by homologous series which contain a decreasing amount of alkyl substitution as the temperature of combustion increases.

— Early diagenesis in recent sediments may be responsible for the aromatization of some naturally occurring cyclic compounds (e.g. steroids,

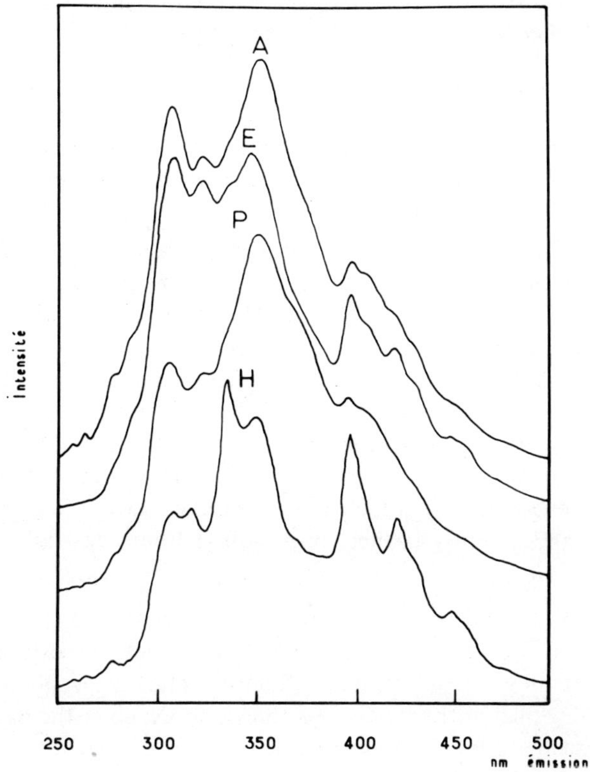

FIGURE 5 Synchronous fluorescence spectra ($\Delta\lambda = 23$ nm) of aromatic hydrocarbons (Côte Bleue, station 4) showing seasonal variations. (A: autumn; H: winter; P: spring; E: summer).

TABLE III

Seasonal variations of the concentration ratios of some selected PAH. (Côte Bleue—Station 4).

	Autumn	Winter	Spring	Summer
Phenanthrene/-1 methyl Phe.	0.06	6.8	0.06	0.06
Phenanthrene/C_1Phe	0.03	1.8	0.03	0.04
Phenanthrene/C_2Phe	0.09	3.1	0.1	0.12
Phenanthrene/Fluoranthene	0.43	0.8	1.5	0.66
Phenanthrene/Pyrene	0.6	1.3	1.5	1
Fluoranthene/C_1Phe	0.07	8.3	0.02	0.06
Pyrene/C_1Phe	0.05	5.2	0.02	0.04
Pyrene/Fluoranthene	0.71	0.62	1	0.66

triterpenoids,[11]) with one or two specific compounds which are directly related to the skeletal structures of the precursors.

For these three sources, relative distribution patterns are different and quantitative measurements of various alkylated and non-alkylated PAH can be useful in the identification of the possible source of PAH, except for the fact that the ability to distinguish between natural and anthropogenic combustion is limited.[8]

— *Zone C* (Côte Bleue): absolute concentrations of selected PAH are the highest for samples taken at Station L and the lowest for samples taken at Station 5 (Table II). Samples H4 and H5 show the same PAH pattern which is different from those of PAH found at Station L. In the case of Stations H-4 and H-5 the PAH mixtures are characterized by the predominence of unsubstituted species (phenanthrene, fluoranthene, pyrene, $C_{18}H_{12}$ and $C_{20}H_{12}$ compounds) over the alkylated species and the relative distribution of the PAH is indicative of combustion (ratios of the phenanthrene to the C_1 or C_2 phenanthrenes, ratio of $C_{16}H_{10}$ isomers: fluoranthene and pyrene to their mono-alkyl homologs, are always $>$to 1).

For Station L the PAH mixture shows both alkylated and non alkylated species and its origin may be direct deposition of non-combusted petroleum (e.g. refinery effluents). The ratios of phenanthrene to the C_1 or C_2 phenanthrenes is less than one which is the case when the source is direct fossil fuel contamination.

For Station 4, several samples were taken at each season of 1979. Samples exhibited lower hydrocarbon concentrations in summer, winter and spring than in autumn.[1] Gas chromatograms and UV fluorescence spectra (Figure 5) show that the composition of the aromatic fraction is not the same for each season. The major difference in the aromatic fraction compositions between winter and the other seasons is seen in the phenanthrene series concentrations. This fact has previously been noticed by Boehm *et al.*[12] in the study of hydrocarbon pollutants in the Georges bank region.

In winter, a combustion source for the PAH seems to be possible whereas for the other seasons a direct fossil fuel contamination seems to be evident. Why do such seasonal variations take place?

Until now we could only make some hypothesis to explain the experimental data including:

a) Combustion source for the PAH in winter and (b) petroleum origin of these PAH during the other seasons (as seen by a deficiency of unsubstituted species, especially phenanthrene). It seems reasonable to assume that natural mechanisms exist that can modify a PAH homolog distribution after the mixture has been deposited in the sediment. Such mechanisms could include differential water solubility of the higher alkyl homologs V/S the unsubstituted species; preferential bioaccumulation of higher homologs, preferential biodegradation of lower homologs, preferential vaporization of lower homologs or preferential accumulation of some specific compounds by organisms.

All these PAH homologs distribution modification mechanisms have yet to be demonstrated. Studies on the seasonal hydrocarbon (PAH) variations will be of great interest on a longer time scale.

— *Zone M* (Monaco F and G): The PAH distribution is similar to the one found for Station H4 and H5 (Zone C) (Figure 3 and 4) with lower amounts of phenanthrene and C_1 C_2 phenanthrenes. A pyrolytic source for this PAH seems to be justified (Fluoranthene/C_1 Fluor. >3) Fluoranthene, Pyrene and Perylene concentrations (sample F) are higher than those found for sample H4.

We analysed a core (15cm length) from Station F and we did not find increasing concentrations of perylene, phenanthrene, fluoranthene, pyrene or chrysene with increasing depth in the core. Based on this it would seem that these PAH cannot result from an in situ diagenesis.

— *Zone E*: We only have determined concentrations of five PAH. Phenanthrene series, fluoranthrene and pyrene and low values (4–10 ng/g) have been found for the concentrations of these compounds. Their relative distribution, ratio of phenanthrene to the C_1 or C_2 phenanthrenes less than one, seems to indicate a petroleum source (Figure 4) for them.

For this station we have begun seasonal samplings in order to know if the composition of the aromatic fraction is the same for each season.

CONCLUSION

Two main sources of polycyclic aromatic hydrocarbons in Mediterranean coastal sediments have been found: combustion source and petroleum source.

Seasonal variations seem to be important and can lead to very different patterns in the distribution of polycyclic aromatic hydrocarbons. The exact reasons for these modifications remain unexplained and further research will be necessary.

References

1. G. Mille, M. Guiliano, A. Arlhac and J. Chouteau, *Vie Marine* **2,** 81 (1980).
2. J. W. Farrington and B. W. Tripp, *Marine Chemistry in the Coastal Environment*, ed. by M. Church, Am. Chem. Soc. Symp, Ser., No. 8, 267 (1975).

3. J. M. Teal, K. Burns and J. W. Farrington, *J. Fish. Res. Bd. Can.* **35,** 510 (1978).
4. J. W. Farrington and B. W. Tripp, *Geochim. Cosmochim. Acta* **41,** 1627 (1977).
5. W. Giger and C. Schaffner, *Anal. Chem.* **50,** 243 (1978).
6. S. G. Wakeham, C. Schaffner and W. Giger, *Geochim. Cosmochim. Acta* **44,** 403 (1980).
7. W. Giger and M. Blumer, *Anal. Chem.* **46,** 1663 (1974).
8. W. W. Youngblood and M. Blumer, *Geochim. Cosmochim. Acta* **39,** 1303 (1975).
9. R. E. Laflamme and R. A. Hites, *Geochim. Cosmochim. Acta* **42,** 289 (1978).
10. R. A. Hites, R. E. Laflamme, J. G. Windsor, Jr., J. W. Farrington and W. G. Deuser. *Geochim. Cosmochim. Acta* **44,** 873 (1980).
11. C. Spyckerelle, *Constituants Aromatiques de Sédiments.* Thèse Sciences. Université Louis Pasteur, Strasbourg, France. (1975).
12. P. D. Boehm, W. G. Steinhauer, D. L. Fiest, N. Mosesman, J. E. Barak and G. H. Perry, *Proceedings of the 1979 Conference on Prevention and Control of Oil Pollution*, pp. 333–341. American Petroleum Institute (1979).

Dissolved and Particulate Hydrocarbons in Water from a Spring Sampling of the Var River Estuary (S. France)†

K. A. BURNS AND J.-P. VILLENEUVE

International Laboratory of Marine Radioactivity, International Atomic Energy Agency, c/o Musée Océanographique, Principality of Monaco

(*Received November 27, 1981*)

To address questions of water quality monitoring and to investigate processes controlling the biogeochemical cycling of organics in estuarine and marine systems, we have used a resin adsorption apparatus with in-line glass fiber filter for sampling "dissolved" and "particulate" organics. High resolution gas chromatographic analyses of water samples collected from the Var River and coastal Mediterranean show rivers are important sources of soluble pesticides and some PCB's. Percentage particle association was related to particle loading as predicted by sorption models. Total concentrations of PCB's resulted from river input superimposed on more widespread distributions consistent with atmospheric transport models. Petroleum products discharged to rivers and the sea surface plus atmospheric inputs of combustion products add to levels of biogenic sources making distributions highly variable. Distributions of hydrocarbons between physical-chemical phases showed high percentage particle association of high boiling relatively insoluble components, while "dissolved" material contained a larger proportion of low boiling unsaturated components at fairly constant concentration. This partitioning suggests different input sources or differential fractionation of components from the same general sources. We also noted the increase in relative importance in new unidentified halogenated hydrocarbon contaminants in coastal systems.

KEY WORDS: Hydrocarbons, chlorinated hydrocarbons, coastal pollution, seawater sampling.

---

†Presented at the Workshop on the Chemistry and Analysis of Hydrocarbons in the Environment, Barcelona, November 1981.

INTRODUCTION

As summarized by Meyers and Gunnerson[1] levels of non-volatile hydrocarbons in seawater are generally in the low μg/l concentration range. Samples may contain both biogenic and petroleum derived material. Measurement of such low levels and distinction between natural and pollutant compounds present extreme analytical difficulty. Efforts to develop methods for measurement of both halogenated and non-halogenated hydrocarbons which are precise and sensitive enough for detecting long term trends in oceanic levels have not yet yielded satisfactory results.[2] To address questions of water quality monitoring and to investigate processes controlling the biogeochemical cycling of organics in marine systems, we use a resin adsorption apparatus with in-line glass fiber filter for sampling "dissolved" and "particulate" organics. We present here results from a spring sampling of the Var River Estuary in Southern France.

METHODS

Water was collected with a metal bucket hand dipped off an inflatable boat deployed from a small research vessel. Water was taken well upwind of the ship and poured into 20 l dark glass carboys and sealed with aluminium foil. The surface sample depicted in Figure 2 was collected by repeated dippings of a Garrett-type stainless steel screen[3] and drained through a glass funnel into a dark 2 l glass bottle containing 50 ml glass distilled hexane. Buckets, carboys, screen and all other gear were precleaned and rinsed with solvents before use. Samples were transported back to the laboratory and contents of each carboy passed through the extraction apparatus. The apparatus consisted of copper tubing, a stainless steel filter-holder housing a precombusted 0.3 μ Gelman type AE glass fiber filter and Teflon seals, a glass column packed with 50 ml precleaned Amberlite XAD-2 resin held in place with glass wool plugs. Water was drawn out of the carboy and through the system with vacuum at approximately 300 ml/min. After concentrating the organics, columns were leached with boiling acetonitrile. Dissolved lipids were partitioned into hexane. Filters were soxhlet extracted with hexane. The surface sample was extracted with three further aliquots of hexane in a separatory funnel. All hexane extracts were dried with anhydrous sodium sulfate, reduced to 1 ml in volume and passed through columns of partially deactivated silica gel over alumina. Elution with hexane followed by 10% and 20% toluene in hexane produced saturated and unsaturated hydrocarbon extracts. These were again reduced in volume by rotary evaporation followed by evaporation in a stream of pure N_2 and analyzed by high resolution gas chromatography.

Petroleum was distinguished from biogenic hydrocarbons according to criteria discussed by Farrington *et al.*,[4,5] Burns and Teal[6] and Wakeham and Farrington.[7] Further details on methodology for seawater sampling and analysis can be found in Burns and Smith[8] and Elder.[9] Samples from the Var Estuary were collected 2 March, 1981 before the major onset of spring runoff. The large volume samples depicted in Figure 1 were collected in September, 1981 approximately 2 km off Monaco. For these samples 100 l of near-surface water was processed from a station with approximately 250 m water depth.

RESULTS AND DISCUSSION

Results of analyses of 20 l samples for chlorinated hydrocarbons are shown in Table I. Those for non-chlorinated hydrocarbons are shown in Table II. The data show that PCB's and total hydrocarbons were adsorbed to particles in the Var Estuary compared to more soluble chlorinated pesticides such as lindane. Plots of concentration verses distance show [PCB], [lindane] and [suspended particles] were inversely correlated with distance from the river mouth (and salinity) (Figure 2).

For PCB's the percentage associated with suspended particles changed in proportion to particle loading suggesting that as particles decrease in abundance a larger portion of the PCB's is found in solution. The percentage association of PCB's on particles was consistent with sorption models predicting a theoretical distribution coefficient in the range of 10^4 to 10^5.[10,11] In contrast to PCB's, [total hydrocarbons] and [unresolved hydrocarbons] (URE), were relatively constant with salinity and usually showed greater than 70% particle association. Exceptions were samples showing a large biogenic input (Station 4) or a large component of relatively fresh fuel oil contamination (Station 6). [HCB] (hexachlorobenzene) showed no consistent trends. pp'DDE was only detectable in river samples of low salinity.

The regressions taken from the data plotted in Figure 2 show that rivers are highly significant sources of particulate matter in coastal waters making material available for binding organic pollutants such as petroleum hydrocarbons and PCB's. The Var River was a significant source of lindane and pp'DDE confirming the view that river inputs are major influences on the distributions of persistent pesticides in Mediterranean ecosystems. The low significance of [PCB] with distance from the river shows that river inputs may be important in localized areas but these inputs are superimposed on widespread distributions consistent

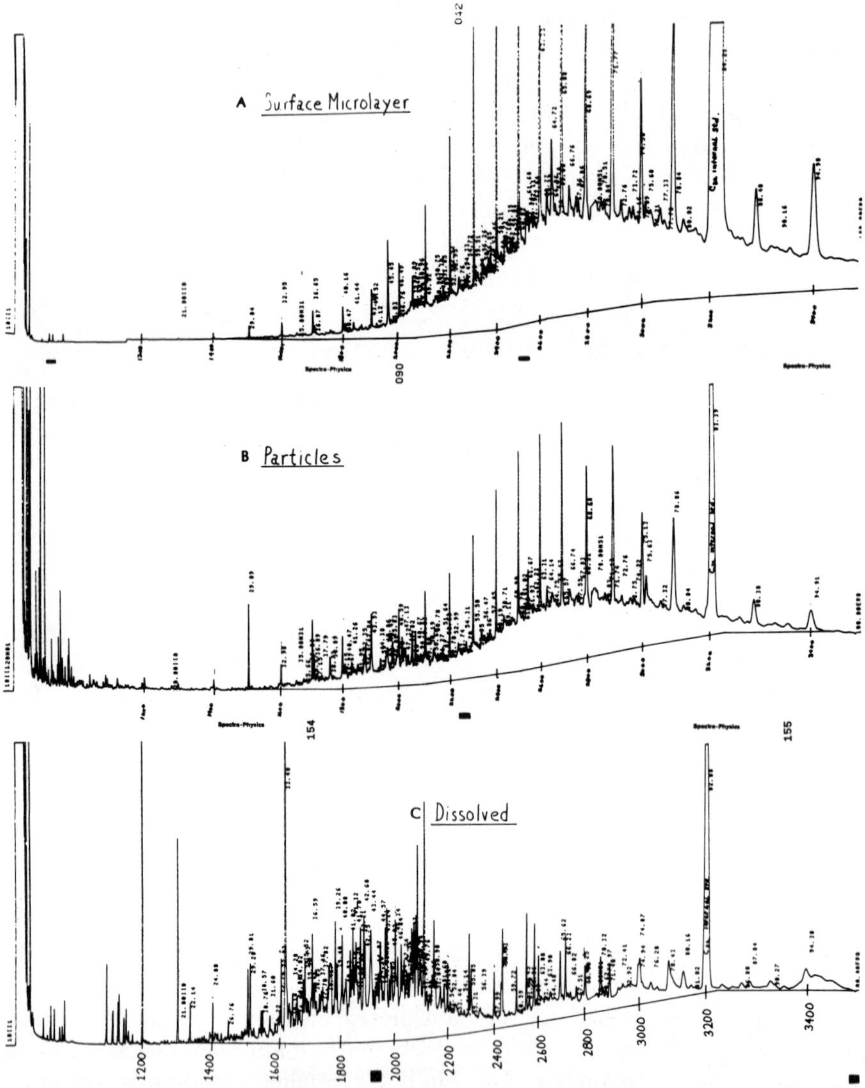

FIGURE 1 F.I.D. gas chromatograms of large volume water samples collected 2 km off Monaco depicting the hydrocarbon composition of various physical-chemical phases. A. Surface microlayer, B. Particles, C. Dissolved. Numbers at the bottom are retention indices of *n*-alkane standards.

TABLE I

Chlorinated hydrocarbons in Var Estuary water. †ng/l=(10^{-9} g/l), ‡pg/l=(10^{-12} g/l).

			Particulate				Dissolved				
Stn.	Salinity ‰	Particles mg/l	PCB †	Lind †	HCB ‡	pp'DDE ‡	PCB †	Lind †	HCB ‡	pp'DDE ‡	% PCB on particles
0	0.081	457	2.25	0.01	25.0	7.0	1.35	1.48	46.5	17.5	63
1	0.463	1,164	2.55	0.02	8.5	21.5	1.75	2.41	38.0	12.5	60
2	9.478	949	1.38	0.02	25.0	2.7	0.22	0.08[b]	13.0	N.D.	86
3	12.905	135	0.80	0.02	26.5	8.0	1.60	0.39	N.D.	N.D.	35
4	14.122	133	0.66	0.02	15.3	N.D.	0.94	0.32	N.D.	N.D.	40
5	24.019	77	0.40	N.D.	37.0	N.D.	1.30	0.39	N.D.	N.D.	22
6	34.634	73	1.00	0.02	5.5	N.D.	55.0[a]	0.24	72.0	N.D.	2

Values excluded from the regressions:
[a]Sample assumed to be contaminated as long term monitoring results show levels of PCB's in the Mediterranean off Monaco average 1.3 ng/l total.
[b]Unusually low recovery in dissolved sample probably due to loss during analysis.
N.D. = not detectable.

TABLE II

Non-halogenated hydrocarbons in Var Estuary Water. †ng/l, ‡μg/l.

		Dissolved			Particulate			% on particles	
Stn.	Miles from river	Biogenic †	Petrol. Peaks ‡	U.R.E. ‡	Biogenic †	Petrol. Peaks ‡	U.R.E. ‡	Total	U.R.E.
0	0	170	0.02	0.26	24	0.17	0.94	71	78
1	0.2	21	0.09	0.19	34	0.13	0.44	66	70
2	0.4	<1	N.D.	0.17	46	0.73	0.64	85	82
3	0.6	<1	N.D.	0.11	66	0.29	0.60	88	84
4	0.8	502	0.05	0.29	753	0.78	0.50	60	63
5	1.0	22	N.D.	0.06	299	0.36	0.34	90	85
6	1.2	2	0.47	0.68	752	0.88	0.36	52	35

Major biogenics were at retention index 1518 on SE-52 g.c. columns and occurred mostly in F_2 and except for Stn. 0 were mostly associated with particles. Other biogenics at 1500, 1700, 3020, 3012, 3040. Particle hydrocarbons were very degraded oil residues C_{14} to C_{35} *n*-alkane boiling range with biogenics on top. Dissolved hydrocarbons were lower boiling oil residues C_{12} to C_{32}. Stn. 6 showed fresh fuel oil contamination.

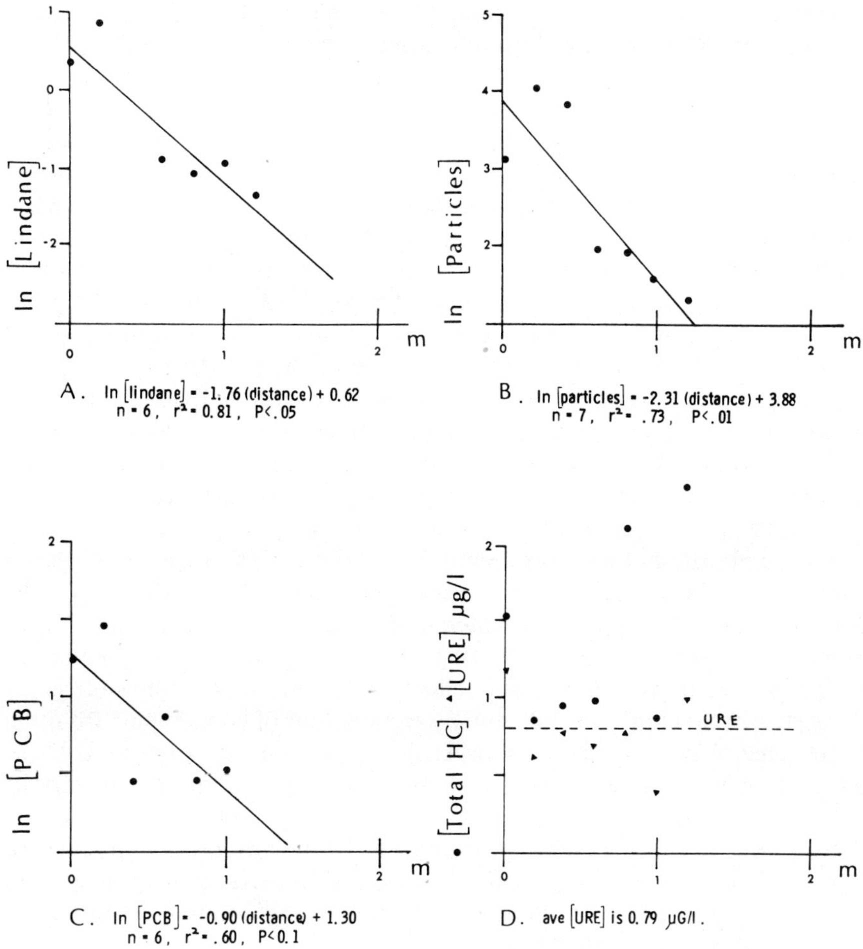

FIGURE 2 Plots of ln of concentration verses distance from River mouth of the Var Estuary. A. Lindane, B. Suspended particles, C. PCB's and D. linear plots of (Total hydrocarbon) and (URE). Data taken from tables.

with atmospheric transport of PCB's. Petroleum products discharged to rivers and sea surface and atmospheric inputs of combustion products are superimposed on levels of biogenic hydrocarbons making distributions highly variable in total concentration. Other as yet unidentified halogenated hydrocarbons were present in these samples and we note the increasing importance of new pollutants in coastal waters.

These and subsequent analyses of Mediterranean water using large volume samples show consistent patterns of distribution of hydrocarbon fractions between ecosystem components. As shown in Figure 1 the surface microlayer shows fresh tar inputs resulting in residues of high boiling point composition with the *n*-alkanes clearly visible over the background of unresolved material. Concentrations fluctuate in response to intermittent inputs of hydrocarbons to the sea surface. Tar-like residues are seen by chemical analysis even in the absence of any visible surface contamination. Near-surface particulate matter is a mixture of adsorbed tar-like components and some biogenics. "Dissolved" phases contain the more soluble (more aromatic) lower boiling residues at more constant concentration than other phases. Large fluctuations in biogenic components result from seasonal injections from plankton communities.[12] Biogenics are seen on top of background pollutant components. Biogenics in our samples totalled 0 to 7% of the hydrocarbon load. Brown and Huffman[13] correlated total hydrocarbon concentration in Mediterranean seawater with total extractable organics when [TEO] exceeded 20 μg/l. Scatter at lower concentrations emphasizes the importance of biogenic hydrocarbon inputs to TEO in coastal and oceanic systems.

The compositional patterns seen in the various physical-chemical phases were consistent with those seen in previous studies of the Mediterranean,[14,15] coastal waters off California[16] and the North Atlantic.[17] We hypothesize that the composition of pollutant hydrocarbons in various phases reflects either different input sources as suggested by Boehm[17] or differential fractionation of components from the same general sources as seen in enclosed ecosystem experiments.[18] In either case it is the distribution of hydrocarbon components in various phases which controls their input to sediments and biota in marine ecosystems. Thus our continuing studies of hydrocarbon fractionation are aimed at elucidating biogeochemical transport processes in marine systems and the development of methods for monitoring long term trends in oceanic levels of selected organic pollutants.

Acknowledgements

The International Laboratory of Marine Radioactivity operates under a tripartite agreement between the International Atomic Energy Agency, the Government of the Principality of Monaco and the Oceanographic Institute at Monaco. Salinity and suspended particulate measurements were done by staff of the Station Zoologique, Villefranche sur Mer, France who gratuitously provided us with time aboard R. V. *Korotneff*. Partial support for this work was provided by the Division of Marine Sciences, UNESCO.

References

1. E. P. Meyers and C. G. Gunnerson, *Hydrocarbons in the Ocean*. Special report of Marine Ecosystem Analysis Program (N.O.A.A., Boulder Colorado, U.S.A. 1976).

2. Intergovernmental Oceanographic Commission. *Global Investigation of Pollution in the Marine Environment.* Group of Experts on Methods, Standards, and Intercalibration. (U.N.E.S.C.O., Paris, 1980).
3. W. D. Garrett, *Limnol. and Oceanog.* **10,** 602 (1965).
4. J. W. Farrington, J. M. Teal, J. G. Quinn, T. Wade and K. A. Burns, *Bull. Environ. Contamin. Toxicol.* **10,** 129 (1973).
5. J. W. Farrington, J. M. Teal and P. L. Parker, in *Strategies for Marine Pollution Monitoring.* Ed. by E. D. Goldberg (J. Wiley, N.Y. 1976).
6. K. A. Burns and J. M. Teal, *Deep Sea Res.* **20,** 207 (1973).
7. S. G. Wakeham and J. W. Farrington, in *Petroleum in the Marine Environment.* Ed. by L. Petrakis and F. T. Weiss (American Chemical Society, Wash. D.C., 1980).
8. K. A. Burns and J. L. Smith, *Aust. J. Marine and Freshwater Res.* **31,** 251 (1980).
9. D. Elder, *Mar. Pollut. Bull.* **7,** 63 (1976).
10. S. P. Pavlou, in *Analytical Techniques in Environmental Chemsitry.* Ed. by J. Albaiges (Pergamon Press, N.Y., 1979).
11. C. T. Chiou, L. J. Peters and V. H. Freed, *Science* **206,** 831 (1979).
12. M. Goutx and A. Saliot, *Mar. Chem.* **8,** 299 (1980).
13. R. A. Brown and H. L. Huffman, *Mar. Pollut. Bull.* **10,** 291 (1979).
14. J. C. Marty and A. Saliot, *Deep Sea Res.* **23,** 863 (1976).
15. R. J. Morris, *Mar. Pollut. Bull.* **5,** 105 (1974).
16. R. W. Risebrough, B. W. de Lappe, W. Walker, A. M. Springer, M. Firestone-Gillis, J. Lane, W. Sistek, E. F. Letterman, J. C. Schropshire, R. Vinc and A. S. Newton, in *Analytical Techniques in Environmental Chemistry.* Ed. by J. Albaiges (Pergamon Press, N.Y. 1980).
17. P. D. Boehm, *Mar. Chem.* **9,** 255 (1980).
18. T. L. Wade and J. G. Quinn, *Mar. Environ. Res.* **3,** 15 (1980).

Petroleum Contamination of Cockburn Sound, Western Australia†

R. ALEXANDER, M. CUMBERS, R. KAGI, M. OFFER, and R. TAYLOR

Petroleum Geochemistry Group, School of Applied Chemistry, Western Australian Institute of Technology, Kent Street, Bentley, 6102, Western Australia

(*Received November 27, 1981*)

Levels of petroleum in refinery effluent streams have been monitored by ultraviolet fluorescence spectroscopy (UVF) and capillary gas chromatography (GC) over a period of one year. Levels in the receiving waters were measured under three different wind patterns by UVF and on one occasion, by both UVF and GC of the saturate fraction. It was concluded that the bulk of the petroleum in the effluent stream is rapidly lost to the atmosphere by evaporation. The hydrocarbons in mullet captured near the outfall and adjudged by a tasting panel to be tainted with petroleum were examined and compared with those from untainted mullet captured in a comparatively pristine area. The tainted mullet contained approximately 15 mg/Kg of aromatics, principally substituted naphthalenes, strongly suggesting that this is the source of the tainting. Preliminary results from analysis of saturate and aromatic fractions of mussels collected 250 m from the outfall are reported and compared with those of mussels from a less contaminated area.

KEY WORDS: Refinery effluents, oil pollution, fish tainting, water analysis.

INTRODUCTION

Cockburn Soundurn Sound is located on the west coast of south-western Australia, immediately to the south of the capital, Perth, a city of just under one million people. Generations of Western Australians have enjoyed the Sound for recreational activities such as swimming, fishing,

†Presented at the Workshop on the Chemistry and Analysis of Hydrocarbons in the Environment, Barcelona, November 1981.

and family outings, and it supports a small professional fishery. From the middle 1950's, heavy industry has been building up on the eastern shores of the Sound, taking advantage of the naturally protected waters for port facilities. Significant developments from the point of view of the present study include the BP oil refinery (1955), the sewerage primary treatment plant at Woodmans Point (1966), the CSBP superphosphate plant (1968), and the KNC plant manufacturing ammonia, nitric acid and ammonium nitrate (1968). The first major signs of deterioration of the Sound appeared in the late 1960's with the commencement of the decline of the *Posidonia* seagrass meadows, so that by the mid 1970's approximately 80% of the original area had died: periodic algal blooms began to appear in the early 1970's. In 1976–1979 a three-year environmental study was carried out to provide a basis for future management of the Sound. This study[1] demonstrated that Cockburn Sound behaves much like a tidal lake, and as a result, steps are in hand to greatly reduce emissions into the Sound waters, particularly of nutrients, heavy metals, and bacteria from sewerage. The present study of petroleum contamination in the Cockburn Sound developed out of a preliminary survey carried out in the context of the major study in 1979. In broad terms, our study which commenced in 1980 had three main objectives. Firstly, to determine the petroleum burden introduced to the Sound in the effluent from the BP refinery. Secondly, to examine the fate of the petroleum in the receiving waters. Thirdly, the impact of the effluent upon the Sound biota was to be investigated. In particular, sensory evaluation trials were to be carried out to establish if fish from the Sound were tainted, and levels of petrogenic material in selected biota were to be determined. The study aimed to provide information about the impact of the present levels of emissions, to determine if reduction to a lower level of emissions was desirable.

The BP refinery at Kwinana was first opened in 1955. It is the only oil refinery in Western Australia, and its capacity of approximately 5·3 million tonnes per year of crude oil and 120,000 tonnes per year of lube oil makes it among the largest in the country. The refinery processes a variety of up to about twenty crudes, principal among which are Iranian, Barrow Island, Kuwait, Murban and Handil. The organisation of water flows in the refinery is outlined in Figure 1. Significant features are that, excepting for tanker ballast water which flows into Circular Separators, all refinery process waters and similar oily waters flow via a sewage system into API-1, a gravity separator, where the bulk of the oil is separated. The effluent water then passes to Sewer Box 33 where it is mixed with a large volume of cleaner water from the Circular Separators and the resulting diluted stream emerges at South Outfall. Prior to entering the Sound, this water is vigorously mixed with cooling water, principally from BP No. 1 (catalytic

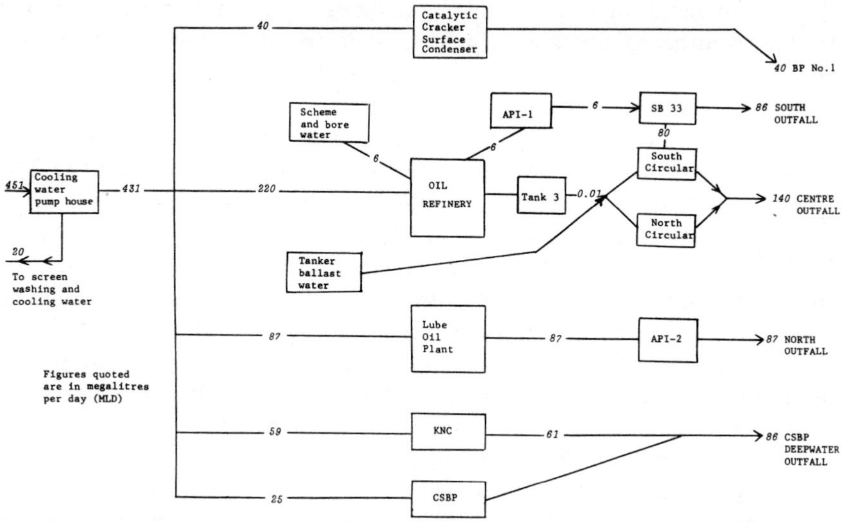

FIGURE 1

cracker). Caustic washes containing mercaptans and phenolics are collected in Tank 3 and are released continuously into the forebay of the Circular Separators, and thence into the Sound via the Centre and South Outfalls. All oil skimmings from separation units are collected in Tank 4, so this may therefore be expected to reflect the composition of the oil in recent effluent. The performance of API-1 has been improved since the initial Cockburn Sound Study with the installation of an oil retention baffle in the forebay and other modifications. Further, a testing station comprising a V-notch weir to measure flows and a sump to induce turbulence to enable collection of a homogeneous sample for analysis is to be constructed at the exit point of API-1. Because of uncertainty in the water flows into Sewer Box 33, considerable uncertainty presently exists in the determination of the petroleum burden to the Sound. These developments will enable the performance of the unit to be monitored directly. BPRK management has a policy of gradual improvement of the performance of API-1, and several options are presently being considered by the company. A number of process modifications are also to be implemented shortly which will greatly reduce the quantities of spent caustic waste, with its contained mercaptans and phenols, which passes into the Sound in the effluents.

Much of the work reported on analysis of waters contaminated with petroleum has involved studies of point sources arising from foundering tankers, oil well blowouts, or similar episodes.[2,3,4] In these situations, it may be possible, although technically challenging, to fully characterise the petroleum source, to observe its behaviour in the water column under the effects of weathering, and to monitor its levels in biota. A refinery effluent stream is a quite different source of petroleum contamination in that a variable oil burden of continuously variable composition, well-mixed with a stream of warm effluent waters, is released into the receiving waters. The fate and effects of oil contained in such an effluent stream is likely to be quite different from that of bulk oil released into receiving waters as in a typical spillage incident. In such circumstances it is possible to regulate levels of oil release by legislation; however, with the limited knowledge of subtle effects of chronic exposure to low levels of oil, the setting of permitted discharge levels is a difficult matter. On the one hand standards should not be set so low as to incur unnecessary costs. On the other hand, standards should not be set so high as to fail to protect the environment to the extent desired by the community.

A further challenge in assessing a situation involving petroleum contamination of a water body is that the processes of weathering modify the nature of the contaminant at a rate which varies greatly with water temperature, wind, sea state, and other environmental factors.[5,6,7] The physical changes which occur are: spreading or extension over the water surface; movement or drifting; evaporation of more volatile constituents; dissolution and dilution of more water-soluble components; emulsification and dispersion; adsorption onto particles; absorption into organisms; and sedimentation. Evaporation and solution are competing processes which leave a similar residue—the higher molecular weight, less volatile, less-soluble components of the oil. The effects of these two processes upon the marine ecosystem are however quite different, as the more soluble low molecular weight material, particularly the aromatics, is more acutely toxic than other hydrocarbons, and when dissolved it is available for uptake by marine organisms. A number of studies[5,6] have indicated that loss of volatile material from a surface film is a rapid process—half-lives of minutes to hours are indicated for compounds in the range C_6–C_{10}, rising to hours and days for material in the range C_{15}–C_{20}. In the weathering of a surface film, indications are that evaporation might occur two to four orders of magnitude faster than dissolution. The vigorous mixing of API-1 effluent with large volumes of water, whilst reducing the formation of an oil slick, may have the effect of greatly increasing levels of finely dispersed and dissolved petroleum, and as a consequence any toxic effects of the effluent may be enhanced. On the other hand, the vigorous mixing would probably also promote evaporation at the mixing point and certainly increase oxygen levels, thereby possibly enhancing the potential for biological degradation.

Chemical degradation processes which typically involve oxygen and often are photoinduced are known to occur to modify petroleum and result in its removal from the water column. Broadly speaking, biota have two effects on petroleum. The first is physical absorption, leading to incorporation into the body tissues or excretion perhaps in a modified form with altered properties, for example faecal pellets which may pass into the sediments where further changes may occur.[5] The second process is biodegradation wherein microorganisms, principally bacteria, yeasts, and filamentous fungi ingest and metabolise the hydrocarbons.[7] More soluble hydrocarbons are biodegraded more rapidly than less soluble hydrocarbons of the same type, and the order of decreasing rate of biodegradation is normal alkanes, branched alkanes, cycloalkanes and aromatics.[6] Certain polycyclic compounds are more resistant to biodegradation. These include the polynuclear aromatics,[8,9] methyl-substituted dibenzthiophenes,[2] triterpanes,[10,11] and sterane-type hydrocarbons.[10] Conditions which favour high rates of biodegradation are warm oxygenated water, high levels of nitrogen and phosphorous nutrients, and continued exposure to petroleum in order that integrated colonies of petroleum-degrading microorganisms are developed.[7] One might therefore expect that biodegradation could presently be playing a major role in controlling levels of hydrocarbons in Cockburn Sound.

Rigorous assessment of petroleum contamination of a water body requires unequivocal identification of petrogenic material and measurement of the levels of this material in the system under examination. In some circumstances this task can be complicated where there is a possibility of hydrocarbon contamination from natural marine seepage or from terrestrial anthropogenic sources. In the case of Cockburn Sound, marine seepages are a most unlikely possibility, and the bulk of the anthropogenic material in the Sound is unlikely to arise from sources other than the BPRK operation, although shipping, pleasure craft, and the sewerage outfall would all be expected to make some contribution. The major complicating factor in the Cockburn Sound situation, apart from the variation in the composition of the contamination source, is the presence of biogenic organics in the water, biota, and sediments: the high levels of nutrients in the warm Sound waters has produced high levels of microorganisms[1] which will not be reduced until the planned reduction of nitrogen input, principally from KNC and the sewerage treatment works, is implemented. Seawater and marine organisms contain a wide range of biogenic hydrocarbons.[12,13,14] There are substantial differences between species: values reported range from 20 to 4800 ppm dry weight. Alkenes are major hydrocarbon components in algae, and polyenes with either straight chains, branches chains, or isoprenoid structures are also prominent. Straight chain alkanes, especially the odd carbon-number series n-C_{15}, n-C_{17}, n-C_{19}, n-C_{21}, are present in marine phytoplankton. Equal amounts of odd carbon-number and even carbon-number alkanes between C_{25} and C_{32} have been reported in bacteria. Branched alkanes are also found in marine organisms: pristane is the most abundant alkane in some fish; however, phytane, its C_{20} analogue, is not commonly found in most biota.

A number of approaches are available for demonstrating contamination of a location with petroleum. One is to carry out measurements, such as ultraviolet fluorescence spectroscopy (UVF), infrared spectroscopy (IR), or gas chromatography (GC), on samples from the subject location. The values obtained can then be compared with those obtained from samples from a pristine, uncontaminated area, and the differences may then be conditionally attributed to pollution. While measurements of these gross parameters may give strong indications of levels of contamination, they suffer from the major difficulty that the analyst does not really know what species are being measured. There are also other deficiencies in this approach: for example, in the UV fluorescence technique, unidentified fluorescent components of the oil, which may perform in the water column quite differently from other components in the oil, are used as a surrogate for the whole oil; again, the IR and GC techniques are subject to interference from biogenic material, and random contamination of samples may not be readily identified. If allowance is made for these deficiencies, these techniques can be very useful, particularly the highly sensitive and rapid UV fluorescence procedure, especially when it is applied in a total fluorescence mode.[15] Gas partitioning[16,17] and HPLC-UVF[18] are particularly useful for investigating more volatile components of petroleum and polynuclear aromatics respectively. The development of high resolution capillary gas chromatography columns of 10^5 theoretical plates and gas chromatograph-mass spectrometer units coupled to a computer with a data system (GC-MS-DS), both afforded quantum jumps in analytical power. When used together they represent the most powerful forensic tool available to the petroleum analyst.

A major task in a study of petroleum contamination is the analysis of petrogenic material in the presence of biogenic material. In the maturation processes of diagenesis and catagenesis, organic material undergoes changes to give compounds which are not found in recent biogenic material, and hence can be used to distinguish petrogenic hydrocarbons in the samples. Gross differences which occur may be summarised as follows:

—Odd-even preference (OEP) of *n*-alkanes.[18] In mature crude oils, the OEP is close to unity in the range C_{10}–C_{20}. Recent biogenic material typically contains predominantly odd numbered *n*-alkanes, especially C_{15}, C_{17}, C_{19} and C_{21}.

— The unresolved complex mixture (UCM).[18] Mature petroleum usually contains a vast number of isomeric and homologous branched and cyclic alkanes. This gives rise to a broad envelope in the gas chromatogram which is particularly evident in weathered samples and contrasts with the alkane fraction of biogenic material which typically comprises a series of individual peaks with baseline resolution.

— Pristane-phytane ratios.[13] Whereas biogenic samples typically contain substantial amounts of pristane, phytane, the C_{20} homologue is noticeably less abundant.

— Aromatics.[2,8,9,19] Aromatic compounds such as methyl-substituted naphthalenes, phenanthrenes, and dibenzthiophenes are prominent components of crude oils and are not typically present in hydrocarbons produced by biota. Polynuclear aromatics are very resistant to biodegradation and this group therefore is a very useful class of market compounds for petroleum contamination.

During maturation, more subtle changes also occur which lead to structural rearrangements and changes in configuration at chiral centres in a variety of classes of compounds.[20] Biogenic acyclic isoprenoid alkanes retain the stereochemistry of the phytol from which they are derived. Material which has undergone maturation shows a loss of this stereospecifity, and by high resolution capillary gas chromatography, the diastereomers can be resolved into a number of peaks. For example, pristane and phytane both appear as a doublet, the front peak of which is due to fossil isomers. As has been pointed out by Gassman,[21] this method has two considerable advantages: it is very sensitive in that it focusses on compounds which are abundant both in biogenic material and in petroleum, and it does not require the use of an expensive GC-MS-DS system. The power of the molecular geochemistry of classes of compounds such as the steranes, diasteranes, and triterpenes applied to environmental problems is presently finding wide use, particularly in fingerprinting petroleum residues so they can be attributed to a particular source.[10,11,22,23]

In the present paper we report the results of our work on petroleum levels in BPRK effluent streams, the levels in the receiving waters, and the impact upon a species of fish, the yelloweye mullet (*Aldrichetta foresteri*) with particular reference to tainting of flesh. Preliminary results of studies of the impact of the effluents upon the mussel (*Mytilus edulis*) are also presented.

SAMPLE COLLECTION

Collection of fish for sensory evaluation and chemical analysis

Fish were collected by netting or trawling on a number of occasions and at a variety of locations: samples described as outfall were collected within a 100 m radius of the BP effluent outfall. Those described as Warnbro Sound were captured outside Cockburn Sound, approximately 10 Km. to the south. Fillets were obtained from a number of fish, wrapped in clean aluminium foil, and frozen. Individual fillets were used for sensory evaluation, but in the work reported in this paper, aggregates of fillets from three individual fish were used for chemical analysis.

Collection of Mussels

On 15 July, 1981, samples of mussels were collected from a bottom site in 2·5 m of water, approximately 250 m southwest of the outfall. On the same day, a control sample was collected from the far northwest extremity of the Sound at a depth of 50–100 cm on Beacon Head Dolphin.

Collection of water samples

Effluent waters Water samples were collected periodically from outfall streams and from the combined effluent stream. The combined effluent stream is not homogeneous, values to the south usually being much higher than those to the north. The precise sampling locations varied with changes in tides and sea states, and, on some occasions, point samples were taken whereas on other occasions samples were collected by traversing the effluent stream. Averaging the values quoted in Table 2 should however give a crude estimate of the petroleum outfall as a whole. At the time of each water sampling, a sample of oil from Tank 4 was also collected for use as a standard.

Cockburn Sound waters Collections of Cockburn Sound waters were carried out on three occasions: 31 March, 15 July and 16 October, 1981. At each site, seawater (3.3 l) was collected at a depth of 50 cm by use of a 3.8 l brown glass bottle in a cradle and buoy sampling device. A number of samples were also collected by divers 30 cm from the bottom. Sample sites were positioned on a time/boat speed basis for those between the BP outfall and the James Point Beacon; others were located by compass triangulation. In order to minimise the risk of contamination, the boat was driven towards a sample site and the motors were stopped and raised so that the boat drifted onto the station. Water samples were also collected at Beacon Head Dolphin at the northwest extremity of the Sound.

SAMPLE TREATMENT

Treatment of water samples Samples of 3.3 l were collected in carefully cleaned 3.8 l brown glass bottles. Dichloromethane (50 ml) was added immediately, then, on return to the laboratory, the water was extracted with further quantities of dichloromethane (2 × 30 ml). The extracts were separated with a pipette, dried with anhydrous sodium sulphate and the dichloromethane was then removed by careful evaporation to a volume of 10 ml. Fluorescence measurements were carried out on this sample, then the volume was reduced to approximately 1 ml and selected samples were fractionated by chromatography on a 6 cm × 0.6 cm silicic acid column. A saturated fraction was obtained by eluting with pentane and an aromatic fraction was obtained by eluting with 5% ether in pentane. The aromatic fraction was concentrated to 20 μl by careful evaporation of the solvent, then it was further purified by chromatography on a Size A Merck Silica 60 pre-packed liquid chromatography column. The fraction eluted with 15% dichloromethane in hexane at 2 ml/min. was collected as aromatic compounds. This material was concentrated to approximately 100–400 μl by evaporation in a stream of nitrogen at 40°C, then set aside for capillary gas chromatography and GC-MS-DS studies.

Treatment of biota samples—fish and mussels. Frozen fish samples were selected for analysis on the basis of the results from sensory evaluation. Mussels were washed with water, shucked, drained to constant weight, and frozen. Flesh (100 g) and KOH (100 g) were weighed directly into a 500 ml RB flask then 70% methanol in water (200 ml) was added and this mixture was refluxed for 12 hours under nitrogen, then cooled. The digest was then extracted with pentane (2 × 200 ml) and the extracts were transferred to a 500 ml RB flask. The pentane was carefully evaporated and KOH (50 g) and 70% methanol in water (100 ml) was added and the mixture was refluxed for a further 4 hours. The digest was then extracted with pentane (2 × 100 ml), the pentane extracts were washed with water (5 × 500 ml), 1% HCl (200 ml), and again with water (2 × 500 ml), then evaporated to approximately 5 ml. The residue was applied to a 6 cm × 1 cm column of silicic acid with a short plug of anhydrous sodium sulphate on top to remove traces of water. The saturate fraction was eluted with pentane and the aromatic fraction with 5% ether in pentane.

The fractions were concentrated to approximately 100 μl by careful evaporation of the solvent. The aromatic fraction was further purified by liquid chromatography in the manner described above for the water extracts. In one trial, 2,3-dimethylnaphthalene (5 mg) was added to the mixture prior to digestion; subsequent GLC showed that 80% of this material was retained through the treatment procedures.

ANALYTICAL METHODS

General Prior to use, clean glassware was washed with distilled water, dried, and rinsed with purified dichloromethane. Solvents were prepared by distillation of commercial analytical reagent material, followed by analysis for contamination by capillary gas chromatography. Fluka 100-mesh silicic acid was washed with ether, then pentane, dried at 100°C, then activated by heating at 200°C. The column was eluted with pentane prior to use. Blanks were routinely carried out on all procedures. At each time of sampling, a standard oil sample was collected from Tank 4 which accumulates the skimmings from all of the oil and water separation units. This tank is routinely pumped out at least once a day; the oil returning to the refinery process stream.

Ultraviolet fluorescence Ultraviolet fluorescence determinations were carried out using an Hitachi 650-10M fluorescence spectrophotometer with an excitation wavelength of 310 nm and emissions were measured at 374 nm. Under these conditions, the standard Tank 4 oil showed a linear response in the range of 0–100 μg/ml. The response ratio of chrysene to each standard sample from Tank 4 was determined. In one trial experiment, Tank 4 oil was added to seawater and extracted as described in the procedure. The fluorescence measurements on the extract were within 10% of those of the standard oil. Values during the study ranged between 3.5 and 10. With water samples of 3.3 l, the limit of detection of the method is approximately 1 μg or a concentration of 0.3 μg/l in the water sample.

Liquid chromatography Liquid chromatography was carried out using a Varian 8500 dual pump LC system equipped with a Variscan variable wavelength UV detector set at a wavelength of 276 nm. Preparative chromatography to prepare samples for further analysis was carried out using a Merck Lobar Grobe A (240-10) Lichroprep Si 60 40–63 μm column. The solvent used was hexane-dichloromethane (7:1) at 2 ml/min.

Capillary gas chromatography Capillary gas chromatography was carried out using a Varian Series 1400 gas chromatograph equipped with an SGE Grob-type splitless injector system, a detector modified to reduce dead volume and an HP 3390 integrator recorder. A 25 m × 0.2 mm ID, SP-2100, fused silica column was used with hydrogen as carrier gas at a linear velocity of 35 cm/sec. Two temperature programs were used: in most cases the regime was 60° to 280° at 4°/min., however, when resolution of isoprenoid diastereomers was required, the program used was 60° to 100° at 2°/min., 100° to 160° at 1°/min. then 160° to 280° at 4°/min. Levels of hydrocarbons were estimated by comparing the levels of benchmark peaks in the C-10 to C-30 region with those in the Tank 4 standard; alternatively, where this was not appropriate, quantitation was carried out using detector response calculations. Injection reproducibility was found to be better than ±2%; injector fractionation was insignificant below C-28, but losses increased to approximately 20% at C-32 and 60% at C-36. Gas chromatograms of samples containing very low levels of material showed peaks at retention times less than that of C-12 due to volatile contaminants derived from the silicic acid.

Capillary gas chromatography—mass spectrometry GC-MS-DS was carried out using a Hewlett Packard 5985B GC-MS-DS system. Chromatography conditions were identical to those described above. The mass spectrometer was operated in the E.I. mode at 70 eV. GC-MS for sterane and terpane analysis was carried out using the HP SIM program; all other measurements were carried out using the standard program.

RESULTS AND DISCUSSION

Levels of petroleum in the effluent streams Results obtained for analysis by UV fluorescence of petroleum in the effluent streams are presented in Table I. There is clearly considerable variation from time to time in the levels in each stream as one would expect with changes in refinery operations. Using the values from this table, it is possible to estimate the contribution to the petroleum burden of the Sound from each of the four treatment units as shown in Table II. From these values it is apparent that API-1 (South Outfall) is usually the major source of petroleum to the Sound, and further, the wide variation of values might be construed as reflecting upon the effectiveness of the unit. The discrepancy between estimates of the petroleum burden to the Sound based on outfall concentrations and estimated water flows for the four outfall pipes on the one hand, and those for the combined effluent stream on the other hand is

TABLE I

Petroleum burdens ($\mu g/l$) of BP refinery outfalls. These were determined by UV fluorescence and are quoted in terms of the Tank 4 Standard. In order to convert these values into terms of chrysene equivalence, divide by the chrysene equivalence factor. Refinery operational statistics are given in detail in Appendix 1.

Outfall	14 Jan. 1981	25 June 1981	8 July 1981	15 Sept. 1981	2 Oct. 1981	8 Oct. 1981	16 Oct. 1981
North	15	30		60		15	
				50			
Centre	40	70		260		20	880
				250			
South	10,000	3,000	9,000	2,000	705	300	1,550
			8,500	2,600			
BP No. 1:							
(Cat cracker)	8	30		10		5	
				10			
						(20 m)(70 m)[a]	(5 m)
Combined effluent			220	250	484	17 16	1,390
stream			150	250	25	55 35	790
			270			27	
			360			60	
Chrysene equivalence factor (divide by this number)	NA	10	10	7	3.5	3.5	4

[a]These samples were collected at points in the combined effluent stream moving from North to South in each case. Averaging the values should give a crude approximation of the petroleum burden of the stream as a whole.

TABLE II

Estimates of contribution of each of the BPRK treatment units to the petroleum burden of Cockburn Sound

Unit		Cat. cracker cooling water	API-2	Circular separators	API-1[a]	Combined effluent stream
Outfall		BP No. 1	North	Centre and South	South	—
Volume of flow (megalitres per day)		40	87	220	6	353
Date						
14/1/81	μg/l	8	15	40	142,000	
	Kg/d	0.32	1.3	8.8	857	
25/6/81	μg/l	30	30	70	43,000	
	Kg/d	1.2	2.6	15.4	252	
8/7/81	μg/l				120,000[b]	250
	Kg/d				750[b]	88
15/9/81	μg/l	10	55	255	33,000	250
	Kg/d	0.4	5	55	178	88
2/10/81	μg/l				9,400	250
	Kg/d				56	88
10/10/81	μg/l		15	20	4,000	35
	Kg/d		1.3	4	24	12
16/10/81	μg/l			880	10,500	1090
	Kg/d			194	63	384

[a]Estimated using API conc. = 1/6(86 × South Outfall Conc) − (80 × Centre Outfall Conc).
[b]Estimated using average values for Centre Outfall concentrations.

L

perplexing, but it may be accounted for by a variety of factors. The difficulty in obtaining a representative sample of the combined effluent stream is manifest from the values presented in Table II. However, the variability observed underlines the importance of the need to obtain more accurate data on the water flows in the refinery and on the performance of API-1, as is proposed by BPRK management. The values obtained on 16th of October are interesting, and merit special comment as they illustrate the difficulties associated both with control of refinery effluents and their monitoring. When the effluent from Circular Separators was checked early in the morning by BPRK staff, it was quite clean; however, at the time of sampling we noted that it was obviously contaminated with petroleum and the analytical results confirmed this observation. As is shown in Table III, there is clearly a discrepancy between the petroleum levels obtained by UVF and those obtained by GLC for samples collected at the outfalls and in the combined effluent stream: the GLC values being higher by a factor of 2.5 to 3.5. The fact that the saturate fractions from both Centre Outfall and South Outfall were virtually identical by GLC strongly suggests that a pulse of petroleum was passing through Circular Separators and being released via these two outfalls at the time of sampling, and further, that this material had a much lower level of fluorescent material than that contained in the Tank 4 standard collected at the time of sampling.

Levels of petroleum in Cockburn Sound waters Samples of Cockburn Sound waters were collected on 31 March, 1981, 15 July, 1981 and 16 October, 1981. The results of analysis of these samples by UVF, using a Tank 4 standard collected on the same day, are presented in Figure 2. One cannot help but be struck by the apparent rapid diminution of the petroleum levels as the effluent plume moves out into the Sound. There are two major limitations inherent in the UVF method. The principal one is the fact that one uses unidentified fluorescent components of the oil, which may of course perform quite differently in the water column to other components of the oil, as a surrogate for the whole oil. There is also the possibility that refinery effluent hydrocarbons could vary in fluorescent properties over time. We therefore elected to analyse the fractions obtained from a suite of water samples by capillary GLC: selected chromatograms of saturate fractions are presented in Figure 3. Unfortunately, this exercise was carried out using samples collected on 16 October, 1981, when there was clearly a major discontinuity in the dominance of API-1 effluents over those from other treatment units. On inspection of the traces, one cannot however help but be struck by the apparent rapidity with which hydrocarbons of increasing molecular weight

TABLE III

Comparison of UVF and GLC values ($\mu g/l$) in water samples

Date	Station number	UVF (tank 4)	GLC Saturates	GLC Aromatics
31 March 1981	1	60	40	15
	2 top	20	25	10
	2 bottom	12	6	2
15 July 1981	Combined effluent stream	322	225	55
	10	32	27	4
	13 bottom	9	20	Not detectable
16 October 1981	South	1,550	5,000	700
	Centre	880	2,200	
	Combined effluent stream (south)	790	1,700	350
	1	23	18	4
	2	5	4	
	3	5	4	
	5	2	1	
	12	9	6	
	13 bottom	6	8	

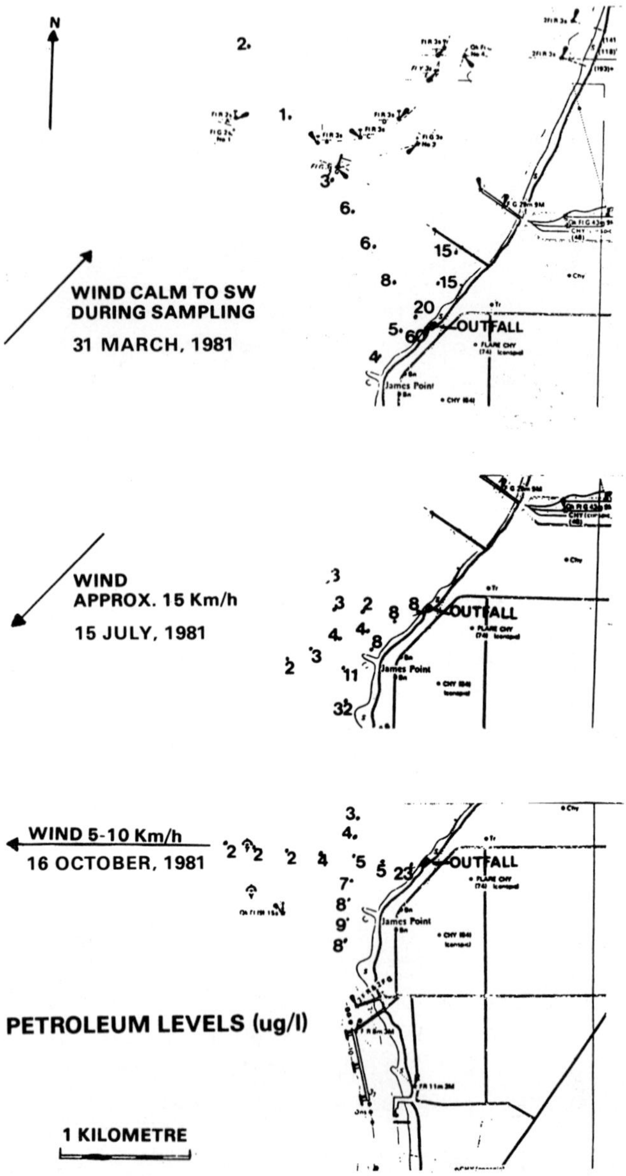

FIGURE 2 Petroleum levels in Cockburn Sound waters determined by UVF.

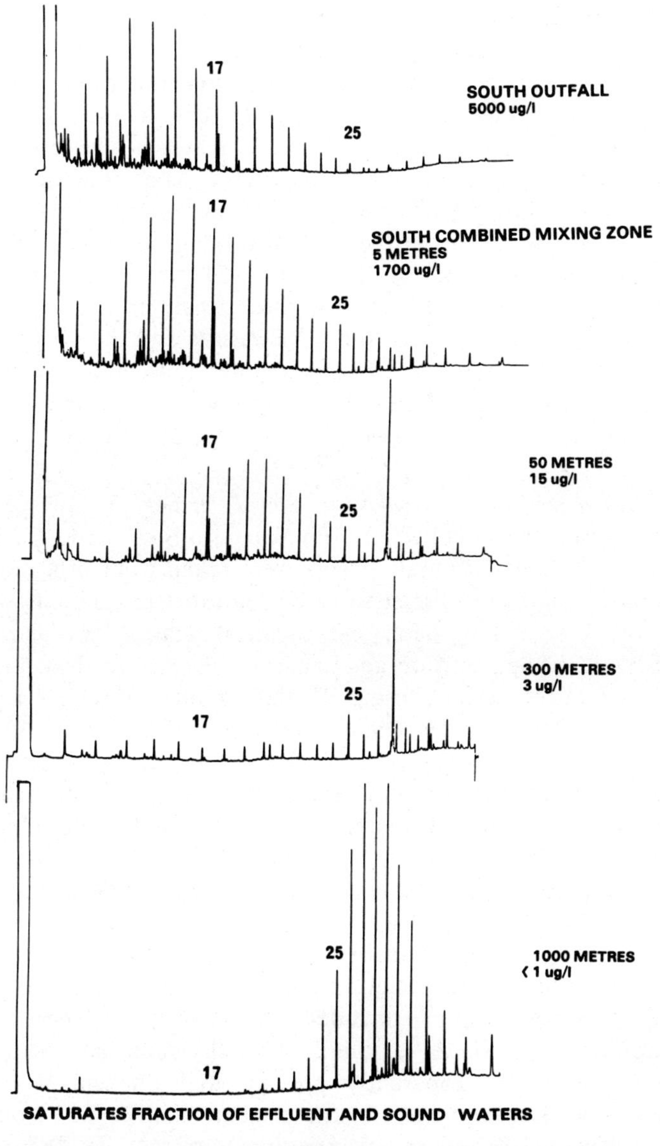

FIGURE 3 Capillary GC's of saturates fractions of effluent and Cockburn Sound waters collected on 16 October, 1981.

are depleted with increasing distance from the outfall. The appearance of the *n*-alkanes in each chromatogram as a smooth envelope with no significant OEP suggests that the saturate fraction of the extracts is comprised largely of petrogenic material. There is also no evidence that the loss of hydrocarbon is due to weathering processes other than evaporation. Indeed, the pristane/n-C_{17} and phytane/$n-C_{18}$ ratios are either constant or else they actually decrease, which is the reverse of the trend expected if biodegradation were playing a major role, and suggests instead that there has been a change in the initial ratios of these components in the petroleum in the effluent leaving the outfalls. Fractions of selected samples collected on 15 July, 1981 were also analysed by capillary GLC, and the results of all these analyses are presented in Table III. The agreement between the UVF results and the GLC results is surprising, considering the fact that probably quite different processes affect the fluorescent material and the alkanes which give rise to the gas chromatograms. It is however, rather reassuring to an analyst.

Levels of hydrocarbons in Cockburn Sound mullet Yelloweye mullet (*Aldrichetta foresteri*) were collected from Warnbro Sound and from within a 100 m radius of BPRK Outfall. Fillets were taken from fish and frozen: individual fillets were used for sensory evaluation trials, details of which will be reported at a later date, and chemical analysis was carried out both on individual fillets and on aggregates of fillets from three individual fish. Only the latter results, using Outfall mullet adjudged as highly tainted with petroleum by the testing panel, are presented in this paper. Two sequential caustic digests were carried out to remove ester material, followed by a preliminary clean-up using a silicic acid column which gave a saturate fraction eluted with pentane. The aromatic fraction eluted with ether in pentane 1:20 was then further purified by HPLC using a Merck Silica 60 column which had been calibrated using standard compounds. The GLC traces shown in Figure 4 of the aromatic fractions of Outfall mullet, effluent water, and Warnbro mullet leave little doubt that the Outfall mullet contains substantial amounts of aromatic material; readily identifiable components include methylnaphthalenes (MN), dimethylnaphthalenes (DMN), and trimethylnaphthalenes (TMN). Phenanthrenes also are apparently present, and one might expect if benzenoid aromatics were present in the original sample, they would have been lost or depleted in the sample treatment procedures. Traces of the gas chromatograms of the saturate fractions of the Outfall mullet, and Warnbro mullet are shown in Figure 5. At first blush, one might conclude that these simply reflect the trend shown with the aromatic fractions. That is, gross contamination of the Outfall sample with petrogenic material.

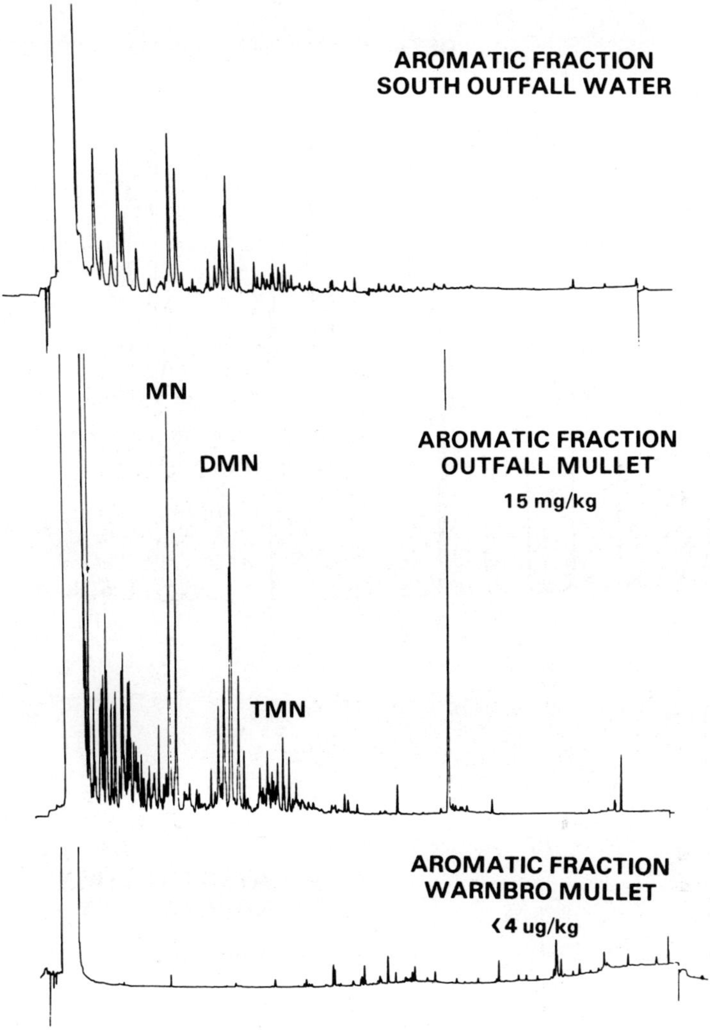

FIGURE 4 Capillary GC's of aromatic fractions of South Outfall water, and mullet extracts.

Closer inspection, however, provides some illuminating insights. The increased level of odd-numbered *n*-alkanes over the largely petrogenic even-numbered homologues is a clue, but the true nature of the situation is revealed in the insets to Figure 5. By far the greatest proportion of pristane in this mullet aggregate sample is biogenic in origin: in the high resolution GC of the Outfall mullet, the peak due to pristane (A) shows

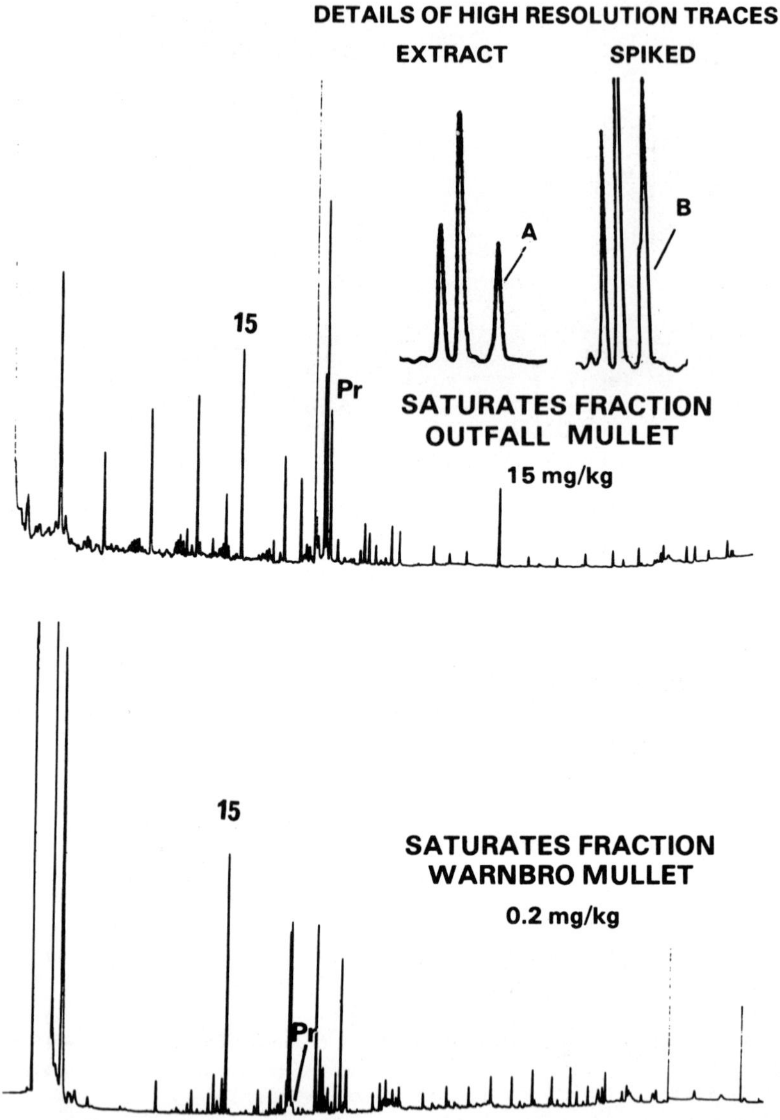

FIGURE 5 Capillary GC's of saturates fractions of mullet samples. Insets are high resolution chromatograms of the pristine peak of the extract (left) and the extract spiked with Tank 4 standard (right).

only a small inflection at the front due to the petrogenic diastereomers. In the trace of the same sample spiked with aromatics from Tank 4, the inflection in the peak B is much more pronounced. The high level of biogenic hydrocarbons present in Cockburn Sound mullet suggests that the nutrient-rich effluents from CSBP, KNC, and WPTP are greatly increasing the production of biogenic hydrocarbons in the Sound, and these are appearing in the fish. The enhanced levels of aromatic components in the fish suggests that these more soluble compounds are being absorbed by fish swimming in the warm effluent waters, and that it is these aromatic compounds which are giving rise to the tainting problem found with fish captured in the vicinity of the outfall..

Levels of hydrocarbons in Cockburn Sound mussels At the time of water sampling, on 15 July, 1981, samples of mussels were collected from Beacon Head and from a bottom site approximately 250 m southwest of the outfall in water approximately 2.5 m deep. The organic extracts were fractionated into saturate and aromatic fractions by column chromatography using silicic acid as described above for the mullet samples.

Saturate fractions Capillary gas chromatograms of saturate fractions of mussels from the Beacon Head and the BPRK outfall are shown in Figure 6. These chromatograms were obtained under identical conditions and are therefore directly comparable. They show little resemblance to those of the water samples shown in Figure 3. The chromatogram of the outfall mussels shows the typical UCM pattern characteristic of weathered petroleum, with peaks possibly due to biogenic hydrocarbons superimposed on the broad envelopes. In order to establish that the UCM was in fact hydrocarbons, the mass spectrum was examined at over a dozen points: at each point the mass spectrum showed only a pattern typical of alkanes. Estimation of the concentration of saturates in the mussels on the basis of GC detector response gave values of 50 mg/Kg for the outfall sample and 5 mg/Kg for the Beacon Head sample.

Aromatic fractions The aromatic fraction obtained from column chromatography on silicic acid was further purified by HPLC on silica gel. Capillary gas chromatograms of the purified aromatic fractions of the outfall and Beacon Head samples are shown in Figure 7. In the trace of the outfall samples, some petroleum aromatics are clearly recognisable, in particular the methyl-, dimethyl- and trimethyl-naphthalenes; however the trace is again characterised by a broad UCM. The chromatogram for the Beacon Head mussels is quite different showing no UCM, and examination of the sample by GC-MS (detection limit approximately

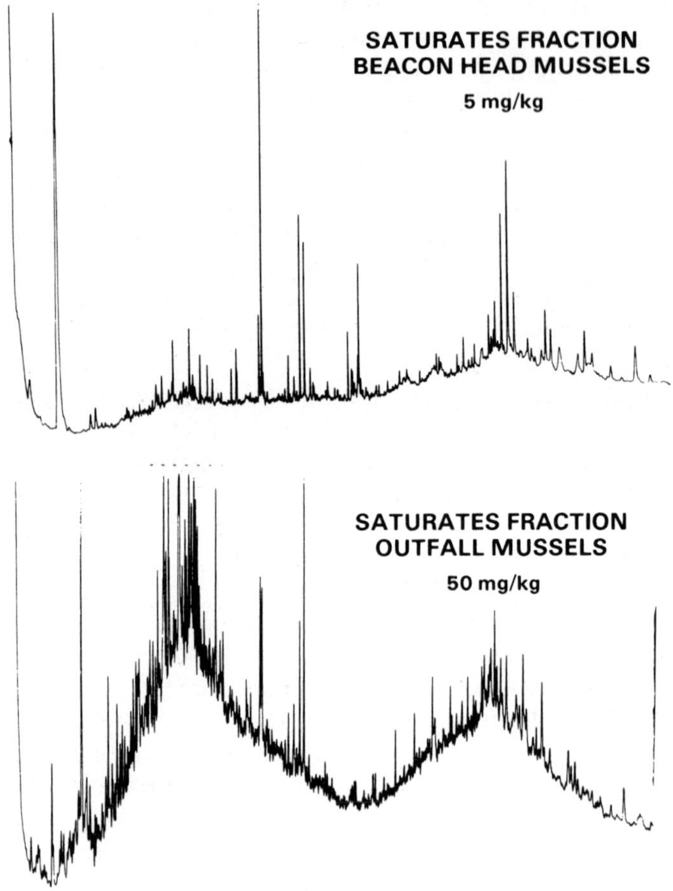

FIGURE 6 Capillary GC's of saturates fractions of mussels from Beacon Head and the BPRK outfall.

4 μg/Kg) identified no aromatic compounds. On the basis of detector response, the chromatogram for the outfall mussels indicates 5 mg/Kg of material.

CONCLUSION

The results of this study focus attention on the API-1 treatment unit of BPRK as the major source of petrogenic hydrocarbons in Cockburn Sound, and indicate that any attempt to improve the effluent quality

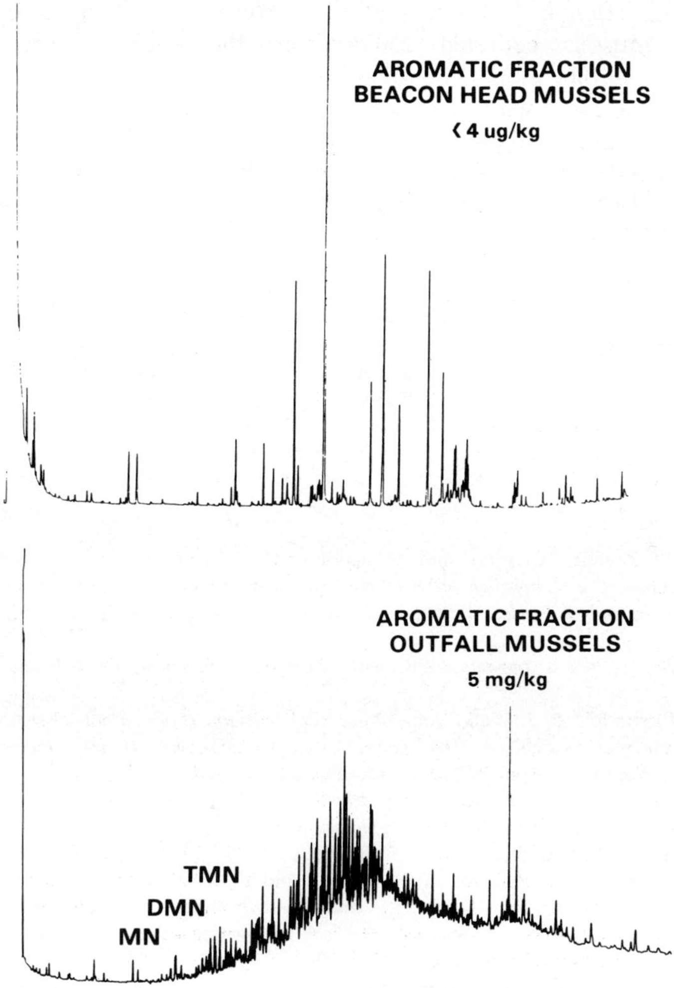

FIGURE 7 Capillary GC's of aromatic fractions of mussels collected at Beacon Head and BPRK outfall.

should be directed at this source. Further, a large proportion of the more volatile components of petroleum released into the Sound is quite rapidly lost by evaporation into the atmosphere; levels usually approaching those of the background in the region 500–1000 m from the outfall. Petroleum-tainted fish captured in the immediate vicinity of the outfall contain petrogenic aromatics similar to those found in the effluent waters. The strong suggestion is therefore that these aromatics are the cause of the tainting. Mussels collected 250 m from the outfall appear to be accumulating the more persistent branched and cyclic alkanes and aromatics and more work is in hand on detailed analysis of these samples.

Cockburn Sound represents an interesting and complex problem in environmental science which is still unfolding as industry continues to upgrade its effluent disposal practices.

References

1. Report No. 2 of the Cockburn Sound Environmental Study 1976–1979, Western Australian Department of Conservation and Environment, October, 1979.
2. F. Berthou, Y. Gourmelun, Y. Dreano and M. Friocourt, *J. Chrom.* **203,** 279 (1981).
3. R. J. Law, *Mar. Poll. Bull.* **9,** 321 (1978).
4. G. W. Dimock, J. L. Lake, C. B. Norwood, R. D. Bowen, E. J. Hoffman, B. Kyle and J. G. Quinn, *Environ. Sci. Technol.* **14,** 1472 (1980).
5. R. F. Lee, "Processes Affecting the Fate of Oil in the Sea" in *Marine Environmental Pollution:* Vol. 1, Hydrocarbons, Edited by R. F. Geyer, Elsevier, Amsterdam, pp. 337–351 (1980).
6. C. D. McCauliffe, "Dispersal and Attraction of Oil Discharged on a Water Surface" in *Proceedings of a Symposium on Fate and Effects of Petroleum Hydrocarbons in Marine Ecosystems and Organisms.* Edited by D. A. Wolfe, Pergamon Press, Oxford, pp. 19–35 (1977).
7. FAO Report No. 6: *Impact of Oil on the Marine Environment*, FAO, Rome pp. 22–41 (1977).
8. R. E. Cripps and R. J. Watkinson, "*Polycyclic Aromatic Hydrocarbons*: Metabolism and Environmental Aspects" in *Developments in Biodegradation of Hydrocarbons—1.* Edited by R. J. Watkinson, Applied Science, London, pp. 113–134 (1978).
9. J. M. Neff, *Polycyclic Aromatic Hydrocarbons in the Marine Environment—Sources, Fates, and Biological Effects.* Applied Science, London (1979).
10. J. Albaiges and P. Albrecht, *Intern. J. Environ. Chem.* **6,** 171 (1979).
11. M. M. Quirk, R. L. Patience, J. R. Maxwell and R. A. Wheatley, "Recognition of the Sources of Isoprenoid Alkanes in Recent Environments" in *Proceedings of the International Congress on Analytical Methods in Environmental Chemistry.* Edited by J. Albaiges, Pergamon, Oxford, pp. 23–32 (1980).
12. A. Saliot, "Natural Hydrocarbons in Sea Water" in *Marine Organic Chemistry.* Edited by E. K. Duursma and R. Dawson. Elsevier, Amsterdam, pp. 327–374 (1981).
13. C. B. Koons and P. H. Monaghan, "Inputs of Hydrocarbons from Seeps and Recent Biogenic Sources" in *Proceedings of the Symposium on Sources, Effects and Sinks of Hydrocarbons in the Aquatic Environment.* Washington, D.C., August 1976. Edited by F. T. Weiss, American Institute of Biological Sciences, pp. 85–109.
14. E. D. S. Corner and R. P. Harris, "Hydrocarbons in marine zooplankton and fish: Part 1" and K. J. Whittle and P. R. Mackie, "Hydrocarbons in marine zooplankton and fish: Part 2" in Effects of Pollutants in Aquatic Organisms. Edited by A. P. M. Lockwood, Cambridge University Press, Cambridge, pp. 71–105 (1976).
15. P. John and I. Soutar, *Chem. in Brit.* **14,** 278 (1981).
16. C. D. McCauliffe, "The Multiple Gas-Phase Equilibration Method and its Application to Environmental Studies" in *Petroleum in the Marine Environment.* Edited by L. Petrakis and F. T. Weiss, A.C.S. Washington, D.C., pp. 192–218 (1980).

17. J. S. Warner, R. M. Riggin and T. M. Engel, "Recent Advances in the Determination of Aromatic Hydrocarbons in Zooplankton and Macrofauna" in *Petroleum in the Marine Environment*. Edited by L. Petrakis and F. T. Weiss, A.C.S. Washington, D.C., pp. 87–103 (1980).
18. C. S. Johnston, "Sources of Hydrocarbons in the Marine Environment" in *Oily Water Discharges*. Edited by C. S. Johnston and R. J. Morris, Applied Science, London, pp. 41–62 (1980).
19. J. W. Farrington, "An Overview of the Biogeochemistry of Fossil Fuel Hydrocarbons in the Marine Environment" in *Petroleum in the Marine Environment*. Edited by L. Petrakis and F. T. Weiss, A.C.S. Washington, D.C., pp. 1–22 (1980).
20. A. S. MacKenzie, R. L. Patience and J. R. Maxwell, "Molecular Changes and the Maturation of Sedimentary Organic Matter" in *Origin and Chemistry of Petroleum*. Edited by G. Atkinson and J. J. Zuckerman, Pergamon, Oxford, pp. 1–32 (1981).
21. G. Gassman, *Mar. Poll. Bull.* **12,** 78 (1981).
22. M. Dastillung and P. Albrecht, *Mar. Poll. Bull.* **7,** 13 (1976).
23. J. Albaiges "Fingerprinting Petroleum Pollutants in the Mediterranean Sea" in *Analytical Techniques in Environmental Chemistry*. Edited by J. Albaiges, Pergamon, Oxford, pp. 69–81 (1980).

Polycyclic Aromatic Hydrocarbons in Exhaust Particles Emitted from Gasoline and Diesel Automobile Engines†

R. HAGEMANN, H. VIRELIZIER, D. GAUDIN and A. PESNEAU

Section d'Etudes et d'Analyse Isotopique et Nucléaire, Centre d'Etudes Nucleaires de Saclay, 91191 Gif sur Yvette Cedex, France

(*Received November 27, 1981*)

This study concerns the polycyclic aromatic hydrocarbons (PAH's) emission from gasoline and diesel automobile engines.

The sampling procedure, the extraction and the analytical method are presented. Using the gas chromatography–mass spectrometry technique it is possible to detect and quantify easily about 15 PAH's. Data obtained on exhaust particles collected from gasoline and diesel automobile engines are presented and discussed.

KEY WORDS: Automobile-engines, GC–MS, analysis, polycyclic aromatic hydrocarbons (PAH's).

INTRODUCTION

Analytical methods for the separation, identification and quantitative determination of polycyclic aromatic hydrocarbons (PAH's) in a wide range of products have been under investigation for more than twenty years, as many of these compounds present mutagenic and carcinogenic properties.

PAH's are formed in small amounts during combustion of aliphatic and aromatic hydrocarbons both in domestic heating systems and in gasoline or diesel automobile engines. The purpose of this study concerns the

†Presented at the Workshop on the Chemistry and Analysis of Hydrocarbons in the Environment, Barcelona, November 1981.

characterization of the emission of PAH's both by gasoline and diesel automobile engines.

SAMPLING

The PAH's emitted by automobile engines are in part fixed on the particles which are present in the exhaust gas and in part in the gas itself. Two sampling methods are currently used. The first uses a glass cooler in front of a microglass fiber filter in which the PAH's are collected both from the condensed phase and from the trapped particles, this sampling method certainly presents a high collecting efficiency.

The second method, used in this study, is the one recommended by the EPA in the United States in which only the particles are considered. They are collected on filters mounted in a constant volume sampler. A diagram of the system is shown on Figure 1, its main advantage is that the exhaust gas is diluted and cooled by air, as it is for automobiles in normal use, and it may be expected that reactions which may occur in the hot gas are reduced.

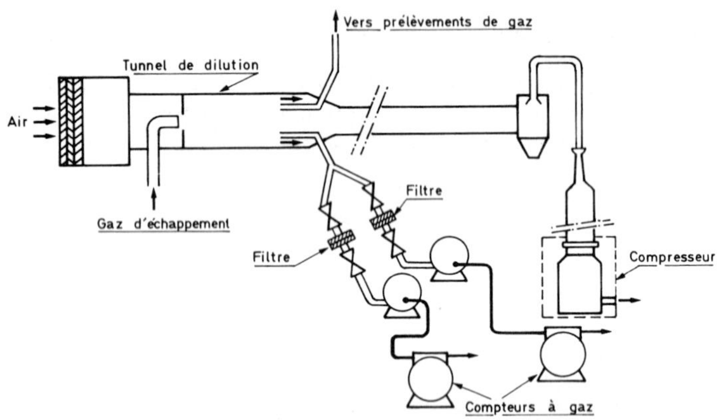

FIGURE 1 Schematic configuration of a Constant Volume Sampler (CVS).

The collected particles are those emitted during a well defined functioning cycle of the engine, the cycle which has been used is the so called U.S. FTP 75 cycle. Two different constant volume samplers have been used, one for gasoline engines, the other for diesel engines. The teflon filters on which the particles are collected have a diameter of 48 mm and a porosity of 1 μm.

EXTRACTION

The filter is first put into a mixture of 8 ml of DMSO (dimethyl sulfoxyde CH_3SOCH_3) and 2 ml of cyclohexane for about 12 hours, then submitted to ultrasonic stirring at room temperature for about 15 minutes. After filtration of the solution, 10 ml of distilled water and 20 ml of cyclohexane are added. This step allows the separation of the PAH's, soluble in cyclohexane, from alcohols, aldehydes and acids which are soluble in water. Before the above treatment 450 ng of 9 chloro anthracene and 1 μg of triphenylbenzene are added on the filter. These two compounds are used as internal standards. The yield for the extraction and purification steps have been measured equal to 50% for coronene.

The cyclohexane solution is evaporated under a stream of nitrogen to reduce its volume to 0.5 ml. This extract is stored in the dark at $-10°C$ before analysis.

ANALYTICAL METHOD

Identification and quantitative determination of PAH's are performed using the gas chromatography–mass spectrometry (GC–MS) technique. This method is preferred to the high performance liquid chromatography technique which has also been tested in the laboratory, and it has already been used by several authors.[1–4]

A Riber 1010 quadripole mass spectrometer connected to a GIRDEL 3000 chromatograph and to a computer system is used for the analysis.

The chromatographic column is a CP tm Sil 5, glass capillary column from Chrompack, the length is 25 metres the diameter 0.25 mm. The glass injector is of the Ros type.

The temperature of the injector and of the glass connecting line to the mass spectrometer is 300°C. The temperature of the column is programmed from 180°C up to 300°C at a rate of 5°C or 10°C per minute.

The identification of each compound (or group of non separated isomers) is possible from the mass spectrum which is obtained at each top of the chromatographic peaks. The mass spectra of the PAH's are very characteristic with an important molecular peak and almost no fragmentation. Two mass spectra are presented as examples on Figures 2 and 3. With one or two nanograms injected in the chromatographic column it is possible to identify a given PAH from its mass spectrum. The chromatographic column used allows the separation of about 15 PAH's, the isomers (a) and (e) of benzopyrene are separated as it is shown on Figure 4 which represents a currentogram obtained from an extract. The

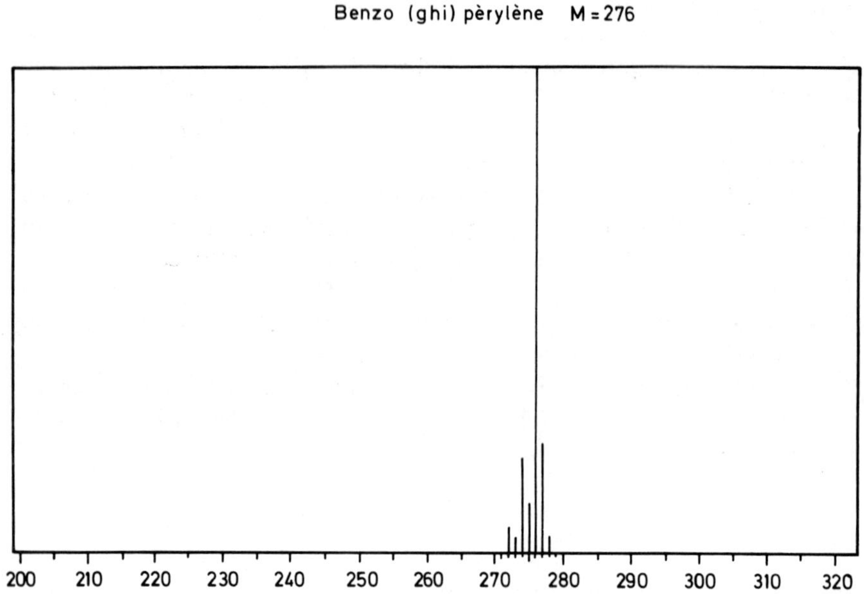

FIGURE 2 Mass Spectrum of benzo (ghi) perylene.

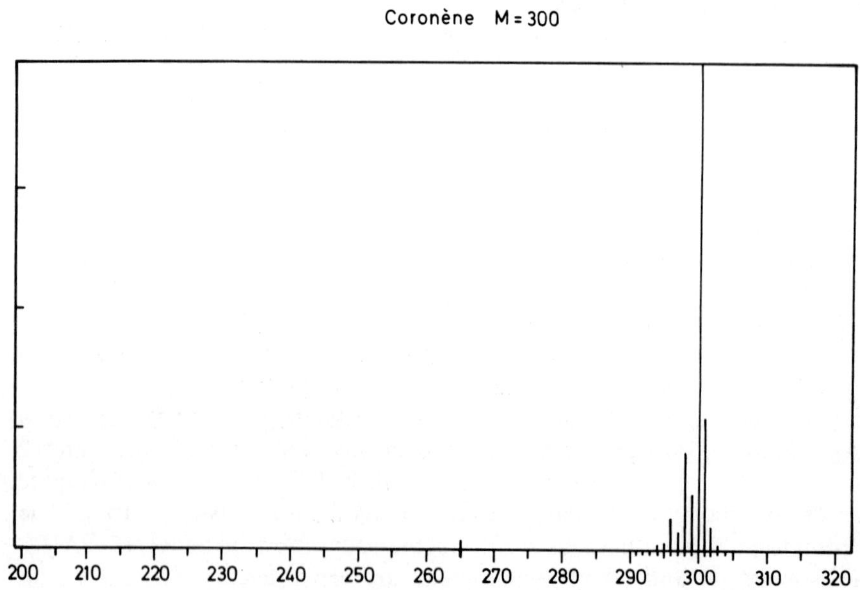

FIGURE 3 Mass spectrum of coronene

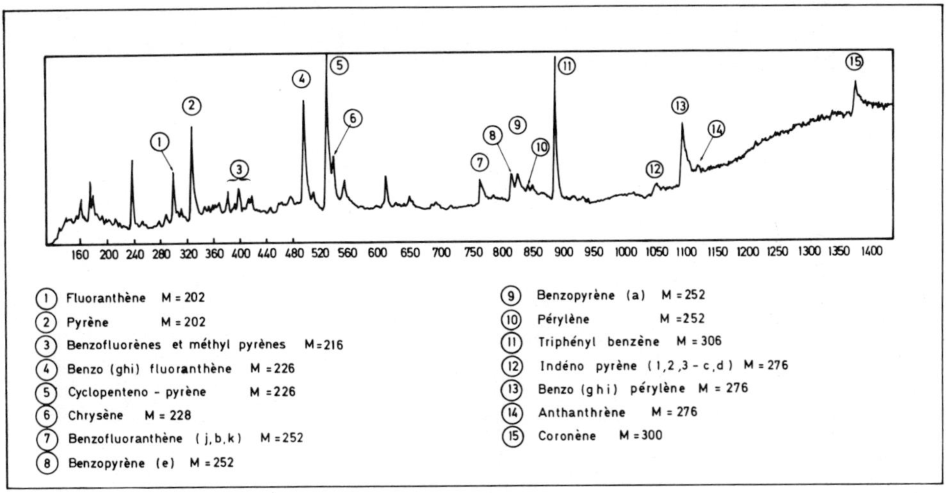

FIGURE 4 Example of currentogram obtained from an extract (gasoline engine).

identified PAH's are listed on the figure. The explored mass range extends from m/z:150 up to m/z:320. The integration time is 1 millisecond for each mass.

The quantitative determinations are performed using mass fragmentography, or multiple ion detection (MID). Figure 5 is an example of such a mass fragmentogram obtained in setting the mass spectrometer on the m/z values: 202, 216, 226, 228, and 212 for the 9 chloroanthracene which is used as internal standard for the PAH's listed on this figure. The amounts of each PAH are deduced from the relative intensities of each peak to the peak of the internal standard compared to these ratios measured on a calibration curve. The integration time on each m/z value is 20 ms.

Two calibration curves are shown on Figures 6 and 7. As it can be seen on Figure 7 it is not possible to avoid some non reversible adsorption on the chromatographic column. In the experimental conditions which are used the sensitivity limits range from 3 nanograms for fluoranthene to 50 nanograms for coronene present on the filter. For each PAH reference solution was prepared from the corresponding pure compound, however the non availability of some PAH's led to evaluate:

a) benzofluorenes, methyl pyrenes and dimethyl phenanthrenes in using a pyrene solution,

b) benzo (ghi) fluoranthene, cyclopentenopyrene and trimethyl phenanthrenes in using a chrysene solution,

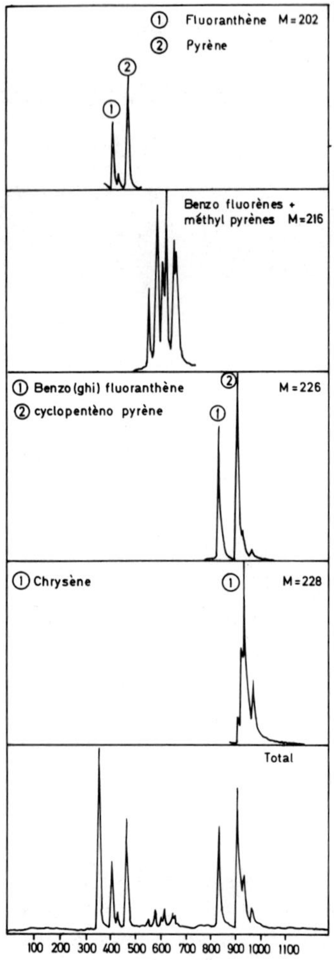

FIGURE 5 Multiple ion detection for quantitation of PAH's.

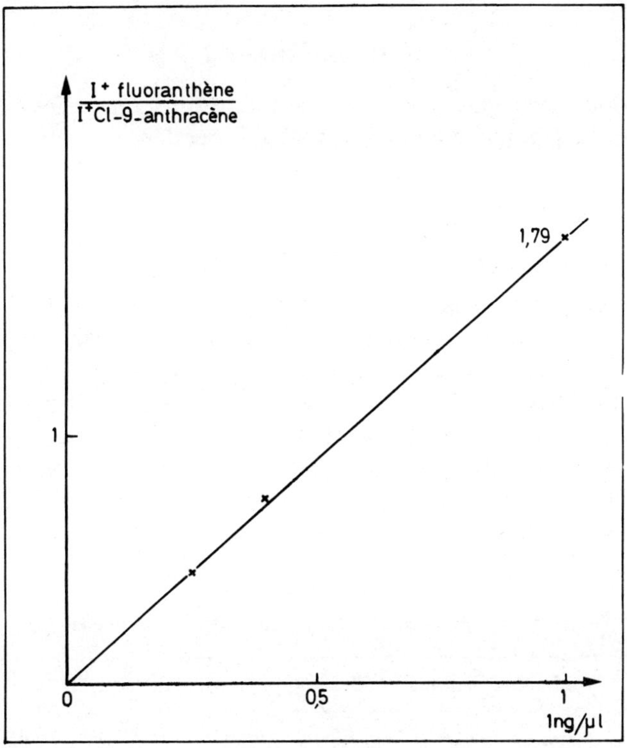

FIGURE 6 Calibration curve for fluoranthene.

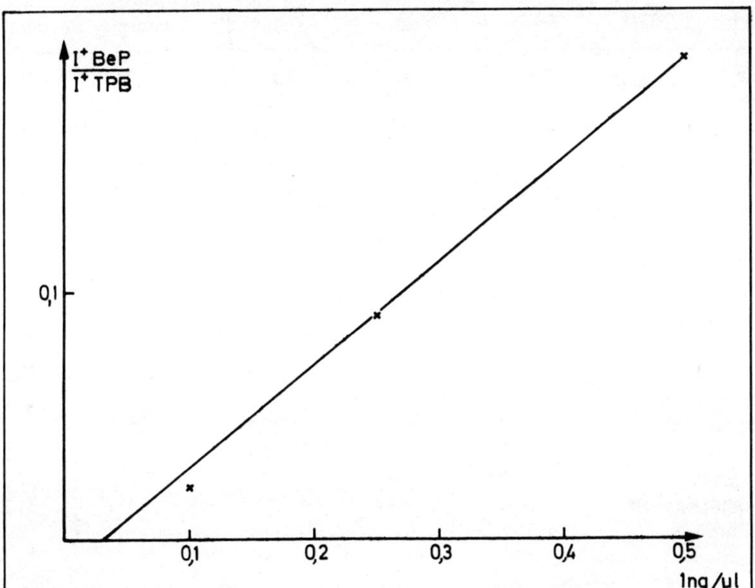

FIGURE 7 Calibration curve for benzopyrene (e).

c) benzofluoranthene (jbk) and anthanthrene in using respectively solutions of benzopyrene (e) and benzo (ghi) perylene.

EXPERIMENTAL RESULTS

1. Gasoline engines

The emission of the PAH's from four different gasoline engines have been measured. The results are presented in Table I. It must be pointed out that each vehicle was equipped with a catalyst system (oxydation catalyst). For each analysed PAH the observed variations from vehicle to vehicle are within a factor of two to three but it is important to emphasize the considerable variations which are observed for the same vehicle equipped successively with a new and a used catalyst system, in the latter case the total amount of emitted PAH's is higher by a factor of 40.

TABLE I

Amounts of PAH's emitted from four different gasoline engines

Amounts of emitted PAH's in μg per mile (cycle US-FTP 75)					
				Vehicle 4	
Detected PAH	Vehicle 1	Vehicle 2	Vehicle 3	New catalyst	Used catalyst
Benzene	8,840	5,370	10,200	5,200	32,000
Toluene	18,740	13,700	10,500	5,700	55,000
Fluoranthene	0.09	0.13	0.22	0.09	4.2
Pyrene	0.18	0.20	0.37	0.15	7.3
Chrysene	0.88	0.17	0.26	0.20	6.9
Benzofluorenes + methyl pyrenes	<0.11	0.06	0.10	<0.05	4.2
Benzo (ghi) fluoranthene	0.15	0.14	0.25	0.12	22.1
Cyclopentenopyrene	0.24	0.37	0.70	<0.05	30.3
Benzofluoranthenes (jbk)	0.53	0.55	0.90	0.43	12.8
Benzopyrene (e)	0.88	0.72	1.51	0.56	15.0
Benzopyrene (a)	0.70	0.72	1.90	0.50	21.8
Perylene	<0.22	0.33	1.61	0.10	2.7
Indenopyrene	1.41	1.03	2.07	1.0	17.0
Benzo (ghi) perylene	2.16	1.34	2.15	1.0	33.2
Anthanthrene	<0.22	0.15	0.15	<0.10	2.75
Coronene	3.53	1.16	1.82	1.5	56.1
Total PAH	11.30	7.07	14.0	5.86	236

2. Diesel engines

The emission of the PAH's from two different diesel engines have been measured, the results are presented in Table II.

These data show that the total amount of PAH's emitted from the diesel engine equipped with an exhaust gas recycling system is about ten times higher than the amount emitted from the other diesel engine.

TABLE II

Amounts of PAH's emitted from two different diesel engines

Amounts of emitted PAH's in μg per mile (cycle US-FTP 75)		
Detected PAH	Vehicle 1 equipped with an exhaust gas recycling system	Vehicle 2
Methyl phenanthrenes (4 isomers)	326	25
Fluoranthene	110	7.1
Pyrene	47	5.5
Chrysene + benzo (a) anthracene	27	5.8
Benzofluorenes + methyl pyrenes	9.8	3.6
Dimethyl phenanthrenes (9 isomers)	119	13.3
Benzo (ghi) fluoranthene	42.4	3.1
Trimethyl phenanthrenes (4 isomers)	74.1	14
Benzofluoranthenes (jbk)	15.9	9.6
Benzopyrene (e)	12.3	7.4
Benzopyrene (a)	4.4	3.4
Benzo (ghi) perylene	<4	<4
Coronene	<4	<4
Total PAH	788	98

CONCLUSIONS

After an extraction with DMSO and a single purification with cyclohexane it is possible to perform, very easily, quantitative determinations of about 15 PAH's fixed on the particles present in the exhaust gas of automobiles using the GC–MS technique. The procedure used to collect the particles depends on the working conditions of the CVS tunnel

TABLE III

Comparison of the amounts of PAH's emitted from gasoline and diesel automobile engines

Amounts of emitted PAH's in μg per mile (cycle US-FTP 75)			
Detected PAH	Vehicle 3 gasoline engine (+new catalyst)	Vehicle 4 gasoline engine (+used catalyst)	Vehicle 1 diesel engine (with EGR system)
Fluoranthene	0.22	4.2	110
Pyrene	0.37	7.3	47
Chrysene	0.26	6.9	27
Benzo (ghi) fluoranthene	0.25	22.1	42.4
Benzo fluoranthenes (jbk)	0.90	12.8	15.9
Benzopyrene (e)	1.5	15	12.3
Benzopyrene (a)	1.9	21.8	4.4

but the measured amounts of PAH's depend also from the conditioning and adjustments of the engines and the state of the catalyst when such a system is set up on the exhaust gas pipe. However to make a first comparison of the PAH's emissions from gasoline and diesel engines, some data obtained during this study are collected in Table III. From these data one observes:

1) The emission of the light PAH's is higher from diesel engines.

2) The differences between the measured amounts tend to diminish with the increase of the molecular weight of the analysed PAH's.

3) The measured amount of benzopyrene (a) is of the same order of magnitude for a gasoline engine and a diesel engine.

4) It is important to emphasize that the collecting method advocated by the EPA, that is to say, the sampling of only the particles in the exhaust gas, favours probably the heaviest PAH's to the detriment of the lightest.

Acknowledgements

This work was supported by the Institute de Recherche des Transports under Contract No. 79025.

References

1. U. Stenberg, T. Alsberg, L. Blomberg and T. Wannhan, *Polynuclear Aromatic Hydrocarbons* (Ann Arbor Science Publishers, 1979) pp. 313–326.
2. G. Grimmer and H. Bohnke, *Z. Anal. Chem.* **261,** 310 (1972).
3. G. Grimmer, H. Bohnke and A. Glaser, *Zbl. Bakt. Hyg. I Abt Orig.* **B164,** 218 (1977).
4. D. T. Kaschani, *Erdgas*, **32,** 12 (1979).

Part III
INDEX

Author Index

Subject Index